KT-376-118

BASIC ELECTRONICS FOR SCIENTISTS

BASIC ELECTRONICS FOR SCIENTISTS

Third Edition

James J. Brophy
Institute of Gas Technology

INTERNATIONAL STUDENT EDITION

McGRAW-HILL KOGAKUSHA, LTD.

Tokyo Auckland Düsseldorf Johannesburg London Mexico
New Delhi Panama São Paulo Singapore Sydney

This book was set in Times New Roman.
The editors were C. Robert Zappa and Michael Gardner;
the cover was designed by Pencils Portfolio, Inc.;
the production supervisor was Leroy A. Young.
New drawings were done by Oxford Illustrators Limited.

Library of Congress Cataloging in Publication Data

Brophy, James John, date
    Basic electronics for scientists.

    Includes bibliographies and index.
    1. Electronics. I. Title.
TK7815.B74   1977   621.381   76–18715
ISBN 0-07-008107-7

BASIC ELECTRONICS FOR SCIENTISTS

INTERNATIONAL STUDENT EDITION

Exclusive rights by McGraw-Hill Kogokusha, Ltd. for
manufacture and export. This book cannot be re-exported
from the country to which it is consigned by McGraw-Hill.
II

Copyright © 1977, 1972, 1966 by McGraw-Hill, Inc.
All rights reserved. No part of this publication may be
reproduced, stored in a retrieval system, or transmitted,
in any form or by any means, electronic, mechanical,
photocopying, recording, or otherwise, without the prior
written permission of the publisher.

TOSHO PRINTING CO., LTD. TOKYO, JAPAN

Dedicated to Muriel

Contents

Preface

Electronic measurement and control pervade all corners of science and engineering. The great power and versatility of electronic devices and, consequently, their widespread application make it imperative that science and engineering students obtain a working familiarity with electronics. Yet this familiarity need not be as intensive as that achieved in the training of electrical engineers.

This text is written to provide undergraduate science or engineering students with a basic understanding of electronic devices and circuits. This understanding should be sufficient to appreciate the operation and characteristics of the many electronic instruments they will use in their professional careers. The analysis of circuits, rather than their design, is emphasized since even as professionals they will design only the simplest of circuits.

The physical basis of electronic circuits is stressed, but emphasis on mathematical complexity has been studiously avoided. This has the effect of highlighting applications of electronic circuits in measurements and instruments without reducing the level of sophistication. It is assumed that the students possess a general acquaintance with electricity and electrical properties of materials as covered in beginning physics courses. Circuit theory is treated by starting with direct currents, however, so that parts of the early chapters may be used as refresher material.

Since this text is written from the point of view of an experimentalist, concurrent laboratory experience is highly recommended. Almost all of the circuits analyzed give actual component values and can easily form the basis for suitable experiments. No attempt has been made to suggest such experiments in detail because of the diversity of equipment likely to be found in different university laboratories. On the other hand, the problems have been chosen to demand quantitative answers and therefore may substitute in some measure for laboratory work where a separate laboratory course is not possible.

This third edition has been modified in several areas, while retaining the emphasis and thrust of the original text. The treatment of semiconductor devices is pointed toward understanding and use of integrated circuits since they are so versatile and directly applied in practice. Vacuum tube circuits are discussed separately in an appendix because many electronic instruments stil! require an appreciation of their characteristics. Additional emphasis is placed on digital electronics in view of the ever-expanding scope of digital instrumentation and control.

For many years I have felt that a working familiarity with electronics contributes immeasurably to a professional scientific or engineering career. If this text makes it possible for others to attain such familiarity, I will be well satisfied. I am deeply indebted to many colleagues who, through their publications, have provided much background material for this book. I must also express my sincere thanks to those who kindly read and criticized the manuscript, and to those instructors and students who have offered suggestions resulting from their use of previous editions. Lastly, only through the collaboration and encouragement by Muriel, my wife, has this project been completed.

James J. Brophy

BASIC ELECTRONICS FOR SCIENTISTS

DIRECT-CURRENT CIRCUITS

The operation of any electronic device, be it as complicated as a television receiver or as simple as a flashlight, can be understood by determining the magnitude and direction of electric currents in all parts of its functional unit, the circuit. In fact, it is not possible to appreciate how any given circuit functions without a detailed knowledge of the currents in its individual components.

Even the most complicated circuits can be examined in easy stages by first considering each part separately and subsequently noting how the various subcircuits fit together. Therefore, circuit analysis starts by treating elementary configurations under the simplest possible conditions. Circuits in which the currents are steady and do not vary with time are called direct current circuits. Such dc circuits, which are considered in this chapter, are important and relatively simple to understand.

INTRODUCTORY CONCEPTS

Current, Voltage, and Resistance

The motion of electric charges (for example, electrons in a conducting material) constitutes an *electric current*. Specifically, the current I is the time rate at which charge Q passes a given point so that

$$I = \frac{dQ}{dt} \tag{1-1}$$

Electric current is measured in coulombs per second, which is termed an *ampere* (A) in honor of the French scientist Andre Marie Ampère.

A number of materials, most notably metals such as copper and silver, contain many free electrons which move in response to electric fields originating from other, external, electric charges and, accordingly, such materials are capable of carrying an electric current. Each free electron in a metal wire carrying current is accelerated by the electric field until it loses its velocity as a result of a collision within the metal. It then accelerates again until it suffers a subsequent collision. The energy required to effect the successive accelerations and move an electron from one point to another is called the electric *potential difference* between the two points. The potential difference V is measured in terms of work per unit charge, or

$$V = \frac{W}{Q} \tag{1-2}$$

Because potential difference is used so frequently in electric circuit analysis, work per unit charge is termed a *volt* (V) in recognition of the early worker in electricity, Alessandro Volta.

The resistance that each free electron encounters as a result of multiple collisions when moving in a conductor depends upon a material property called *resistivity*, ρ. The range of resistivities typical of several metals and alloys at room temperatures is shown in Table 1-1. In addition to resistivity, the shape of the

Table 1-1. Resistivities of metals and alloys

Material	Resistivity, $10^{-8}\ \Omega \cdot m$
Aluminum	2.6
Brass	6
Carbon	3.5×10^2
Constantan (Cu 60, Ni 40)	50
Copper	1.7
Manganin (Cu 84, Mn 12, Ni 4)	44
Nichrome	100
Silver	1.5
Tungsten	5.6

conductor is also important, so that the *resistance*, R, of a wire L meters long and A square meters in cross sectional area is given by

$$R = \rho \frac{L}{A} \tag{1-3}$$

According to this expression, the resistance of long thin wire is greater than that of a short, thick wire of the same material.

The unit of resistance is labeled the *ohm*, after Georg Simon Ohm, who first discovered the relationship between current, voltage, and resistance discussed in the next section. The commonly accepted symbol for the resistance of a conductor in ohms is the Greek letter omega, Ω. It often proves convenient to describe the ability to conduct current in terms of the reciprocal of resistance, or *conductance*, which is measured in terms of reciprocal ohms, or *mhos*.

Ohm's Law

In order to maintain a large current in a conductor, more energy, hence a greater potential difference, is required than in the case of a small current in the same conductor. The constant of proportionality between current and potential difference is just the resistance of the conductor, or

$$V = RI \tag{1-4}$$

This equation is known as *Ohm's law*. According to Ohm's law, whenever a conductor of resistance R carries a current I, a potential difference, or *voltage V*, must be present across the ends of the conductor. This relation is fundamental to electric circuit analysis and is used repeatedly in subsequent sections.

Joule's Law

The kinetic energy of the electrons in a conductor, which results from acceleration by the electric field, is dissipated in inelastic collisions within the conductor and converted to heat energy. Consequently the temperature of a conductor carrying a current must increase slightly, and it is apparent that electric power is expended in forcing a current through the resistance of the conductor.

The power P that must be supplied to the conductor is given by

$$P = \frac{dW}{dt} = V\frac{dQ}{dt} = VI \tag{1-5}$$

where the definitions of potential difference, Eq. (1-2), and current, Eq. (1-1), have been used. This expression may be written in terms of the resistance of the conductor using Ohm's law. The result,

$$P = I^2 R \tag{1-6}$$

is known as *Joule's law,* after Sir James Prescott Joule, who discovered experimentally that the rate of development of heat in a resistance is proportional to the square of the current.

According to Joule's law, electric power is dissipated in a conductor whenever it carries an electric current. This effect is put to use in incandescent lamps, where a thin metal filament is heated to white heat by the current, and also in electric fuses, in which the conductor melts when the current exceeds a predetermined value. On the other hand, the size of wires, and therefore their resistance, is selected so that the power loss is small and the temperature rise negligible when the wire is carrying less than the maximum design current. The joule heat in a conductor is commonly spoken of as the "*I*-squared-*R*" loss. As usual, the unit of power, according to Eq. (1-5), is a joule per second, which is called a *watt* (W), in honor of James Watt, developer of the steam engine.

CIRCUIT ELEMENTS

Resistors

An electrical component very frequently used in electronic circuits is the *resistor,* which is a circuit element having a specified value of resistance. Resistance values commonly encountered range from a few ohms to thousands of ohms, or *kilohms* (abbreviated as kΩ), and even millions of ohms, or *megohms* (abbreviated as MΩ). Lumped resistances that resistors introduce into a circuit are large compared with those of wires and contacts between wires. According to Ohm's law, a potential difference develops across the resistor as a result of current in it at the place in the circuit where the resistor is inserted. The conventional symbol for a resistor in a circuit diagram is a zigzag line, as illustrated in Fig. 1-1.

Some resistors are constructed from a long, very fine wire wound on an insulating support. Resistance values can be increased by decreasing the cross-sectional area of the wire and by increasing its length, as Eq. (1-3) shows, and by selecting wire materials having a large resistivity (see Table 1-1). Such *wire-wound* resistors commonly employ metal-alloy wires which have resistivities relatively independent of temperature. Typical materials are manganin and constantan. Wire-wound resistors are used where it may be necessary to dissipate sufficient

Figure 1-1 Conventional circuit symbols for fixed (top) and variable resistors.

joule heat for the temperature of the resistor to rise significantly. The resistance of wire-wound resistors can be determined quite precisely by choosing the proper wire length, so wire-wound resistors are also useful in applications where accurate resistance values are desired.

Thin-film resistors are made by depositing a thin film of a metal on a cylindrical insulating support. High-resistance values are a consequence of the thinness of the film. Because of the difficulty in producing uniform films, it is not possible to control resistance values as precisely as in the case of wire-wound resistors. However, thin-film resistors are free of troublesome inductance effects common in wire-wound units (Chap. 2), and this is important in high-frequency circuits. Thin-film resistors fabricated from nonmetallic materials, particularly finely divided carbon granules, are also common. Carbon itself has very high resistivity, as do the points of contact between the granules. In fact, it is possible to achieve such high resistance values with carbon granules that in many situations it is unnecessary to employ thin films at all and the resistance element is a simple rod of pressed carbon granules. Such units are known as *composition* resistors.

Both thin-film and composition resistors are provided with insulation and wire leads to facilitate inserting them into circuits. It is common practice to provide colored markings which denote the resistance value of each unit according to a universal *resistor color code*. In addition, the physical size of the resistor is a rough indication of the maximum permissible power the unit is capable of dissipating without appreciable increase in temperature caused by joule heating. Thus, for example, common resistor power ratings are 1 W, $\frac{1}{2}$ W, and $\frac{1}{4}$ W, although other values are used as well. Examples of typical thin-film and composition resistors are shown in Fig. 1-2.

It is often necessary to vary the resistance of a resistor while it is permanently connected in a circuit. Such *variable resistors* employ a mechanical slider or arm which rides over the resistance element, thus selecting the length of the element included in the circuit. Both wire-wound and composition resistance elements are commonly made circular so that the position of the slider may be adjusted by rotation of a shaft. The circuit symbols for variable resistors are of two types, as in Fig. 1-1, depending upon whether two or three terminals are provided for external

Figure 1-2 Typical composition resistors. (*Allen-Bradley Co.*)

connections. A variable resistor having two terminals is called a *rheostat*, while one with three is known as a *potentiometer*. Obviously a potentiometer, with its terminals at each end of the resistance element and a third terminal attached to the slider, can be used as a rheostat if one of the resistance-element terminals is ignored.

Batteries

According to Joule's law, electric energy is dissipated in any conductor when it carries a current. In simple dc circuits the source of this energy, which must be supplied in order to maintain the current, is often a chemical *battery*. Other sources of dc electric power are considered in a later chapter. In a battery, chemical energy is converted into electric energy, and the chemical reactions maintain a potential difference between the battery terminals whether or not a current is present. This potential difference is commonly referred to as an *electromotive force*, abbreviated *emf*, in order to distinguish it from the potential difference which appears across a resistance in accordance with Ohm's law. As a battery continues to supply the energy necessary to maintain current in a circuit, the chemical constituents eventually become depleted and the battery is said to be *discharged*. Depending upon the particular chemical nature of the battery, it may be possible to *charge* it, that is, return it to its original chemical composition, by passing a current between its terminals in a direction opposed to the internal emf. The symbol for a battery in circuit diagrams, Fig. 1-3, consists of a short heavy line parallel to a longer thin line. It is always assumed, if not explicitly indicated, that the longer line represents the higher, or positive, terminal of the internal emf. Since the internal emf is a potential difference, its unit is the volt.

The carbon-zinc battery is by far the most common, and least expensive, source of electrical energy. Although it is conventionally referred to as a *dry cell*, it actually consists of a moist paste of zinc chloride, ammonium chloride, and manganese dioxide (called the *electrolyte*) contained between a zinc electrode and a carbon electrode. The zinc and carbon electrodes serve as the terminals of the battery. The operation of such a cell is briefly as follows. At the zinc electrode, zinc atoms are dissolved into solution as doubly charged zinc ions. The zinc electrode becomes negatively charged because each zinc atom leaves behind two electrons. At the carbon electrode, ammonium ions reacting with manganese dioxide withdraw electrons from the carbon, and thus it becomes charged positively. If the negative zinc electrode is connected externally through a circuit to the positive carbon electrode, electrons can flow between them to complete the chemical reaction.

Figure 1-3 Conventional circuit symbol for a battery.

Notice that in order for the chemical reaction to continue, zinc ions must move away from the negative electrode and the reaction products near the positive terminal must likewise move away from the carbon electrode. Thus, current is carried internally to the battery by means of ions moving in the electrolyte, and this is a source of internal resistance. Current in the internal resistance has the effect of reducing the terminal voltage of the battery. The terminal voltage of the dry cell slowly decreases with use as the internal resistance increases because of depletion of the manganese dioxide. The internal resistance eventually becomes so large that the battery is useless.

If the dry cell is left idle for some time before it is completely discharged, the internal resistance gradually reduces because of internal diffusion of the ions. On the other hand, if a dry cell is allowed to age for extended periods (more than one year) internal ionic diffusion increases the internal resistance so much that the cell becomes inoperative, even though it may never have been used. The emf of a freshly prepared dry cell is 1.5 V. Higher voltages are conventionally obtained by connecting a number of individual units (Fig. 1-4); in fact the term *battery* originated from just such assemblies. Dry-cell batteries of 1.5, 9, 22.5, 45, 67.5, and 90 V are most commonly available.

The familiar lead-acid *storage battery* used in automobiles is an example of a battery that can be repeatedly recharged. The positive electrode of a fully charged storage battery is a porous coat of lead dioxide on a grid of metallic lead. The negative electrode is metallic lead, and both electrodes are immersed in a liquid sulfuric acid electrolyte at a specific gravity of about 1.3. During discharge the lead dioxide is converted to lead sulfate, which is poorly soluble and clings to the positive plate. This reaction withdraws electrons from the electrode, thus charging it positively. At the negative electrode, sulfate ions from solution produce lead sulfate and release electrons. Again the lead sulfate adheres to the electrode and at discharge both electrodes are nearly entirely converted to lead sulfate. The loss of sulfate ions from solution during discharge reduces the specific gravity to about 1.16, so that the condition of the battery may be determined by measuring the specific gravity of the electrolyte.

These chemical reactions are easily reversible, and current directed into the positive terminal acts to return the electrodes to their original chemical composition. Charging requires an external source to furnish electric energy, after which the battery again can supply energy during discharge. Thus, the storage battery may be said to store electric energy in chemical form. In addition, the internal resistance of the lead-acid battery is very low and the battery is capable of delivering currents of several hundred amperes for short times. The fully charged cell has

Figure 1-4 Four batteries connected in series.

Figure 1-5 Discharge curve of carbon-zinc dry cell compared with that of mercury battery.

Figure 1-6 Typical modern batteries. (*Union Carbide Co.*)

an emf of about 2.1 V, and commercial units are available as 6-, 12-, and 24-V batteries. It is important to maintain an idle storage battery fully charged, for otherwise the electrodes slowly become converted to a sulfate which cannot be returned to the original chemical composition by a charging current. In this condition, the electric energy capacity of the battery is reduced.

The internal resistance of the recently developed *mercury battery* does not change appreciably during discharge. This means that the terminal voltage remains essentially constant throughout the useful life. It then falls precipitously when the battery is exhausted, as illustrated in Fig. 1-5. The constant-voltage characteristic of mercury batteries is important in those electronic applications where the proper operation of a circuit depends critically upon the battery voltage. Such situations are not uncommon in vacuum tube and transistor circuits. In addition, the constant-voltage feature means that the mercury battery is useful as a voltage standard in electrical measurement circuits. The mercury battery has a zinc amalgam for one electrode and mercuric oxide and carbon for the other. The chemical reactions at the electrodes are somewhat similar to those of the dry cell, and the potential developed is 1.35 V.

Other recent battery types include the *alkaline* battery and the *nickel-cadmium* battery. The alkaline battery is chemically quite similar to the dry cell, but has a strongly basic electrolyte between the electrodes. This, together with a modified electrode structure, lowers internal resistance, increases energy capacity, and improves shelf life. The nickel-cadmium battery can be repeatedly recharged like the lead storage battery, but is completely sealed, since gas evolution during charging acts as a self-regulating mechanism to prevent the buildup of a large gas pressure. This feature, and the fact that a liquid electrolyte is not required, compensates for the high cost of this battery. Typical modern batteries are illustrated in Fig. 1-6.

SIMPLE CIRCUITS

Series Circuits

If several electric components, such as resistors, are connected so that the current is the same in every one, the components are said to be in a *series* circuit. Consider the simple series circuit comprising the battery and three resistors illustrated in Fig. 1-7a. The current I results in a potential difference between the terminals of each resistor which is given by Ohm's law. That is,

$$V_1 = R_1 I \qquad V_2 = R_2 I \qquad \text{and} \qquad V_3 = R_3 I \qquad (1\text{-}7)$$

Clearly, the sum of these voltages is equal to the battery emf, or

$$V = V_1 + V_2 + V_3 \qquad (1\text{-}8)$$

Equation (1-8) is a simple example of a principle of electronic circuits which is considered in greater detail in the next section. The equation states that the algebraic sum of the potential differences around any complete circuit is equal to

Figure 1-7 (a) Simple series circuit and (b) its equivalent.

zero. Note the polarity distinction between the potential difference at the terminals of a resistor compared with that of a source of emf such as a battery: the current direction is *into* the positive terminal of a resistance while it is *out of* the positive terminal of an emf source. Since, the potential decreases in the direction of the current through a resistance, the potential difference is commonly referred to as the *IR drop* across the resistor.

If the *IR* drops of Eq. (1-7) are inserted into Eq. (1-8), the result is

$$V = IR_1 + IR_2 + IR_3 = I(R_1 + R_2 + R_3)$$

Thus, the current in the series circuit is

$$I = \frac{V}{R_1 + R_2 + R_3} = \frac{V}{R_{eq}} \qquad (1\text{-}9)$$

where the equivalent resistance R_{eq} is defined as

$$R_{eq} = R_1 + R_2 + R_3 \qquad (1\text{-}10)$$

Evidently the equivalent resistance of any number of resistors connected in series equals the sum of their individual resistances. Insofar as the current is concerned,

Output

Figure 1-8 Potential-divider circuit.

the circuit of Fig. 1-7b containing the one single resistor R_{eq} is equivalent to that of Fig. 1-7a, which has three resistors.

A useful circuit based on the series connection of resistors is the *potential divider*, Fig. 1-8, in which the junction between each pair of resistors is connected to a terminal of a *multiple-tap* selector switch. By positioning the switch on each of its various taps it is possible to present a given fraction of the battery voltage V at the output terminals. The division of the potential V among the various taps depends upon the magnitudes of the resistances in the potential divider. Obviously, if the series resistors are replaced by a potentiometer, the output voltage may be set at any desired fraction of V. This is the principle of the volume control in radio and television receivers.

Parallel Circuits

Another way of connecting electric components, such as resistors, is shown in Fig. 1-9. Here the potential difference across each resistor in the circuit is the same; this form of connection is called a *parallel circuit*. The current in each resistor is given by Ohm's law as

$$I_1 = \frac{V}{R_1} \qquad I_2 = \frac{V}{R_2} \qquad \text{and} \qquad I_3 = \frac{V}{R_3} \tag{1-11}$$

In this case, the sum of the currents equals the battery current,

$$I = I_1 + I_2 + I_3 \tag{1-12}$$

Substituting for the currents from Eq. (1-11), this becomes

$$I = \frac{V}{R_1} + \frac{V}{R_2} + \frac{V}{R_3} = V\left(\frac{1}{R_1} + \frac{1}{R_2} + \frac{1}{R_3}\right) \tag{1-13}$$

Now, in order to determine the equivalent resistance for parallel resistors, we define R_{eq}, using Ohm's law, as

$$V = IR_{eq} \tag{1-14}$$

Inserting Eq. (1-14) into Eq. (1-13)

$$I = \frac{V}{R_{eq}} = V\left(\frac{1}{R_1} + \frac{1}{R_2} + \frac{1}{R_3}\right) \tag{1-15}$$

Figure 1-9 Parallel-connected resistors.

So that

$$\frac{1}{R_{eq}} = \frac{1}{R_1} + \frac{1}{R_2} + \frac{1}{R_3} \tag{1-16}$$

which states that for any number of resistors in parallel the reciprocal of the equivalent resistance equals the sum of the reciprocals of the individual resistances.

Networks

Network connections of series and parallel resistances can be analyzed by successive applications of Eqs. (1-10) and (1-16). Consider, for example, the network of Fig. 1-10a with the resistance values as marked on the circuit diagram. The parallel combination of R_5 and R_6, each of which is 10 Ω, may be replaced by a 5-Ω resistor since according to Eq. (1-16)

$$\frac{1}{R_{eq}} = \frac{1}{R_5} + \frac{1}{R_6} = \frac{1}{10} + \frac{1}{10} = \frac{2}{10}$$

$$R_{eq} = 5\ \Omega \tag{1-17}$$

Therefore the network is reduced to that shown in Fig. 1-10b. Next, the combination of R_{eq} with R_4 (= 10 Ω) is, by Eq. (1-10),

$$R'_{eq} = R_{eq} + R_4 = 5 + 10 = 15\ \Omega \tag{1-18}$$

and the network is now Fig. 1-10c. R'_{eq} and R_3 are in parallel, so that their equivalent is

$$R''_{eq} = \frac{R'_{eq} R_3}{R'_{eq} + R_3} = \frac{15 \times 15}{15 + 15} = \frac{225}{30} = 7.5\ \Omega \tag{1-19}$$

Lastly, the series combination of R''_{eq}, R_1, and R_2 is simply

$$R_T = R''_{eq} + R_1 + R_2 = 7.5 + 5 + 5 = 17.5\ \Omega \tag{1-20}$$

and the entire network of Fig. 1-10a may now be replaced by its simple equivalent, Fig. 1-10e, where R_T represents the resistance of the entire network. The current in the battery is therefore

$$I = \frac{V}{R_T} = \frac{35}{17.5} = 2\ \text{A} \tag{1-21}$$

Suppose it is desired to determine the current I_3 in R_3. This is accomplished by first calculating the potential difference V_3 between points b and c in the circuit diagram. The IR drop across R_1 is $IR_1 = 2 \times 5 = 10$ V, and a similar value applies to the IR drop across R_2. According to Eq. (1-7)

$$V = V_1 + V_3 + V_2 \tag{1-22}$$

Figure 1-10 Network reduction by series and parallel equivalents.

Thus

$$V_3 = V - V_1 - V_2 = 35 - 10 - 10 = 15 \text{ V} \tag{1-23}$$

The current in R_3 is therefore

$$I_3 = \frac{V_3}{R_3} = \frac{15}{15} = 1 \text{ A} \tag{1-24}$$

By similar reasoning it is possible to determine the current in each resistor.

CIRCUIT ANALYSIS

Kirchhoff's Rules

It is not possible to reduce many of the networks important in electronics to simple series-parallel combinations, so that more powerful analytical methods must be used. Two simple extensions of Eqs. (1-8) and (1-12), known as *Kirchhoff's rules*, are most helpful in this connection. Consider first the simple parallel circuit, Fig. 1-9, redrawn as in Fig. 1-11 to illustrate the idea of a *branch point*, or *node*, of a circuit. A node is the point at which three (or more) conductors are joined. Kirchhoff's first rule is that the algebraic sum of the currents at any node is zero. Symbolically

$$\Sigma I = 0 \tag{1-25}$$

Note that Eq. (1-25) is essentially a statement of continuity of current; it may also be looked upon as a result of the conservation of electric charge.

Kirchhoff's second rule has already been applied implicitly in using Eq. (1-25) to calculate I_3 in Fig. 1-10a. It states that the algebraic sum of the potential differences around any complete loop of a network is zero. Symbolically

$$\Sigma V = 0 \tag{1-26}$$

A loop of a network is understood to be any closed path such as *abcda* in Fig. 1-10a which returns to the same point. Other examples of complete loops in the same network are *befgcb* and *daefgd*. Equation (1-26) is a consequence of the conservation of energy.

In applying Kirchhoff's rules to any network the first step is to assign a current of arbitrary direction to each of the resistances in the network. The polarity of the potential difference across each resistor is then marked on the circuit diagram using the convention already noted that the current enters the positive terminal of a resistance. The polarity of emf sources are, of course, specified in advance from the circuit diagram itself. Kirchhoff's rules are then applied to the various nodes and circuit loops to obtain a sufficient number of simultaneous equations to solve for the total number of unknown currents.

Figure 1-11 Nodes in a simple parallel circuit.

It is true that if a network contains m nodes and n unknown currents there are $m - 1$ independent equations which result from Eq. (1-25). Similarly, there are $n - (m - 1) = n - m + 1$ independent equations derived from Eq. (1-26). The total number of independent equations obtained from Kirchhoff's rules applied to any network is therefore $(m - 1) + n - (m - 1) = n$. This is just the number of unknown currents and the network solution is therefore completely determined. It is generally possible to write down more node and loop equations than are needed but only n of them are truly independent.

The solution of these independent equations often results in certain of the currents being negative. This means that the original arbitrarily assigned current direction is, in fact, incorrect and the current is actually in the opposite direction. Thus, it is not necessary to know the true current direction in advance. Once the various currents have been calculated, the IR drops in any portion of the circuit are determined using Ohm's law.

The technique of applying Kirchhoff's rules to a network can best be illustrated with a few examples. Consider first the simple parallel-resistor circuit of Fig. 1-12. The current direction in each resistor has been arbitrarily selected and the polarity of the IR drops marked in accordance with these assigned directions. Note that this network has only two nodes, one at b and the other at e. Therefore, there is only $2 - 1 = 1$ independent node equation. Considering the branch point at b, Eq. (1-25) yields

$$I - I_1 + I_2 = 0 \tag{1-27}$$

Notice at branch point e the current equation is

$$-I + I_1 - I_2 = 0 \tag{1-28}$$

Clearly Eq. (1-28) is simply the negative of (1-27) and the two relations are therefore not independent. Either equation may be used in the solution of the network.

Consider now the loop $abef$. According to Eq. (1-26)

$$V - I_1 R_1 = 0 \tag{1-29}$$

Similarly, around the loop $abcdef$

$$V + I_2 R_2 = 0 \tag{1-30}$$

Since there are three unknown currents, there must be $3 - 2 + 1 = 2$ independent loop equations and these are (1-29) and (1-30). Note, however, that around loop $bcde$

$$I_1 R_1 + I_2 R_2 = 0 \tag{1-31}$$

Figure 1-12

This is not an independent relation, as may be shown by subtracting Eq. (1-29) from (1-30). The result is Eq. (1-31). Thus these three loop equations are not independent, and any two may be used in solving the network.

Choose Eqs. (1-27), (1-29), and (1-31) as the three independent equations to solve for the three unknown currents. The solution is accomplished by first solving Eq. (1-29) for I_1,

$$I_1 = \frac{V}{R_1} \tag{1-32}$$

Next, I_2 is determined from (1-31),

$$I_2 = -\frac{I_1 R_1}{R_2} \tag{1-33}$$

Substituting (1-33) into (1-27),

$$I - I_1 - \frac{I_1 R_1}{R_2} = 0 \tag{1-34}$$

Substituting from (1-32) for I_1,

$$I - \frac{V}{R_1} - \frac{V R_1}{R_1 R_2} = I - V\left(\frac{1}{R_1} + \frac{1}{R_2}\right) = 0 \tag{1-35}$$

The current I is therefore

$$I = V\left(\frac{1}{R_1} + \frac{1}{R_2}\right) \tag{1-36}$$

which is quite equivalent to the solution corresponding to Eq. (1-13) arrived at by considering parallel resistors.

Finally, I_2 is determined by substituting for I_1 in Eq. (1-33),

$$I_2 = -\frac{V R_1}{R_1 R_2} \tag{1-37}$$

or

$$I_2 = -\frac{V}{R_2} \tag{1-38}$$

According to the minus sign in Eq. (1-38), this current is actually in the opposite direction to that assumed in Fig. 1-12. Correspondingly, the IR drop across R_2 has the opposite polarity from that shown on the circuit diagram.

More complicated networks require more than three equations, and it is usually desirable to employ the standard method of determinants to solve the set of simultaneous equations. This technique, illustrated in the following section, has the considerable advantage that it is possible to solve directly for only those currents that are of interest. Often, only one or two of the currents in a network are of direct concern, and in this case a complete solution for all the unknowns is superfluous.

Wheatstone Bridge

In this section Kirchhoff's rules are used to analyze the *Wheatstone bridge* circuit illustrated in Fig. 1-13. This extremely useful circuit was developed in 1843 by Charles Wheatstone and is widely used in electrical measurements to determine values of unknown resistances. The manner in which it is used may be understood from an analysis of the circuit. According to Kirchhoff's rule for branch points *a*, *b*, and *d*,

$$I - I_1 - I_2 = 0$$
$$I_1 - I_3 + I_5 = 0 \qquad (1\text{-}39)$$
$$I_3 + I_4 - I = 0$$

Because there are four branch points in the Wheatstone bridge circuit, these three current equations are independent, so the fourth one, which could be written for branch point *c*, is not used.

Applying Kirchhoff's rule to loops *abdefa*, *acba*, and *bcdb*, the equations are

$$-I_1 R_1 - I_3 R_3 + V = 0$$
$$-I_2 R_2 - I_5 R_5 + I_1 R_1 = 0 \qquad (1\text{-}40)$$
$$I_5 R_5 - I_4 R_4 + I_3 R_3 = 0$$

Note carefully the indicated polarities of the various IR drops as they are encountered in traversing each loop. Since there are six unknown currents, $6 - 4 + 1 = 3$ loop equations are needed here and any others are redundant.

Equations (1-39) and (1-40) are six equations in six unknowns. Therefore, in applying the method of determinants to these simultaneous equations it is necessary to evaluate two sixth-order determinants in calculating each current. The total solution involves seven different such determinants. While the evaluation of a sixth-order determinant is straightforward and a number of standard approaches exist to reduce the order before final evaluation, the complete solution

Figure 1-13 Wheatstone bridge.

of seven sixth-order determinants is quite laborious. Therefore, although the solution to the set of Eqs. (1-39) and (1-40) is in principle accomplished, it is useful to seek alternative methods.

The analysis of complex networks can usually be simplified by the use of *loop currents*. This technique, known as *Maxwell's method*, after James Clerk Maxwell, in effect applies both Kirchhoff's rules simultaneously and thereby reduces the number of simultaneous equations needed. The loop currents are drawn around any complete loop as the three illustrated for the case of the Wheatstone bridge in Fig. 1-14. After the polarity of the IR drops are indicated in accordance with the current directions, the usual voltage equations around each loop are written. Thus, with reference to Fig. 1-14

$$V - R_1(I_a - I_b) - R_3(I_a - I_c) = 0$$

$$-R_2 I_b - R_5(I_b - I_c) + R_1(I_a - I_b) = 0 \qquad (1\text{-}41)$$

$$R_3(I_c - I_a) + R_5(I_c - I_b) + R_4 I_c = 0$$

Here again, note the polarity of the IR drops and the current directions. Upon rearranging

$$-(R_1 + R_3)I_a + R_1 I_b + R_3 I_c = -V$$

$$R_1 I_a - (R_1 + R_2 + R_5)I_b + R_5 I_c = 0 \qquad (1\text{-}42)$$

$$-R_3 I_a - R_5 I_b + (R_3 + R_4 + R_5)I_c = 0$$

The solution of Eqs. (1-42) for any current, say I_b, using determinants is found by forming a ratio in which the denominator is the determinant of the coefficients of the currents and the numerator is a similar determinant with the coefficients of the unknown current replaced by the right side of the equation. That is, the solution for I_b is

$$I_b = \frac{\begin{vmatrix} -(R_1 + R_3) & -V & R_3 \\ R_1 & 0 & R_5 \\ -R_3 & 0 & R_3 + R_4 + R_5 \end{vmatrix}}{\begin{vmatrix} -(R_1 + R_3) & R_1 & R_3 \\ R_1 & -(R_1 + R_2 + R_5) & R_5 \\ -R_3 & -R_5 & R_3 + R_4 + R_5 \end{vmatrix}}$$

$$= \frac{VR_5 R_3 + VR_1(R_3 + R_4 + R_5)}{\Delta} \qquad (1\text{-}43)$$

where Δ signifies the denominator. Similarly, I_c is

$$I_c = \frac{\begin{vmatrix} -(R_1 + R_3) & R_1 & -V \\ R_1 & -(R_1 + R_2 + R_5) & 0 \\ -R_3 & -R_5 & 0 \end{vmatrix}}{\Delta}$$

$$= \frac{VR_1 R_5 + VR_3(R_1 + R_2 + R_5)}{\Delta} \qquad (1\text{-}44)$$

Figure 1-14 Loop current analysis of Wheatstone bridge.

Now, the current through R_5, which in Fig. 1-14 is labeled I_5, is

$$I_5 = I_b - I_c$$

$$= \frac{V}{\Delta}(R_5 R_3 + R_1 R_3 + R_1 R_4 + R_1 R_5 - R_1 R_5 - R_1 R_3 - R_2 R_3 - R_5 R_3)$$

$$= \frac{V}{\Delta}(R_1 R_4 - R_2 R_3) \tag{1-45}$$

Equation (1-45) is a most important relation for the Wheatstone bridge. Note that if

$$R_1 R_4 = R_2 R_3$$

or

$$\frac{R_1}{R_2} = \frac{R_3}{R_4} \tag{1-46}$$

then I_5 is zero, independent of the applied voltage. If the resistances in the arms of the bridge obey the ratios indicated in Eq. (1-46), the bridge is said to be *balanced*. Thus, for example, if R_1, R_2, and R_3 are known resistances and I_5 is zero, the value of R_4 may be immediately calculated from the condition for balance, Eq. (1-46).

In the common version of a Wheatstone bridge, resistances R_1 and R_2 are connected to a switch to give decade values of the ratio R_2/R_1, and R_3 is a continuously variable calibrated resistor. Once the bridge is balanced by adjusting R_3, the value of the unknown R_4 is simply $(R_2/R_1)R_3$. The decade values of the ratio (R_2/R_1) may range from 10^{-3}, 10^{-2}, and 10^{-1} to 1, 10, 10^2, and 10^3, so that a very wide range of resistance values can be measured. In practice, a current-indicating instrument is connected in the position of R_5 to indicate balance. Note that this meter need not be calibrated, since it is only used to indicate the balance condition, that is, a zero current.

Equation (1-45) is an example of a case in which useful information concerning the circuit is derived without carrying through the complete solution for all currents. It is often possible to draw the loop currents in such a way that only one current need be determined. The facility of choosing loop currents that minimizes the effort required to solve any given network is attained with experience.

Potentiometer Circuit

A very accurate way to compare two potential differences uses the *potentiometer circuit*, a simple version of which is illustrated in Fig. 1-15. Here an accurately variable precision potentiometer is connected in series with a variable resistance and a battery. The slider of the precision resistor is connected to one external terminal through a current-indicating meter and one end of the resistor is connected to the other output terminal. Suppose a specific value of the current $I = V/(R + R_A)$ is selected by adjusting the variable resistor R_A. Then, the potential V' of the slider on R is simply IR', where R' is the resistance between the end of the resistor and the slider. Since I and R are known, the position of the slider can be calibrated in terms of potential difference in volts. If now an unknown voltage, for example, a battery, is connected to the output terminals, and the slider is adjusted until the current indicated by the meter M is zero, the value of the unknown voltage is equal to IR'.

The potentiometer is a comparison measuring device which determines the value of the unknown potential difference in terms of the voltage of a standard battery. To show how this is accomplished, refer to the practical circuit in Fig. 1-16. Here, V_S is the emf of a standard such as a mercury battery or, more usually, a *Weston cell*, which is a special battery with an extremely stable emf. Suppose now R_A is adjusted until the meter M_1 indicates zero current. This means that

$$IR_S = V_S \tag{1-47}$$

In this circuit the precision variable resistance is composed of nine identical series resistors, each of value R, and a variable resistance R'. The current in the output circuit, indicated by M_2, is adjusted to zero with the unknown voltage source V_u connected to the output terminals. This is done by selecting the proper switch position for the switch and the slider. When the current in M_2 is zero

$$V_u = I(nR + R') \tag{1-48}$$

Figure 1-15 Simple potentiometer circuit for measuring unknown voltage.

Figure 1-16 Practical potentiometer circuit.

where n is the tap number of the selector switch. Substituting for I in Eq. (1-47) from Eq. (1-48)

$$V_u = V_S \frac{nR + R'}{R_S} \tag{1-49}$$

According to this expression, the unknown voltage is determined entirely in terms of the standard emf and the resistances of the potentiometer circuit. Note, particularly, that neither the current I nor the battery voltage V need be known.

The accuracy of the potentiometer circuit depends upon the precision with which the various resistors are constructed and the mechanical stability of the slider on the variable resistance. Accuracy can be improved by combining fixed resistors selected by a tap switch together with the continuously variable resistor, as in Fig. 1-16. This is so because the potential drop across the variable resistor is only one-tenth of that, say, in the circuit of Fig. 1-15, and voltage errors caused by mechanical irregularities in the slider are reduced by the same factor.

A great virtue of the potentiometer circuit is that no current flows in the measuring circuit at balance. This means that the unknown potential is measured under effectively open-circuit conditions and the measurement is unperturbed by internal IR drops. Actually, a small current may be present, depending upon the sensitivity of the null indicator, and sensitive electronic amplifiers are often used in place of the meter to minimize the null current.

EQUIVALENT CIRCUITS

Thévenin's Theorem

Many times the analysis of electronic circuits is facilitated by replacing all or part of a network by an *equivalent circuit* which, for certain purposes, has the same characteristics as the original. An example of this possibility has already been discussed in connection with series and parallel combination of resistors. There,

Figure 1-17 (a) Two-terminal network and (b) its Thévenin equivalent circuit.

(a)   (b)

an entire network of resistances was replaced by a single equivalent resistance in order to calculate the current. In other situations, particularly in connection with vacuum tube and transistor circuits, equivalent circuits may be used to represent the behavior of electronic devices.

One of the most useful equivalent circuits is the one that results from *Thévenin's theorem*, which states that any network of resistors and batteries having two output terminals may be replaced by the series combination of a resistor and a battery, as illustrated in Fig. 1-17. The form of the Thévenin equivalent circuit shows immediately how the proper values of V_{eq} and R_{eq} can be determined without knowing the actual configuration of the network itself. The equivalent emf is just the potential at the output terminals when the load current is zero, that is, the *open-circuit* voltage. The equivalent resistance is then the ratio of V_{eq} to the load current when $R_L = 0$, or to the *short-circuit* current.

Note also that R_{eq} is equal to the load resistance for which the voltage across the load is one-half of V_{eq}. This is useful in situations where the short-circuit current cannot be easily determined. The form of the Thévenin equivalent circuit, Fig. 1-17b, also shows that R_{eq} is the resistance of the network between the output terminals when V_{eq} is considered to be replaced by a short circuit. This way of determining R_{eq} analytically is useful when the configuration of the network is known since it usually involves only simple network reductions. Which of these alternative methods is employed to determine the equivalent emf and the equivalent internal resistance depends only upon which is easiest in any given situation.

Consider, for example, the Thévenin equivalent of the simple circuit in Fig. 1-18. The equivalent emf is just

$$V_{eq} = \frac{VR_2}{R_1 + R_2} \tag{1-50}$$

Figure 1-18

Replacing the battery with a short circuit, the resistance between the terminals is just the parallel combination of R_1 and R_2. Accordingly,

$$R_{eq} = \frac{R_1 R_2}{R_1 + R_2} \tag{1-51}$$

The load current is then

$$I_L = \frac{V_{eq}}{R_L + R_{eq}} = \frac{VR_2}{R_1 R_2 + R_L(R_1 + R_2)} \tag{1-52}$$

The right side of Eq. (1-52) may be derived by direct analysis of Fig. 1-18.

To illustrate the power of the equivalent-circuit method, consider the Wheatstone bridge circuit of Fig. 1-19a. The current through R_5 is analyzed by replacing the remainder of the circuit with its Thévenin equivalent. Next, replacing the battery with a short circuit puts R_3 in parallel with R_1 and this combination in series with the parallel combination of R_2 and R_4 across the output terminals, as illustrated in Fig. 1-19b. Therefore, R_{eq} in the equivalent circuit, Fig. 1-19c, is

$$R_{eq} = \frac{R_1 R_3}{R_1 + R_3} + \frac{R_2 R_4}{R_2 + R_4} \tag{1-53}$$

The open-circuit voltage at the output terminals is simply the potential difference between the junction of R_1 and R_3 and the junction of R_2 and R_4. This potential difference is found by subtracting the IR drop across R_2 from the IR drop across R_1. Therefore the equivalent battery is

$$V_{eq} = \frac{VR_1}{R_1 + R_3} - \frac{VR_2}{R_2 + R_4} \tag{1-54}$$

Figure 1-19 (*a*) Conventional Wheatstone bridge circuit; (*b*) after replacing battery with a short circuit in order to calculate R_{eq}; (*c*) the Thévenin equivalent.

Finally, according to the equivalent circuit, Fig. 1-19c, the current through I_5 is

$$I_5 = \frac{V_{eq}}{R_{eq} + R_5} \qquad (1\text{-}55)$$

The ease and rapidity with which this result is obtained should be compared with that necessary using Kirchhoff's rules. Note that the balance condition, Eq. (1-46), follows immediately from Eqs. (1-54) and (1-55), since $I_5 = 0$ at balance.

Norton's Theorem

A second form of equivalent circuit useful in situations as, for example, transistor circuits where current sources, rather than emfs, are of major interest, is one given by *Norton's theorem*. Norton's theorem states that any network of batteries and resistors having two output terminals can be replaced by the parallel combination of a current source I_{eq} and a resistance R_{eq}. The current source I_{eq} is the short-circuit current in the output terminals of a circuit, while the resistance R_{eq} is the same as for Thévenin's theorem.

Norton's equivalent circuit is shown in Fig. 1-20, where the triangle represents the current source I_{eq}. No simple electric component acts as a current source the way a battery acts as a voltage source. Nevertheless, the idea of a current source is conceptually very useful in circuit analysis.

Since it is possible to represent any network by either the Thévenin or the Norton equivalent circuit, it must be possible to convert from one equivalent circuit to the other. Referring to Fig. 1-21a and b, the short-circuit current through the load is V_{eq}/R_{eq} for the Thévenin equivalent circuit, and is equal to I_{eq} for the Norton equivalent circuit. For both circuits to represent the same network, it must be true that

$$I_{eq} = \frac{V_{eq}}{R_{eq}} \qquad (1\text{-}56)$$

Thus, it is a simple matter to convert from one equivalent circuit to another. Which circuit is used to represent any given network is entirely a matter of choice and convenience.

Maximum Power Transfer

In many electronic circuits, such as a radio transmitter or phonograph amplifier, it is important to transfer efficiently the maximum amount of electric power from the circuit to the load, which may be an antenna or loudspeaker. Therefore, it is of

Figure 1-20 Norton equivalent circuit.

Figure 1-21 Relationship between (a) Thévenin equivalent circuit and (b) Norton equivalent circuit.

interest to determine circuit conditions for which it is possible to achieve the maximum power transfer. Suppose the network is represented by its Thévenin equivalent circuit shown in Fig. 1-22 and that the load connected to the output terminals is represented by the resistance R_L. The subscripts on the equivalent battery and equivalent resistance have been eliminated for convenience.

Following Joule's law, the power delivered to the load resistance is

$$P = I^2 R_L = \left(\frac{V}{R + R_L}\right)^2 R_L$$

$$= \frac{V^2 / R_L}{(1 + R/R_L)^2} \tag{1-57}$$

According to Eq. (1-55), the power in the load is zero if the load resistance is very small and is also zero when the load resistance is very large. Thus, there must be an optimum load resistance for which the power in R_L is a maximum.

To find the condition for *maximum power transfer*, differentiate Eq. (1-57) with respect to R_L and equate the result to zero,

$$\frac{dP_L}{dR_L} = \frac{V^2}{R_L} \frac{2R/R_L^2}{(1 + R/R_L)^3} + \frac{V^2}{(1 + R/R_L)^2} \frac{-1}{R_L^2} = 0 \tag{1-58}$$

$$\frac{2R}{R_L} = 1 + \frac{R}{R_L} \tag{1-59}$$

so that

$$R_L = R \tag{1-60}$$

This means that maximum power is delivered to the load when the load resistance is equal to the internal resistance of the network delivering the power. When the

$$P_B = I^2 R + I^2 R_L \qquad P = I^2 R_L$$

Figure 1-22 Thévenin's equivalent circuit used to examine maximum power transfer to load resistance R_L.

load resistance is equal to the internal resistance of the network, the load is said to be *matched* to the circuit.

Since the equivalent circuit of Fig. 1-22 represents any network, the result applies equally well to all circuits. The use of the equivalent-circuit concept in this connection has thus made it possible to prove a very general result quite easily.

ELECTRICAL MEASUREMENTS

D'Arsonval Meter

By far the most common design for electric current-measuring instruments is the *d'Arsonval meter*, named after its inventor. A multiturn coil of fine wire wound on an aluminum frame is pivoted between the poles of a horseshoe permanent magnet (Fig. 1-23). Two fine spiral springs serve to position the coil and to carry the current to be measured. A pointer attached to the coil indicates the current on a scale as the coil rotates in response to the interaction between the current in the coil and the magnetic field of the magnet. A soft-iron pole piece is fitted between the poles of the magnet so that the sides of the coil move in a radially directed field.

With this design, the deflection of the pointer is directly proportional to the current through the coil. The sensitivity of the meter, that is, the deflection for a given current, is improved by increasing the magnetic field of the magnet, the coil area, and the number of turns on the coil, or by decreasing the torque constant of the springs.

The size of the coil and springs is dictated by mechanical ruggedness, since a large coil suspended by weak springs is subject to damage from mechanical shock and vibration. Also, it is not desirable to increase the number of turns of wire inordinately in an effort to improve sensitivity, since the resistance of the coil is

Figure 1-23 Sketch of the essential features of d'Arsonval meter.

increased thereby. This may adversely affect the operation of the meter, as explained in a later section. The magnetic field is limited to that available from conventional permanent magnets. In spite of these practical limitations, common d'Arsonval meters have full-scale deflections for currents as small as 10^{-3} A (1 *milliampere*, abbreviated mA), or even 50×10^{-6} A (50 *microamperes*, abbreviated μA). Laboratory instruments, which may be shock-mounted and therefore designed for maximum sensitivity, are capable of measuring 10×10^{-12} A (10 *picoamperes*, abbreviated pA).

Ammeters and Voltmeters

The d'Arsonval meter is a current-sensitive device, or *ammeter*. It is often convenient to change the current required for full-scale deflection in order to increase the range of currents over which the meter is useful. This is accomplished by *shunting* a portion of the current around the ammeter with a parallel resistance, as diagrammed in Fig. 1-24. Note that the internal resistance of the ammeter's coil R_m is indicated explicitly. Following Kirchhoff's rules, $I = I_m + I_s$ and $I_m R_m = I_s R_s$, so that the current to be determined is

$$ I = I_m + \frac{I_m R_m}{R_s} = I_m\left(1 + \frac{R_m}{R_s}\right) \tag{1-61} $$

If, for example, the shunt resistance is one-ninth of the meter resistance, $1 + R_m/R_s = 10$, and the full-scale deflection is extended to ten times the inherent sensitivity of the meter.

It is always necessary to consider the effect of the meter resistance on the circuit. Suppose it is desired to measure the current in R_L of the Thévenin equivalent circuit of Fig. 1-25 using an ammeter having an internal resistance of R_m. With the ammeter connected into the circuit the current is

$$ I = \frac{V}{R + R_L + R_m} \tag{1-62} $$

Unless $R_m \ll R + R_L$, the current indicated by the ammeter is different from the true current. For this reason it is always desirable that the internal resistance of an

Figure 1-24 Increasing range of an ammeter by using shunt resistor in parallel with meter resistance.

Figure 1-25 Effect of ammeter resistance on current in circuit.

ammeter be small. On the other hand, in circumstances where the internal resistance may not be small compared with circuit resistances, it is possible to correct for the disturbing influence of the meter resistance and thus determine the true current.

Since the full-scale deflection of an ammeter may be attributed to the voltage $V_m = R_m I$ across the meter resistance R_m, a d'Arsonval meter also is a *voltmeter*. Here again it is useful to change the range of any given voltmeter by introducing a resistance in series with the meter. Referring to the circuit of Fig. 1-26, the voltage to be measured is

$$V = I_m(R_m + R_s) \tag{1-63}$$

and it is obvious that the series resistance R_s increases the maximum fullscale voltage of a given meter. It is common practice to provide several such series resistance *multipliers* to allow a given meter to be used over a wide range of voltages.

The effect of a voltmeter on the circuit to which it is attached must be considered, just as in the case of an ammeter. This is so because the voltmeter requires a small current to deflect the pointer and this current must be supplied by the circuit. If the meter current is not negligible compared with the normal currents in the circuit, the voltmeter is said to *load* the circuit and a correction must be applied to the indicated meter reading to determine the true voltage in the absence of the disturbing meter influence.

Figure 1-26 Using d'Arsonval meter as voltmeter with series resistor multiplier.

Sensitive d'Arsonval meters are useful as voltmeters since they require only a very small current to achieve full-scale deflection. It is common practice to specify the sensitivity of a voltmeter in terms of the ratio of its internal resistance to the full-scale voltage in units of *ohms per volt*. Note that, according to Eq. (1-63), the ratio of $R_m + R_s$ to the voltage required for full-scale deflection is simply the current sensitivity of the meter, and therefore the two specifications are quite equivalent. For example, a voltmeter which uses a d'Arsonval meter having a full-scale sensitivity of 1 mA is rated at $1/10^{-3} = 1000\ \Omega/V$. This means that the voltmeter has an internal resistance of 100,000 Ω on the 100-V scale, etc. Similarly, a 20,000-Ω/V voltmeter (employing a 50-μA meter) has a resistance of 2 megohms (MΩ) on its 100-V scale.

A convenient technique for determining the resistance of a portion of a network is to measure both the current and the voltage and then apply Ohm's law. There are two distinct ways that the meters can be connected in this *voltmeter-ammeter method* (Fig. 1-27a and b). The choice between the two possibilities depends upon the relative values of the meter resistances and the circuit resistances, as can be shown in the following way. Consider first the circuit in Fig. 1-27a. From Kirchhoff's rules

$$V = AR + AR_A \tag{1-64}$$

where V and A are the meter readings. The unknown resistance is given by

$$R = \frac{V}{A} - R_A \tag{1-65}$$

which shows that the true resistance is smaller than the indicated V/A ratio.

Similarly, in the circuit of Fig. 1-27b, the current A divides between the parallel paths R and R_V so that

$$V = \frac{R_V R}{R_V + R} A \tag{1-66}$$

(a)   (b)

Figure 1-27 Two ways of connecting voltmeter and ammeter to measure resistance R.

Solving for the unknown resistance, the result is, after some rearrangement,

$$R = \frac{V}{A} \frac{1}{1 - (V/A)/R_V} \tag{1-67}$$

According to Eqs. (1-65) and (1-67) the first circuit is most useful when the ammeter resistance is small compared with the unknown (or to V/A), while the second circuit applies when the voltmeter resistance is large compared with the unknown. In either case, the unknown resistance is then given simply by the ratio V/A.

Ohmmeters and Multimeters

A simple extension of the voltmeter-ammeter circuit can be used as an *ohmmeter* in which the meter scale is calibrated directly in ohms. In a typical circuit (Fig. 1-28) the same meter is used successively to measure first the voltage across the unknown resistor and then the current in it. The way the meter scale is calibrated directly in ohms can be understood from the following analysis. First, suppose the terminals a and b in Fig. 1-28 (which are usually connected to *test leads* to facilitate connecting the ohmmeter to the unknown resistor) are shorted together. The voltmeter then measures the battery voltage V.

Next, the test leads are connected to the unknown resistance. If the voltage across R_1, as measured by the meter, is now V_R, Ohm's law gives

$$R_U + R_1 = \frac{V}{V_R/R_1} \tag{1-68}$$

Solving for R_U

$$R_U = R_1 \left(\frac{V}{V_R} - 1 \right) \tag{1-69}$$

According to Eq. (1-69) the unknown resistance can be calculated directly from the two meter readings, but it is more useful to calibrate the meter scale directly in ohms in the following way.

The variable resistance R_2 is used to adjust the meter reading to full-scale when the test leads are first shorted together. This point on the scale is then "zero ohms," and is so marked. Suppose that when the test leads are connected to the unknown resistor, the meter deflects to half-scale. This means $V_R = V/2$ and,

Figure 1-28 A simple ohmmeter circuit.

Figure 1-29 Circuit diagram of a simple volt-ohm-milliammeter (VOM).

according to Eq. (1-69), $R_U = R_1$. Thus, the midpoint on the scale can be marked with the resistance corresponding to R_1. Similarly, note that the quarter-scale reading, $V_R = V/4$, corresponds to $3R_1$, while a zero reading indicates an open circuit, or infinite ohms. The scale of an ohmmeter is clearly nonlinear although it is not difficult to use since it is direct-reading in ohms.

According to Eq. (1-69) the midscale reading of an ohmmeter depends upon R_1, so that by selecting different values for R_1 it is possible to encompass a wide range of unknown resistances. Equation (1-69) assumes that the meter current is negligible, which may not be true on high-resistance ranges where R_1 is large. Therefore, practical ohmmeter circuits are slightly more complicated than the elementary one illustrated by Fig. 1-28, but the principle of operation is identical. Note that an ohmmeter cannot be used to determine resistance values in a circuit which is in operation, for erroneous readings result due to IR drops in the circuit itself.

It is convenient to include the functions of a voltmeter, ammeter, and ohmmeter within one instrument since all three employ the same basic d'Arsonval meter. In such an instrument, commonly termed a *multimeter* or *VOM* (volt-ohmmilliammeter), switches or a number of terminals select the function and range to be used. The circuit in Fig. 1-29 is an example of an elementary instrument which has four voltage ranges, two current ranges, and a single ohms scale. By carefully tracing through the connections on this circuit diagram it is possible to draw the simple functional circuits for each use and in this way subdivide the analysis of the circuit into easy stages.

The VOM circuit of Fig. 1-29 employs a multitap switch with three wipers which are mechanically linked to move together. More elaborate multimeters have considerably more complicated switching arrangements in order to accommodate additional ranges and other functions.

SUGGESTIONS FOR FURTHER READING

A. M. P. Brookes: "Basic Electric Circuits," The Macmillan Company, New York, 1963.
Leigh Page and Norman Ilsley Adams: "Principles of Electricity," D. Van Nostrand Company, Inc., Princeton, N.J., 1931.
M. E. Van Valkenburg: "Network Analysis," Prentice-Hall, Inc., Englewood Cliffs, N.J., 1955.

EXERCISES

1-1 Calculate the resistance of a copper wire 1 m long and 0.5 mm in diameter. Repeat for a similar-sized nichrome wire.
    Ans.: 8.65×10^{-2} Ω; 5.1 Ω

1-2 What is the maximum current that can be in a 1-W 1-MΩ resistor? In a $\frac{1}{2}$-W 10,000-Ω resistor?
    Ans.: 10^{-3} A; 7.07×10^{-3} A

1-3 Given that the maximum-current capability of a flashlight dry cell is 0.5 A, what is the internal resistance of the cell? Compare with the internal resistance of a storage battery, if the maximum current in this case is 500 A.
    Ans.: 3 Ω; 4.2×10^{-3} Ω

1-4 Assume that Fig. 1-7a represents a battery power source connected to a load R_3 by means of copper wires of resistance R_1 and R_2. If $V = 10$ V, the load is 5 Ω, and the wires are 0.5 mm in diameter and 100 m long, calculate the current, the power delivered to the load, the power lost in the wires, and the voltage across the load resistor.
    Ans.: 1.07 A; 5.7 W; 4.9 W; 5.4 V

1-5 Design a voltage divider (Fig. 1-8) in which the output voltages possible are 1.0, 2.0, 5.0, and 10.0 V, if the battery voltage is 10 V and no current is taken by the output terminals.

1-6 How many identical 1-W resistors, and of what resistance value, are needed to yield an equivalent 1000-Ω 10-W resistor? Obtain two different solutions.

Ans.: Ten 10-kΩ resistors in parallel; ten 2.5-kΩ resistors in series-parallel

1-7 Determine the current through each resistor of Fig. 1-10a. Verify that the total I^2R loss in the resistors equals the power delivered by the battery.

Ans.: 2 amp in R_1 and R_2; 1 amp in R_3 and R_4; 0.5 amp in R_5 and R_6

1-8 In Fig. 1-30, if $R_1 = 2\ \Omega$, $R_2 = 5\ \Omega$, $R_3 = 2\ \Omega$, $R_4 = 5\ \Omega$, $R_5 = 10\ \Omega$, and $V = 10$ V, find the total current supplied by the battery.

Ans.: 2.18 A

Figure 1-30

1-9 Find the resistance of the network in Fig. 1-31 between the input terminals. What voltage applied to the input results in a 1-A current in the 4-Ω resistor?

Ans.: 8 Ω; 72 V

Figure 1-31

1-10 Using Kirchhoff's rules, find the current in the 4-Ω resistor in the network in Fig. 1-32.

Ans.: 0.12 A

Figure 1-32

1-11 Find the current in each resistor of the circuit in Fig. 1-33.

Ans.: 1.15 A; 0.883 A; 0.267 A

Figure 1-33

1-12 Note that in the Wheatstone bridge (Fig. 1-13) at balance $I_5 = 0$. Using this condition, compare the voltage drops in corresponding arms of the bridge and thus derive the balance condition, Eq. (1-46).

1-13 Solve the Wheatstone bridge network of Fig. 1-14 by drawing loop currents such that there is only one current in R_5. Use the expression for this current to derive the balance condition.

1-14 In the Wheatstone bridge circuit of Fig. 1-13 take $R_1 = R_2 = 100 \ \Omega$, $R_4 = 100 \ \Omega$, $V = 12$ V, and R_3 to be a calibrated variable resistor. Plot the voltage across the detector, $R_5 = 100 \ \Omega$, as a function of R_3 for the region near balance. Repeat for $R_1 = R_2 = 1000 \ \Omega$. Which configuration is more sensitive?

Ans.: The first

1-15 Experimental values of load current and terminal voltage of the unknown network in Fig. 1-34 are listed in Table 1-2. Determine the Thévenin equivalent network from the data if the voltmeter is 1000 Ω/V used on the 1-V scale. Plot the load power IV and the measured resistance V/I as a function of V and compare the load resistance at the maximum power point with R_{eq}.

Ans.: 67 Ω, 0.5 V, 65 Ω, 55 Ω

Figure 1-34

Table 1-2

I, $-$mA	V, volts
0 ($R_L = \infty$)	0.468
1.0	0.405
2.0	0.342
3.0	0.279
4.0	0.216
5.0	0.153
6.0	0.090
6.4 ($R_L = 0$)	0.064

1-16 Design an *Ayrton shunt* as in Fig. 1-35 for a 50-μA meter which has an internal resistance of 1000 Ω, if the desired current ranges are 10 mA, 100 mA, 1 A, and 10 A.

    Ans.: 4.523 Ω; 0.4523 Ω; 4.523 $\times$ 10^{-2} Ω; 5.025 $\times$ 10^{-3} Ω

    Input   **Figure 1-35**

1-17 Suppose, in the circuit of Fig. 1-25, $R = 1000$ Ω, $R_L = 5000$ Ω, and $R_m = 1000$ Ω. If the indicated current is 1.5 mA, what is the true current when the ammeter is not present?

    Ans.: 1.75 mA

1-18 The voltage in a circuit as measured with a 20,000-Ω/V meter on the 500-V scale is 200 V. On the 100-V scale the reading is 95 V. What is the true voltage?

    Ans.: 278 V

1-19 What are the meter readings in the two versions of the voltmeter-ammeter method of Fig. 1-27 if the voltmeter is 1000 Ω/V, the internal resistance of the ammeter is 100 Ω, the "unknown" resistance is 1000 Ω, and the applied voltage is 10 V? Which of the two versions is more satisfactory?

    Ans.: 10 V, 9.1 $\times$ 10^{-3} A; 9.008 V, 9.92 $\times$ 10^{-3} A; circuit *b* is better

1-20 Draw the scale of an ohmmeter, assuming $R_1 = 10,000$ Ω in Eq. (1-69).

TWO

ALTERNATING CURRENTS

The currents and voltages in most practical electronic circuits are not steady but vary with time. For example, when a circuit is used to measure some physical quantity, such as the temperature of a chemical reaction, the voltage or current in the circuit representing the temperature may vary significantly. Similarly, detection of nuclear disintegrations results in a series of rapid voltage pulses of very short duration. In order to understand such effects it is necessary to study the properties of time-varying currents.

The simplest time-varying current alternates direction periodically and accordingly is called an alternating current, abbreviated ac. Obviously an ac circuit is one in which alternating currents are active, but direct currents may be present as well. Most of the concepts developed for dc circuits in the previous chapter carry over to ac circuits. Two new elements in addition to resistance are important in ac circuits and these are treated in this chapter.

SINUSOIDAL SIGNALS

Frequency, Amplitude, and Phase

The simplest alternating waveform is *sine-wave* voltage or current, which varies sinusoidally with time. A sinusoidal waveform is generated by the variation of the vertical component of a vector rotating counterclockwise with a uniform angular velocity ω as in Fig. 2-1. One complete revolution is termed a *cycle* and the time interval required for one cycle is called the *period T*. The number of cycles per second is the *frequency f*, which is just the reciprocal of the period. The scope of frequencies encountered in electronic circuits is often as small as a few cycles per second, which is called a *hertz* (abbreviated Hz) in honor of Heinrich Hertz, discoverer of radio waves. The range of frequency extends to *kilohertz* (kHz, 10^3 Hz) and *megahertz* (MHz, 10^6 Hz), on up to the *gigahertz* (GHz, 10^9 Hz) range.

Since there are 2π radians in one complete revolution and this requires T seconds, the angular frequency ω is $2\pi f$. If the length of the rotating vector is V_p, the instantaneous value at any time t is just $V_p \sin \omega t$, where V_p is the *maximum* or *peak amplitude* of the sine wave.

Two sinusoidal waveforms that have the same frequency but pass through zero at different times are said to be out of *phase*, and the angle between the two rotating vectors is called the *phase angle*. In Fig. 2-2 the voltage v_2 is *leading* the

Figure 2-1 Generation of sine wave by vertical component of rotating vector.

Figure 2-2 Illustrating phase angle between two sinusoidal voltages.

sine-wave voltage v_1 because it passes through zero first, and the phase difference is the angle ϕ. Note that it is only possible to specify the phase angle between two sine waves if they have the same frequency. A voltage sine wave is completely described by its frequency and amplitude unless it is compared with another signal of the same frequency. In this case the most general equation for the voltage must include the phase angle

$$v = V_p \sin (\omega t + \phi) \tag{2-1}$$

Note that lowercase symbols are used to indicate time-varying voltages (and currents), while capital letters refer to constant values or to dc quantities.

RMS Value

It is often necessary to compare the magnitude of a sine-wave current with a direct current. This is accomplished by comparing the joule heat produced in a resistor by the two currents. That is, the *effective value* of a sinusoidal current is equal to the direct current which produces the same joule heating as the alternating current. To determine this value, the heating effect of an alternating current is calculated by averaging the I^2R losses over a complete cycle. That is, the average power is given by

$$P = \frac{1}{T} \int_0^T i^2R \; dt = \frac{I_p^2R}{T} \int_0^T \sin^2 \omega t \; dt = \frac{I_p^2R}{T} \left[\frac{t}{2} - \frac{\sin 2\omega t}{4\omega} \right]_0^T = \frac{I_p^2R}{2} \tag{2-2}$$

Since the joule heating in a resistor caused by a direct current is equal to I^2R, the effective value of an alternating current I_e is simply

$$I_e^2R = \frac{I_p^2R}{2} \tag{2-3}$$

or

$$I_e = \frac{I_p}{\sqrt{2}} \tag{2-4}$$

According to Eq. (2-4) the effective value of a sine wave is simply its peak value divided by the square root of two. The effective value is commonly referred to as the *root-mean-square* or *rms* value. Voltmeters and ammeters capable of measuring ac signals are almost universally calibrated in terms of rms readings to facilitate comparison with dc meter readings. It is generally understood that ac currents and voltages are characterized by their rms values, unless stated otherwise.

Power Factor

Suppose that the current and voltage in some portion of a circuit are given by

$$i = I_p \sin \omega t$$

$$v = V_p \sin (\omega t + \phi) \tag{2-5}$$

where the phase angle ϕ is introduced to account for the possibility that the current and voltage are not in phase. The instantaneous power p is then

$$p = vi = V_p I_p \sin \omega t \sin (\omega t + \phi) \tag{2-6}$$

According to Eq. (2-6) the instantaneous power in this part of the circuit varies with time and may even become negative, as illustrated by the waveforms in Fig. 2-3. The interpretation of negative power in Eq. (2-6) is that during some portion of a cycle this part of the circuit gives up electrical power to the rest of the circuit. For the remainder of the cycle the circuit delivers power to the part under investigation.

The average power is found by averaging Eq. (2-6) over a complete cycle,

$$P = \frac{1}{T} \int_0^T vi \, dt = \frac{V_p I_p}{T} \int_0^T \sin \omega t \sin (\omega t + \phi) \, dt \tag{2-7}$$

The second factor under the integral is expanded using a standard trigonometric identity

$$P = \frac{V_p I_p}{T} \left(\cos \phi \int_0^T \sin^2 \omega t \, dt + \sin \phi \int_0^T \sin \omega t \, dt \right) \tag{2-8}$$

Both integrals are standard forms and may be evaluated directly to give

$$P = \frac{V_p I_p}{2} \cos \phi \tag{2-9}$$

$$P = VI \cos \phi \tag{2-10}$$

where V and I are rms values.

The meaning of Eq. (2-10) is that the useful power in ac circuits depends not only upon the current and voltage in the circuit but also upon the phase difference between them. The term $\cos \phi$ is called the *power factor* of the circuit. Note that when the phase angle is 90° the power factor is zero and no useful electrical power

Figure 2-3 Instantaneous power in an ac circuit.

is delivered. Therefore, it is possible for the current and voltage to be very large and, consequently, the instantaneous power large, yet the average power to be zero. On the other hand, when the current and voltage are in phase, the power factor is unity and the power is equal to the current times the voltage, as in a dc circuit.

CAPACITANCE AND INDUCTANCE

Capacitive Reactance

Consider two parallel metal plates separated by a narrow air gap and connected to the terminals of a battery. The potential difference of the battery exists between the plates and the positive electric charge on the plate connected to the positive terminal of the battery is attracted to the negative electric charge on the plate connected to the negative terminal. The capacity of the two plates to hold electric charge is proportional to the voltage, so that

$$Q = CV \tag{2-11}$$

where C is a constant of proportionality called the *capacitance*. Capacitance is a geometric factor that depends upon the size, shape, and separation between any two conductors such as the two plates.

Capacitance is important in ac circuits because a voltage which changes with time gives rise to a time-varying charge, and this is equivalent to a current. For example, differentiating both sides of Eq. (2-11) with respect to time and using the definition of current in Eq. (1-1) yields

$$i = \frac{dq}{dt} = C \frac{dv}{dt} \tag{2-12}$$

In particular, if the voltage is sinusoidal the current in the capacitance,

$$i = C \frac{d}{dt} (V_p \sin \omega t) = \omega C V_p \cos \omega t = \omega C V_p \sin \left(\omega t + \frac{\pi}{2} \right) \tag{2-13}$$

is also sinusoidal and leads the voltage with a phase angle of $\pi/2$.

In terms of rms values, Eq. (2-13) can be written

$$V = \frac{1}{\omega C} I \tag{2-14}$$

which shows that the ac current in a capacitance increases with the ac voltage applied quite analogously to the current in a resistance. Indeed, the form of Eq. (2-14) is quite similar to Ohm's law, Eq. (1-4). The proportionality factor between current and voltage, $1/\omega C$, is called the *capacitive reactance*. It is analogous to resistance in dc circuits except, of course, that it decreases with respect to frequency. Note also that capacitive reactance is measured in ohms since it is the ratio of the voltage to the current.

Capacitors

Circuit elements having specific values of capacitance are known as *capacitors*. Most of the capacitors used in electronic circuits consist of two conducting plates separated by a small air gap or thin insulator. The capacitance of such a parallel-plate capacitor is increased by making the area of the plates large and by making the separation between them small. The unit of capacitance is the *farad*, named in honor of Michael Faraday. Actually, values of practical capacitors used in electronic circuits are 10^{-6} farad (1 *microfarad*, abbreviated μF), or even 10^{-12} farad (1 *picofarad*, abbreviated pF).

It turns out that insulating material between the parallel plates increases the capacitance of a capacitor because the insulator, in effect, permits a greater charge on the plates for a given applied voltage. The increased capacitance is accounted for by the *dielectric constant* of the insulator. For example, the dielectric constant of mica is about 6 and that of paper is about 2, so that the capacitances of capacitors made from these materials are greater by factors of 6 and 2, respectively, than a parallel-plate capacitor with air between the plates.

Conventional capacitors are made of two thin metal foils separated by a thin insulator or *dielectric*, such as paper or mica. This sandwich is then rolled or folded into a compact size and covered with an insulating coating. One axial wire lead is attached to each foil. In order to increase the capacitance, it is desirable for the insulator to be as thin as possible. This can only be done at the expense of limiting the maximum voltage that can be applied before the insulator ruptures because of the intense electric field. Another important factor is the resistivity of the insulator. Thin, large-area shapes increase the *leakage* resistance between the plates and thus degrade the capacitor. Mica and paper dielectric capacitors are available in capacitances ranging from 0.001 to 1 μF and can be used in circuits where the maximum voltage is of the order of hundreds of volts.

Ceramic and plastic-film capacitors are also used, generally with metal-film plates deposited directly on the dielectric. Plastic dielectrics have very high resistivity, which means that the leakage resistance is extremely small. The large dielectric constant of many ceramic materials provides large capacitance values in a small package.

In several applications, most notably transistor circuits, very large capacitance values are desirable and leakage resistances are of secondary concern. *Electrolytic* capacitors made of an oxidized metal foil in a conducting paste or solution are used to achieve high capacitance values. The thin oxide film is the dielectric between the metal foil and the solution. Because the film is extremely thin, the capacitance is quite large. Several metals, such as tantalum and aluminum, can be used in electrolytic capacitors, and capacitance values range from 1 to 10^4 μF. The largest capacitances are useful in circuits where the applied voltage does not exceed a few volts, because the oxide dielectric is so thin. Electrolytic capacitors can only be used in circuit situations where the metal foil never becomes negative with respect to the solution. If the foil becomes negative, electrolytic action destroys the film and the capacitor becomes useless. Some typical capacitors are shown in Fig. 2-4.

Figure 2-4 Typical capacitors. (*Sprague Electric Company.*)

It is often convenient to vary the capacitance in a circuit without removing the capacitor. The components described above are *fixed* capacitors, since it is clearly not easy to vary their capacitance. The common *variable* capacitor is made of two sets of interleaved plates, one immobile and the other set attached to a shaft. Rotating the shaft effectively changes the area of the plates, thus changing the capacitance. Because the dielectric is air and it is necessary to make the separation plates relatively large to assure that they do not touch, maximum capacitance values are limited to about 500 pF. In the fully unmeshed state a 500-pF variable capacitor may have a minimum capacitance of 10 pF or so. Mica *trimmer* capacitors use a mica dielectric. The separation between plates is adjusted with a screwdriver. They are commonly used where variation in capacitance is only occasionally necessary. The range of capacitance values is about the same as for air dielectric capacitors.

Conventional symbols for capacitors in circuit diagrams are reminiscent of the parallel-plate construction, as illustrated in Fig. 2-5. The slightly curved plate

(*a*)   (*b*)

Figure 2-5 Conventional circuit symbols for (*a*) fixed and (*b*) variable capacitors.

is the negative terminal in the case of electrolytic capacitors, and the outside plate on foil capacitors. This is sometimes significant in electronic circuits in order to connect the capacitor properly, even though foil capacitors are not sensitive to voltage polarity.

Inductive Reactance

The current in any electric circuit is accompanied by a magnetic field that exists in the region surrounding the current. Changes in the magnetic field arising from a varying current induce an emf in the circuit which is given by

$$v = L \frac{di}{dt} \qquad (2\text{-}15)$$

where the factor L is called the *inductance*. Inductance is a geometric constant that depends upon the size and shape of the circuit and upon magnetic properties of nearby materials. The induced emf acts to oppose the change in current and increases with the rate of change of current, according to Eq. (2-15). This means that the circuit inductance prevents current in the circuit from changing instantaneously.

The inductance of simple circuits is small enough so that the induced emf may usually be ignored. This is true except at the very highest frequencies of interest in electronic circuits, where the rate of change of current becomes exceedingly large. Circuits used at such frequencies are kept as small as possible to minimize inductive effects. By contrast, it is possible to produce electric components that have appreciable inductance, and these prove to be very useful in ac circuits.

Consider the case of a sinusoidal current in such an inductance. The induced emf in the inductance is also sinusoidal, for, according to Eq. (2-15),

$$v = L \frac{d}{dt} (I_p \sin \omega t) = \omega L I_p \cos \omega t = \omega L I_p \sin \left(\omega t + \frac{\pi}{2} \right) \qquad (2\text{-}16)$$

and in this case the current lags the voltage by a phase angle of $\pi/2$. Expressing Eq. (2-16) in terms of rms values

$$V = \omega L I \qquad (2\text{-}17)$$

which shows that the ac current in an inductance increases with applied ac voltage just as in the case of a resistor and in the case of a capacitor. The quantity ωL is called the *inductive reactance* and is measured in ohms. Note that the magnitude of the inductive reactance increases with frequency in contrast to the case of capacitive reactance.

Inductors

Electric components with appreciable inductance are called *inductors*, or *inductances*, and in some applications, *chokes*. They consist of many turns of wire wound adjacent to one another on the same support. In this way every single coil of wire links the magnetic flux produced by the current in all other coils, and the total flux intercepted by all the coils together can be made large. The unit of inductance, is called the *henry* after Joseph Henry, an early American investigator of inductive effects.

High-frequency electronic circuits frequently employ inductances of the order of 10^{-6} henry (*microhenry*, or μH) which may be a helical coil of a few turns on, say, a 1-cm-diameter support. A few tens of turns produce inductance values in the 10^{-3} henry (*millihenry*, or mH) range. Large inductances for use at low frequencies are obtained by winding many hundreds of turns of wire on a core of a ferromagnetic material such as iron. The magnetic properties of these materials are such that the magnetic flux is increased appreciably. In this fashion inductances of several hundred henrys are attained. Several typical inductors are pictured in Fig. 2-6.

The cores of iron-core inductors are laminated in order to interrupt currents induced in the metal core by the changing magnetic flux. This reduces the I^2R

Figure 2-6 Typical inductors used in electronic circuits. (*J. W. Miller Co. and Essex International, Inc.*)

(a)   (b)   **Figure 2-7** Circuit symbols for (a) inductor and (b) iron-core inductor.

losses of these so-called *eddy currents* in the core. Individual laminations are stacked on top of each other and separated with insulating varnish to make a core of desired size. The circuit symbol for an inductor is a helical coil, as shown in Fig. 2-7. Parallel lines along the helix, also illustrated in Fig. 2-7, signify a magnetic core.

Variable inductances can be achieved by moving one portion of the windings relative to the other, but such components are not widely used and most inductors are fixed. In many applications it is necessary to take account of the resistance of the wire in the windings and the capacitance between layers of the winding in determining the effect of a choke in a given ac circuit.

The resistive component in a practical inductor may also include the power lost in the magnetic material as a result of the rapidly changing magnetic field in the core. As mentioned earlier, metallic cores must be laminated to reduce eddy-current losses and the thickness of the lamination determines the maximum useful frequency of the inductor. This technique reaches its practical limit in powdered iron cores in which the iron is present as fine particles. *Ferrite* cores, made of high-resistivity ferromagnetic materials are used at high frequencies because their high resistivity makes eddy-current losses negligible. Such materials are not as useful as iron at low frequencies because magnetic-saturation effects limit the maximum power levels of the inductor.

SIMPLE CIRCUITS

RL Filter

The series combination of a resistor and an inductor is a simple but useful ac circuit. Suppose this *RL filter* is connected to a source of sinusoidal voltage symbolized on the circuit diagram, Fig. 2-8, by a circle containing a one-cycle sine wave. The current and voltage in the circuit are determined in the following way.

Figure 2-8

According to Kirchhoff's rules the sum of the voltages around the loop must be equal to zero at every instant. This means that the source voltage must equal the induced voltage across the inductor plus the IR drop across the resistor, or

$$v_i = Ri + L\frac{di}{dt} \tag{2-18}$$

where i is the current. Assuming the current is sinusoidal,

$$i = I_p \sin \omega t \tag{2-19}$$

the derivative is

$$\frac{di}{dt} = \omega I_p \cos \omega t \tag{2-20}$$

Introducing Eqs. (2-19) and (2-20) into the circuit equation, Eq. (2-18), the voltage is given by

$$v_i = RI_p \sin \omega t + \omega L I_p \cos \omega t \tag{2-21}$$

This expression may be put into a more illustrative form by introducing the phase angle between the voltage and the current with the aid of Fig. 2-9. Expressing the coefficients of $\sin \omega t$ and $\cos \omega t$ in terms of $\cos \phi$ and $\sin \phi$, Eq. (2-21) becomes

$$v_i = I_p\sqrt{R^2 + (\omega L)^2}\,(\cos \phi \sin \omega t + \sin \phi \cos \omega t) \tag{2-22}$$

The expression in parentheses is just the trigonometric identity for $\sin (\omega t + \phi)$, so that

$$v_i = I_p\sqrt{R^2 + (\omega L)^2}\, \sin (\omega t + \phi) \tag{2-23}$$

where

$$\phi = \arctan \frac{\omega L}{R} \tag{2-24}$$

Note that, according to Eq. (2-24), the current lags the voltage, as previously observed in connection with Eq. (2-16). Note also that the inductive reactance and resistance both are important in determining the current in this ac circuit.

Suppose that the voltage drop across R is considered to be an output voltage of the circuit. Then, using Eq. (2-19),

$$v_o = RI_p \sin \omega t \tag{2-25}$$

Figure 2-9

Figure 2-10 Frequency characteristic of an *RL* low-pass filter.

The ratio of the rms output voltage to the rms input voltage is, from Eqs. (2-23) and (2-25)

$$\frac{V_o}{V_i} = \frac{RI_p}{I_p\sqrt{R^2 + (\omega L)^2}} \qquad (2\text{-}26)$$

or,

$$\frac{V_o}{V_i} = \frac{1}{\sqrt{1 + (\omega L/R)^2}} \qquad (2\text{-}27)$$

According to Eq. (2-27), at low frequencies where $\omega L/R \to 0$, the output voltage is equal to the input voltage. At high frequencies the output voltage is smaller than the input, as a plot of Eq. (2-27) in Fig. 2-10 illustrates, and the circuit is known as a *low-pass* filter. Consider the frequency f_0, where

$$\frac{2\pi f_0 L}{R} = \frac{\omega_0 L}{R} = 1 \qquad (2\text{-}28)$$

According to Eq. (2-27) this is where

$$\frac{V_o^2}{V_i^2} = \frac{1}{2} \qquad (2\text{-}29)$$

Since the output power in R is proportional to the voltage squared, f_0 is called the *half-power* frequency. In effect, input signals above this frequency are choked off by the inductance; this is the origin of the term *choke* introduced earlier.

RC **Filter**

An elementary but very useful circuit employing a capacitor and resistor connected in series, is shown in Fig. 2-11. This *RC filter* is connected to a source of sinusoidal voltage

$$v_i = V_p \sin \omega t \qquad (2\text{-}30)$$

Figure 2-11 RC filter.

voltage drops around the loop give

$$v_i = \frac{Q}{C} + Ri \qquad (2\text{-}31)$$

where i is the current. Differentiating each term with respect to time and putting $i = dQ/dt$, Eq. (2-31) becomes, after rearranging,

$$R\frac{di}{dt} + \frac{1}{C}i = \omega V_p \cos \omega t \qquad (2\text{-}32)$$

To solve this circuit differential equation assume that the current is given by

$$i = I_p \sin (\omega t + \phi) \qquad (2\text{-}33)$$

where I_p and ϕ are to be determined. Differentiating Eq. (2-33)

$$\frac{di}{dt} = \omega I_p \cos (\omega t + \phi) \qquad (2\text{-}34)$$

Now Eqs. (2-33) and (2-34) are substituted into the circuit differential equation, Eq. (2-32). The result is

$$R\omega I_p \cos (\omega t + \phi) + \frac{I_p}{C} \sin (\omega t + \phi) = V_p \omega \cos \omega t \qquad (2\text{-}35)$$

and the equation is solved by choosing values for I_p and ϕ that make Eq. (2-35) true for all values of t. This substitution has changed the differential equation into a trigonometric equation.

The values for I_p and ϕ which satisfy Eq. (2-35) are found by first expanding each term using a trigonometric identity. This gives, after dividing through by ωI_p,

$$R(\cos \omega t \cos \phi - \sin \omega t \sin \phi)$$
$$+ \frac{1}{\omega C}(\sin \omega t \cos \phi + \cos \omega t \sin \phi) = \frac{V_p}{I_p} \cos \omega t \quad (2\text{-}36)$$

Collecting terms in $\sin \omega t$ and $\cos \omega t$,

$$\left(R \cos \phi + \frac{1}{\omega C} \sin \phi - \frac{V_p}{I_p}\right) \cos \omega t + \left(\frac{1}{\omega C} \cos \phi - R \sin \phi\right) \sin \omega t = 0$$
$$(2\text{-}37)$$

Now consider that $t = 0$; then $\sin \omega t = 0$ and

$$R \cos \phi + \frac{1}{\omega C} \sin \phi - \frac{V_p}{I_p} = 0 \qquad (2\text{-}38)$$

Similarly, suppose $\omega t = \pi/2$, so that $\cos \omega t = 0$. Then

$$\frac{1}{\omega C} \cos \phi - R \sin \phi = 0 \qquad (2\text{-}39)$$

In order that Eq. (2-37) be satisfied for all values of t, both Eqs. (2-38) and (2-39) must be true. Equation (2-39) may be solved immediately for ϕ,

$$\frac{1}{\omega C} \cos \phi = R \sin \phi$$

$$\tan \phi = \frac{\sin \phi}{\cos \phi} = \frac{1}{R\omega C}$$

so that

$$\phi = \arctan \frac{1}{R\omega C} \qquad (2\text{-}40)$$

The solution for ϕ is used in Eq. (2-38) to solve for I_p. This is done exactly as in the case of the RL filter, Eq. (2-21). Equation (2-38) becomes

$$R \frac{R}{\sqrt{R^2 + (1/\omega C)^2}} + \frac{1}{\omega C} \frac{1/\omega C}{\sqrt{R^2 + (1/\omega C)^2}} - \frac{V_p}{I_p} = 0 \qquad (2\text{-}41)$$

So finally

$$I_p = \frac{V_p}{\sqrt{R^2 + (1/\omega C)^2}} \qquad (2\text{-}42)$$

Thus the current in the circuit is

$$i = \frac{V_p}{\sqrt{R^2 + (1/\omega C)^2}} \sin (\omega t + \phi) \qquad (2\text{-}43)$$

where

$$\phi = \arctan \frac{1}{\omega RC}$$

Note that according to Eq. (2-38), the phase angle is positive. This means that the current leads the voltage, which is characteristic of a capacitive circuit. When $1/\omega RC \to 0$ at high frequencies, the phase angle is zero and the current is in phase with the voltage. At very low frequencies, the phase angle approaches $\pi/2$.

The output voltage across the resistance is

$$v_o = RI_p \sin (\omega t + \phi) \qquad (2\text{-}44)$$

Figure 2-12 Frequency characteristic of an RC high-pass filter.

which means that the ratio of the rms output voltage to the input voltage, found from Eqs. (2-30), (2-42), and (2-44), is then

$$\frac{V_o}{V_i} = \frac{1}{\sqrt{1 + (1/\omega RC)^2}} \tag{2-45}$$

A plot of Eq. (2-45), Fig. 2-12, shows that the output voltage V_o is very small at low frequencies and is equal to the input voltage at high frequencies. Since low frequencies are attenuated while high frequencies are not, this circuit is called an *RC high-pass filter*. The frequency f_0 where

$$2\pi f_0 RC = 1 \tag{2-46}$$

is called the *half-power* frequency, as in the *RL* filter.

Differentiating and Integrating Circuits

Suppose that the series resistance and capacitance in the simple RC filter are small enough to make $\omega RC \ll 1$ over a given frequency range. Under this condition the output voltage is, from Eqs. (2-44) and (2-45),

$$v_o = V_p \omega RC \sin (\omega t + \pi/2)$$

$$= V_p \omega RC \cos \omega t \tag{2-47}$$

But, note that the time derivative of the input signal is

$$\frac{dv_i}{dt} = \omega V_p \cos \omega t \tag{2-48}$$

Therefore, combining Eqs. (2-47) and (2-48),

$$v_o = RC \frac{dv_i}{dt} \tag{2-49}$$

The interpretation of Eq. (2-49) is that when $\omega RC \ll 1$ the RC filter circuit performs the operation of differentiation. That is, the output voltage signal is the time derivative of the input voltage. This useful property is applied extensively in electronic circuits, most notably in electronic computers.

Figure 2-13 *RC* low-pass filter.

Correspondingly, the voltage across the capacitor is, by Eq. (2-12), the integral of the current,

$$v_c = \frac{1}{C} \int I_p \sin\left(\omega t + \phi\right) dt = -\frac{I_p}{\omega C} \cos\left(\omega t + \phi\right) \tag{2-50}$$

Suppose now $\omega RC \gg 1$, which may be accomplished by making R and C very large. Then, $I_p = V_p/R$, $\phi = 0$, and Eq. (2-50) becomes

$$v_c = -\frac{V_p}{RC\omega} \cos \omega t \tag{2-51}$$

Introducing the integral of the input voltage results in

$$v_c = \frac{1}{RC} \int v_i \, dt \tag{2-52}$$

which means that under these conditions the RC circuit performs the operation of integration. The output voltage is only a small fraction of the input voltage when the circuit is used to differentiate or integrate, according to Eqs. (2-49) and (2-52). This is not a serious disadvantage in practical applications, however, since the output signal may be subsequently increased in magnitude using amplifier circuits described in later chapters.

It is permissible to ignore the limits of integration in deriving Eq. (2-52), because we are only concerned with steady-state conditions after any initial transient voltages, which may accompany turning on the input voltage, have died away. The possibility of transient effects in RC circuits is considered in the next section. The RC circuit is a *low-pass* filter when the output voltage signal is taken from the capacitor, Fig. 2-13. The transition frequency from one domain to the other is marked approximately by the half-power frequency, where $\omega_0 RC = 1$, as in the case of the high-pass filter.

TRANSIENT CURRENTS

Time Constant

The study of ac circuits to this point has implicitly assumed that the rms values of the sinusoidal emfs are constant. Transient effects are associated with sudden changes in the voltages applied to ac networks, as, for example, when an emf is first applied. These transient currents dissipate rapidly, leaving steady-state currents in the network which persist as long as the ac emf is applied. Although most often the steady-state currents are of primary concern, in many cases transient effects are important as well. This is particularly so in determining the response of a network to isolated voltage pulses.

Transient currents in ac networks are found by solving the circuit differential equation, taking into account the magnitude of the current at the time the applied voltage is changed. Actually, the sum of the transient current and the steady-state current is the complete solution of the circuit differential equation. It is generally satisfactory to consider the two aspects of ac-circuit analysis separately because the response of circuits is most often of interest in connection either with steady emfs alone or with transient effects alone.

It is convenient to investigate first the transient currents in a simple RC series circuit incorporating a battery and a two-position switch, as shown in Fig. 2-14. Suppose the switch is suddenly connected to terminal 1. The circuit differential equation is, from Kirchhoff's rule,

$$V = Ri + \frac{q}{C} \tag{2-53}$$

Differentiating with respect to t and rearranging,

$$\frac{di}{dt} + \frac{1}{RC}i = 0 \tag{2-54}$$

To solve this circuit differential equation, rewrite it as

$$\frac{di}{i} = -\frac{1}{RC}dt \tag{2-55}$$

Both the left side and the right side of Eq. (2-55) are standard integrals, so that upon integrating,

$$\ln i = -\frac{1}{RC}t + K \tag{2-56}$$

where K is a constant to be evaluated from the initial conditions. Another way of writing Eq. (2-56) is

$$i = \exp\left(-\frac{t}{RC} + K\right) = Ae^{-t/RC} \tag{2-57}$$

Figure 2-14

where A replaces exp K for simplicity. In order to evaluate A, note that the current is equal to V/R when the switch is closed at $t = 0$. Therefore,

$$i = \frac{V}{R} e^{-t/RC} \qquad (2\text{-}58)$$

According to Eq. (2-58), the current in this simple RC circuit decays exponentially when the switch is put in position 1. The quantity RC, which has the dimension of seconds, is called the *time constant*. It determines how rapidly the current decreases.

The voltage across the capacitor at any time is found by integrating the current, according to Eqs. (1-1) and (2-11),

$$v_c = \frac{1}{C} \int_0^t i \, dt = \frac{V}{RC} \int_0^t e^{-t/RC} \, dt$$
$$= V(1 - e^{-t/RC}) \qquad (2\text{-}59)$$

The decay of charging current and the growth of voltage across the capacitor are symmetrical according to Eqs. (2-58) and (2-59), as shown in Fig. 2-15. Note that the voltage starts from zero and increases exponentially to the battery voltage. After a time equal to one time constant the voltage is equal to $1 - e^{-1} = 1 - 0.356 = 63$ percent of the final value. The actual time at which this voltage is attained depends upon the magnitude of the resistance and the capacitance. A long time is required for large values of R and C, and vice versa.

The decay of current upon closing the switch to position 2 as the charge on the capacitor discharges through the resistor is the same as that given by Eq. (2-58). This can be seen by writing the circuit voltage equation for this case

$$\frac{q}{C} + Ri = 0 \qquad (2\text{-}60)$$

Figure 2-15 Charging current and capacitor voltage in an RC circuit.

i, units of V/R; v_c, units of V

t, units of time constants

Differentiating and rearranging,

$$\frac{di}{dt} + \frac{1}{RC} i = 0 \tag{2-61}$$

which is identical to Eq. (2-54). Furthermore, the potential across the capacitor is equal to V when the capacitor is fully charged and this means that the initial current is V/R, just as in the charging current case. The voltage across the capacitor decays exponentially to zero with a time constant of RC, as may be shown by calculating the charge remaining on the capacitor at any time using a procedure similar to that leading to Eq. (2-59).

The simple inductive circuit is analogous to the RC case, as is illustrated by considering the circuit in Fig. 2-16. The circuit voltage equation with the switch in position 1 is

$$V = Ri + L\frac{di}{dt} \tag{2-62}$$

which upon rearranging becomes

$$\frac{di}{dt} + \frac{R}{L} i - \frac{V}{L} = 0 \tag{2-63}$$

This equation is solved exactly like Eq. (2-61) with the result that the current is given by

$$i = \frac{V}{R}(1 - e^{-tL/R}) \tag{2-64}$$

Evidently, the current in an inductive circuit grows exponentially with a time constant L/R. That is, small R and large values of inductance result in long transient currents. Note that, according to Eq. (2-64), the steady-state value is just the dc current V/R.

The decay of current when the switch is put in position 2 is also exponential since the circuit differential equation in this case is simply

$$L\frac{di}{dt} + R_i = 0 \tag{2-65}$$

By analogy with Eq. (2-54), the current is given by

$$i = \frac{V}{R}e^{-tL/R} \tag{2-66}$$

The growth and decay of current in an inductive circuit are symmetrical, according to Eqs. (2-64) and (2-66), exactly as in the corresponding case of the RC circuit.

Figure 2-16

AC Transients

Transient currents also accompany the turning on of ac voltage sources. For example, consider the circuit in Fig. 2-17, for which the circuit differential equation is

$$R\frac{di}{dt} + \frac{1}{C}i = \frac{dv}{dt} \tag{2-67}$$

It is convenient to express the applied voltage as

$$v = V_p \sin(\omega t + \theta) \tag{2-68}$$

where the phase angle θ is introduced in order to choose the instantaneous input voltage amplitude at $t = 0$. Since the full solution of Eq. (2-67) involves a transient current and a steady-state current of frequency $\dot{\omega}$, the total current is

$$i = i_1 + i_2 \tag{2-69}$$

where i_1 is the transient current and i_2 is the steady-state current.

Substituting Eq. (2-69) into the differential equation (2-67) and rearranging, the result is

$$\left(R\frac{di_1}{dt} + \frac{1}{C}i_1\right) + \left(R\frac{di_2}{dt} + \frac{1}{C}i_2 - \frac{dv}{dt}\right) = 0 \tag{2-70}$$

This equation is solved if both parentheses are equal to zero. The first one is just Eq. (2-54), for which the solution previously found is Eq. (2-58). Similarly, the second parentheses is just Eq. (2-32), and the solution is Eq. (2-43). Using these expressions, the total current is written

$$i = Ae^{-t/\tau} + \frac{V_p}{\sqrt{R^2 + (1/\omega C)^2}} \sin(\omega C + \theta + \phi) \tag{2-71}$$

where $\tau = RC$

$\phi = \arctan 1/R\omega C$

and the constant A is determined from the voltage of the source at the instant the switch is closed. Note that Eq. (2-71) consists of a transient part together with a steady-state term, as discussed above.

Figure 2-17

Suppose the applied voltage is at its peak, that is, $v = V_p$, when the switch is closed at $t = 0$. This means we choose $\theta = \pi/2$, according to Eq. (2-68), and the initial current is V_p/R. Thus, Eq. (2-71) for $t = 0$ becomes

$$\frac{V_p}{R} = A + \frac{V_p}{\sqrt{R^2 + (1/\omega C)^2}} \sin\left(\frac{\pi}{2} + \phi\right) \tag{2-72}$$

This is solved for the constant A, which is then substituted into Eq. (2-71). The result is the final solution

$$i = \frac{V_p}{R}\frac{1}{1 + (\omega RC)^2}e^{-t/\tau} + \frac{V_p}{\sqrt{R^2 + (1/\omega C)^2}} \sin\left(\omega t + \frac{\pi}{2} + \phi\right) \tag{2-73}$$

In order for the transient term of Eq. (2-73) to have an appreciable magnitude compared with the steady-state term, the denominator of its coefficient must not be too large. This means that $(\omega RC)^2 < 1$, or

$$\tau < \frac{1}{\omega} \tag{2-74}$$

According to this expression, the time constant of the transient current is smaller than one period of the sine-wave current. Therefore, any transient current is negligible after only a fraction of a cycle of the ac wave, and the transient may therefore be ignored. If, on the other hand, the time constant is much greater than one period, the transient persists over many cycles, but, according to Eq. (2-73), the amplitude is small. This reasoning demonstrates why the steady-state solution alone is sufficiently accurate for most purposes.

Ringing

Sudden changes of voltage in a *RLC* circuit may produce *ringing*, in which a decaying ac current is generated at a frequency characteristic of the circuit, called the *resonant* frequency. The amplitude of the ac current dies away exponentially as in nonresonant circuits. For example, after the switch is closed in the circuit of Fig. 2-18, an ac current may be generated which decreases relatively slowly. The phenomenon is analogous to striking a bell a sharp mechanical blow with a hammer, whence the origin of the term ringing.

Figure 2-18

Resonant circuits are examined in greater detail in the next chapter. To examine ringing in the circuit of Fig. 2-18 start by applying Kirchhoff's rules,

$$V = Ri + \frac{q}{C} + L\frac{di}{dt} \tag{2-75}$$

which becomes, after differentiating,

$$\frac{d^2i}{dt^2} + \frac{R}{L}\frac{di}{dt} + \frac{1}{LC}i = 0 \tag{2-76}$$

To solve this circuit differential equation it is assumed, on the basis of previous analyses, that an ac current is present which has an amplitude that decays exponentially with time. Therefore, the current is written as

$$i = Ae^{-at} \sin(\omega t + \phi) \tag{2-77}$$

where A, a, ω, and ϕ are constants. Equation (2-77) is differentiated once, twice, and all three expressions introduced into the circuit differential equation. After some algebraic manipulation, (2-76) reduces to

$$\left[\left(a^2 - \omega^2 - \frac{Ra}{L} + \frac{1}{LC}\right) \sin(\omega t + \phi) + \left(\frac{R\omega}{L} - 2a\omega\right) \cos(\omega t + \phi)\right] Ae^{-at} = 0 \tag{2-78}$$

In order for (2-77) to be a solution of (2-76) the coefficients of both terms in (2-78) must separately be equal to zero. From the coefficient of the cosine term,

$$a = \frac{R}{2L} \tag{2-79}$$

while the coefficient of the sine term gives an expression for the frequency

$$\omega^2 = \frac{1}{LC} - \frac{R^2}{4L^2} \tag{2-80}$$

Therefore, the current in this circuit is sinusoidal at a frequency given by (2-80) and with an amplitude that decreases with a time constant determined from (2-79). A sketch of the current behavior is illustrated in Fig. 2-19.

It is conceivable that values of the circuit elements are such that

$$\frac{R^2}{4L^2} > \frac{1}{LC} \tag{2-81}$$

in which case the ac frequency calculated from (2-80) is imaginary. The interpretation of this situation may be developed from the mathematical expression for the sine of an imaginary angle,

$$\sin j\omega t = \frac{j}{2}(e^{\omega t} - e^{-\omega t}) \tag{2-82}$$

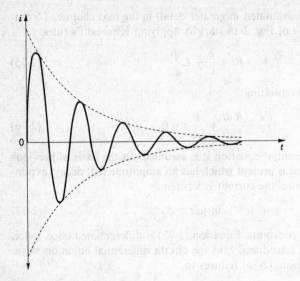

Figure 2-19 Transient current in an *RLC* circuit can be decaying oscillation.

Therefore, Eq. (2-77) reduces to a simple exponential decay with a time constant determined by the *R, C,* and *L* values in the circuit. In effect, according to (2-81), when the circuit resistance is very large, the ac current is damped out by the joule losses in the resistance, and the circuit is said to be *overdamped*. The point of *critical damping* is when the frequency calculated from (2-80) vanishes and the exponential decay is governed solely by (2-79).

COMPLEX WAVEFORMS

Fourier Series

Most steady ac signals of practical interest are not simple sine waves, but often have more complex waveforms. Nevertheless, the analysis based on sine waves can apply because complex waveforms may be represented by summing sine waves of various amplitudes and frequencies. Consider, for example, the sum of the two sine waves shown in Fig. 2-20, one of which is twice the frequency of the other. The resulting waveform is more complex than either constituent. If this periodic voltage is applied to a circuit, the resulting current may be determined by calculating the current caused by each sinusoidal voltage component separately. In this way the effect of any complex waveshape can be investigated.

Any single-valued periodic waveform may be represented by a *Fourier series* which is a summation of a *fundamental* (or lowest) frequency together with its *harmonics* (integral multiples of the fundamental frequency). The mathematical proof of this representation and the procedure by which the proper series is constructed to represent any given waveshape are not of direct concern here. It is interesting, however, to note the harmonic composition of several waves of practical importance to demonstrate the main features of the series representation.

Figure 2-20 Sum of two sine waves is complex waveform.

The Fourier series for a square wave of peak voltage V_p and frequency ω is given by

$$v = \frac{4V_p}{\pi}\left(\sin \omega t + \frac{1}{3}\sin 3\omega t + \frac{1}{5}\sin 5\omega t + \frac{1}{7}\sin 7\omega t + \cdots\right) \qquad (2\text{-}83)$$

The faithfulness of the series representation improves as the number of terms included in Eq. (2-83) is increased, as illustrated graphically in Fig. 2-21. This means that the higher frequencies are necessary in order to reproduce the sharp corners of the square wave. A second example is the sawtooth wave (Fig. 2-22) which is represented by

$$v = \frac{2V_p}{\pi}\left(\sin \omega t - \frac{1}{2}\sin 2\omega t + \frac{1}{3}\sin 3\omega t - \frac{1}{4}\sin 4\omega t + \cdots\right) \qquad (2\text{-}84)$$

Here, again, the high-frequency terms must be included in order to obtain the sharp corner of the sawtooth waveform.

According to Eqs. (2-83) and (2-84), complex waveforms, such as the square wave and the sawtooth wave, are composed of a wide spectrum of frequencies. If the waveform of a given input signal applied to any network is to be preserved at the output, the attenuation of the network must be independent of frequency. For example, a square wave applied to an RC high-pass filter can result in a square-wave output signal only if the fundamental frequency of the square wave is greater than the cutoff frequency of the filter. It is equally important that the relative phases of the harmonics remain constant in traversing the network if the waveform is to be preserved.

On the other hand, the harmonic composition of a wave is often purposefully altered by a network in order to change one waveform into another. Suppose a square wave is applied to an RC differentiating circuit. The output waveform is a series of alternating positive and negative sharp pulses, as sketched in Fig. 2-23b.

$\frac{4}{\pi} (\sin \omega t +$

$+ \frac{1}{3} \sin 3\omega t +$

$+ \frac{1}{5} \sin 5\omega t +$

$+ \frac{1}{7} \sin 7\omega t + \dots)$

Figure 2-21 Many frequency components are present in a square wave. Faithfulness of representation improves as number of harmonics is increased.

In order for the circuit to differentiate a square wave adequately, it is necessary for the highest frequency term in the Fourier series representation to satisfy the approximation $\omega RC \ll 1$ leading to Eq. (2-49). This is usually easily accomplished in practice, and the simple differentiating circuit is widely used to generate pulses from a square wave.

A second example is the integration of a square wave using a simple RC circuit. The result is a triangular wave, Fig. 2-23c. Here the fundamental frequency of the square wave must be large enough that the approximation $\omega RC \gg 1$ leading to Eq. (2-52) is satisfied.

Figure 2-22 Sawtooth wave.

(a)

(b)

(c)

Figure 2-23 (a) Square wave, (b) differentiated square wave, (c) integrated square wave.

Effective Value

The effective value of a complex waveform is defined in exactly the same way as in the case of the sinusoidal wave discussed earlier. Consider, for example, the square wave of period T and maximum value I_p illustrated in Fig. 2-24. The effective current is just

$$I_e^2 = \frac{1}{T} \int_0^T I_p^2 \, dt = \frac{I_p^2}{T} \int_0^T dt = I_p^2 \qquad (2\text{-}85)$$

This means that the effective value of a square wave is equal to the maximum value. Since the joule heat in a resistor depends upon the square of the current, the current direction is immaterial and Eq. (2-85) is just what is expected.

Figure 2-24 Square wave.

Figure 2-25 Sawtooth wave.

Consider next the simple triangular sawtooth waveform illustrated in Fig. 2-25. The current varies as

$$i = \frac{2I_p}{T} t \tag{2-86}$$

over the interval from $-T/2$ to $T/2$. Therefore, the rms value is

$$I_e^2 = \frac{1}{T} \int_{-T/2}^{T/2} \left(\frac{2T_p}{T} t\right)^2 dt = \frac{4I_p^2}{T^3} \int_{-T/2}^{T/2} t^2 \, dt = \frac{I_p^2}{3}$$

$$I_e = \frac{I_p}{\sqrt{3}} \tag{2-87}$$

Another waveform of interest is a series of half-sine waves, Fig. 2-26. The effective value can be found by the procedure used above. This waveform has, however, a dc component, as can be shown by computing the average value,

$$I_{dc} = \frac{2}{T} \int_0^{T/2} I_p \sin \omega t \, dt = \frac{4I_p}{\omega T} = \frac{4I_p}{2\pi}$$

$$I_{dc} = \frac{2}{\pi} I_p \tag{2-88}$$

This dc component contributes to the heating effect of the complete waveform.

For most purposes the rms value of the ac components alone is of interest. It is convenient to consider the Fourier series of complex waveforms. Since the heating effect of each frequency component is independent of the others, the effective value of any complex waveform is given by

$$I_e^2 = I_{dc}^2 + I_1^2 + I_2^2 + I_3^2 + \cdots \tag{2-89}$$

where I_{dc} is the average value and I_1, I_2, I_3, ... are the rms values of each frequency component. It is often sufficient to consider only the fundamental frequency, since it predominates.

The Fourier series of the waveform in Fig. 2-26 is

$$i = \frac{2}{\pi} I_p - \frac{4I_p}{3\pi} \cos 2\omega t - \frac{4I_p}{15\pi} \cos 4\omega t + \cdots \tag{2-90}$$

Figure 2-26 Waveform of half-sinusoids.

Note that the first term corresponds to Eq. (2-89), in conformity with the definition of the dc component. The effective value of the ac components alone is, using Eqs. (2-4) and (2-89),

$$I_{rms}^2 = \frac{I_p^2}{2}\left[\left(\frac{4}{3\pi}\right)^2 + \left(\frac{4}{15\pi}\right)^2 + \cdots\right] \cong \frac{1}{2}\left(\frac{4I_p}{3\pi}\right)^2 \tag{2-91}$$

$$I_{rms} = \frac{2\sqrt{2}\,I_p}{3\pi} \tag{2-92}$$

The preceding results are used in Chap. 4 in connection with further applications of inductance and capacitance filters.

Square-wave Response

It is useful to investigate the response of circuits to square-wave voltages by the transient methods of the previous section. This is possible since a square wave can be considered to be a dc voltage which is switched on and off at the frequency of the wave. Transient-current analysis complements the Fourier-series method and is often simpler. The transient technique yields the output waveform directly and makes it unnecessary to determine the response of the circuit to each of the harmonics.

Consider first the high-pass RC filter of Fig. 2-27 to which is applied the square wave of period T as indicated on the diagram. According to Eq. (2-58) the transient current accompanying each half-cycle of the square wave has the form

$$i = Ae^{-t/\tau} \tag{2-93}$$

where the constant A depends upon the current in the circuit resulting from the previous transient. For example, if $\tau = T/2$, the current has decayed 63 percent of the way at the time the square-wave voltage reverses. This means that the voltage across the capacitor causes a transient current in the reverse direction, and so forth. The output voltage $(= Ri)$ is a succession of exponential transients, as shown in Fig. 2-28. These waveforms correspond to different values of the RC time constant compared with the period of the square wave.

Note that when the circuit time constant is very small compared with the period of the square wave, the output waveform approaches the differential of the

Figure 2-27 Square-wave input signal applied to high-pass filter.

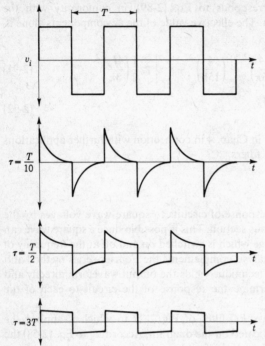

Figure 2-28 Square-wave (top) input to a high-pass filter produces different output waveforms depending upon ratio of period to circuit time constant.

square wave. This is an alternative approach to the study of differentiating circuits discussed in the previous section.

The low-pass filter circuit, Fig. 2-29, may be analyzed in the same fashion. The voltage across the capacitor is an exponential, as discussed earlier, and the output waveform, Fig. 2-30, again depends upon the relative magnitude of the circuit time constant compared with the period. In particular, note that when the circuit time constant is large the capacitor voltage rise is essentially linear during the pulse time. The output voltage is thus just the integral of the input signal, again in conformity with harmonic-circuit analysis.

Figure 2-29 Square-wave input signal applied to a low-pass filter.

Figure 2-30 Square-wave input to low-pass filter produces different output waveforms depending upon ratio of period to circuit time constant.

Oscilloscope

Undoubtedly the most useful instrument for studying complex waveforms is the *cathode-ray oscilloscope*, which is capable of displaying voltage waveforms visually. The heart of the oscilloscope is the cathode-ray tube, in which the position of a thin beam of electrons is electrically controlled to "paint" the waveform on a fluorescent screen. In the cathode-ray tube (CRT) an *electron gun* directs a high-velocity focused beam of electrons onto a glass face-plate covered with fluorescent material that emits light when bombarded by electrons. The beam passes between a pair of horizontal deflecting plates and a pair of vertical deflecting plates on its way to the screen, as sketched in Fig. 2-31. The beam produces a spot of light in the center of the screen when the voltage on the deflection plates is equal to zero. The position of the spot on the screen is easily changed by deflecting the electron beam with voltages applied to either pair of deflecting plates. The beam is deflected as a result of the sidewise force on the electrons caused by the electric field between each pair of deflecting plates. Practical oscilloscopes include a number of auxiliary circuits to increase the versatility of the instrument, as discussed in a subsequent chapter.

The way in which the oscilloscope displays a waveshape can be understood by referring to Fig. 2-32. Suppose a sine-wave voltage of sufficient amplitude to cause

Figure 2-31 Sketch of cathode-ray tube.

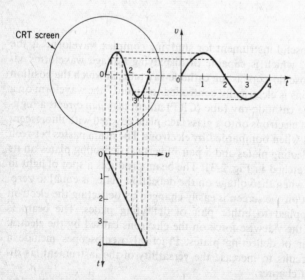

Figure 2-32 Oscilloscope having a sawtooth horizontal-deflection voltage displays waveform of voltage applied to vertical plates.

a peak-to-peak deflection of about one-half the screen diameter is applied to the vertical pair of deflection plates. A sawtooth voltage waveform of the same frequency is applied to the horizontal deflecting plates with sufficient amplitude to cause the beam to sweep one-half the screen diameter in this direction. If both waveforms start from zero at the same instant, the spot begins in the center of the screen. The sawtooth wave on the horizontal plates causes the spot to move uniformly to the right while vertical motion depends upon the sine-wave voltage. The result is that the spot traces out a sine-wave pattern on the screen, as the point-by-point plot in Fig. 2-32 illustrates.

When the sawtooth voltage on the horizontal plates drops to zero at the end of the cycle, the spot rapidly returns to the starting point. The same pattern is traced out on subsequent cycles. At most frequencies of interest the spot moves too rapidly to be seen and the pattern on the screen appears to be a stationary sine wave because of the persistence of vision. The sawtooth voltage and the sine wave must start from the beginning at exactly the same instant on each cycle, for otherwise the spot does not trace out the same pattern every time. Therefore, the sawtooth *sweep* voltage is *synchronized* with the signal on the vertical plates by auxiliary circuits associated with the oscilloscope. It is also common practice to turn off the beam during the fast return of the spot to the beginning, in order to eliminate the *retrace* line from the pattern. The sweep voltage may conveniently be made a submultiple of the signal frequency in order to display more than one cycle of the waveform.

The oscilloscope is also useful in measuring the phase angle between sine waves. This is done by applying one sine wave to the vertical plates and the other to the horizontal plates. The pattern which results can be a straight line, a circle, or an ellipse, depending upon the phase angle, as illustrated in Fig. 2-33 for the corresponding phase angles of 0, 45, and 90°. The manner in which the phase angle is calculated is as follows. Suppose the horizontal voltage is

$$v_H = V_p \sin \omega t \qquad (2\text{-}94)$$

and the vertical sine wave is given by

$$v_V = b \sin (\omega t + \phi) \qquad (2\text{-}95)$$

When $t = 0$, $v_H = 0$, which means the horizontal deflection is zero. The vertical deflection at this point may be labeled a,

$$v_V = b \sin \phi = a \qquad (2\text{-}96)$$

Solving for ϕ,

$$\phi = \arcsin \frac{a}{b} \qquad (2\text{-}97)$$

The ratio a/b can be determined directly from the dimensions of the pattern on the screen, as shown in Fig. 2-34. Note that the pattern must be centered on the horizontal and vertical reference lines in order for Eq. (2-97) to be valid.

Figure 2-33 Lissajous figures for phase differences of 0°, 45°, and 90° between horizontal- and vertical-deflection voltages.

The patterns in Fig. 2-33 are specific examples of so-called *Lissajous figures* which are used to determine the ratio between the frequencies of two sine waves. One sine wave is applied to the vertical deflection plates and the other is applied to the horizontal deflection plates. If the ratio of their frequencies is some integral fraction, such as $\frac{1}{2}$, $\frac{1}{4}$, and $\frac{2}{3}$, the pattern is stationary. The frequency ratio is determined by the number of loops of the pattern touching a vertical line at the edge of the pattern compared with the number of loops touching a horizontal line at the edge of the pattern. The reason for this is that an integral number of sine waves on the horizontal deflection plates are completed in the same time that an integral number of sine waves are completed on the vertical plates. Some examples of common Lissajous figures are given in Fig. 2-35 for different frequency ratios. These patterns are derived in the same fashion used in connection with Figs. 2-32 and 2-33.

Figure 2-34 Phase angle between sine-wave voltages applied to horizontal and vertical plates is determined from ratio a/b.

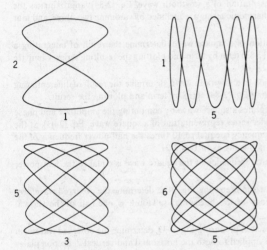

Figure 2-35 Lissajous figures for frequency ratios of 2 : 1, 1 : 5, 5 : 3, and 6 : 5.

Several different patterns can represent the same frequency ratio, depending upon the interval between the times the two sine waves pass through zero. Figure 2-33 is an example of this for a 1 : 1 frequency ratio. Although in principle Lissajous figures can be used to determine the frequency ratio between nonsinusoidal waveforms, in practice the resulting patterns are confusing and difficult to interpret.

SUGGESTIONS FOR FURTHER READING

R. M. Kirchner and G. F. Corcoran: "Alternating Current Circuits," 4th ed., John Wiley & Sons, Inc., New York, 1960.

F. H. Mitchell and F. H. Mitchell, Jr.: "Essentials of Electronics," Addison-Wesley Publishing Company, Inc., Reading, Mass., 1969.

Leigh Page and Norman Isley Adams: "Principles of Electricity," D. Van Nostrand Company, Inc., Princeton, N.J., 1931.

EXERCISES

2-1 Plot the frequency-response characteristic of an RC high-pass filter using Eq. (2-45) for $R = 10^6 \ \Omega$ and $C = 0.01 \ \mu\text{F}$. Cover the frequency interval from two decades below the half-power frequency to two decades above it.

2-2 Develop an expression analogous to Eq. (2-45) for an RC low-pass filter. Plot the frequency-response characteristic for $R = 10^6 \ \Omega$ and $C = 100$ pF.

2-3 Plot the frequency-response characteristic of an RL low-pass filter, Eq. (2-27), if $L = 10$ H and $R = 100 \ \Omega$. Design an RC circuit having the same frequency characteristic.

2-4 Show how the Fourier series representation of a sawtooth wave, Eq. (2-84), approximates the sawtooth more accurately as additional harmonic terms are included by summing two, three, and four terms, as in Fig. 2-21.

2-5 Using the Fourier series representation of a square wave, determine the result of integrating a square wave. Do this by integrating Eq. (2-83) term by term and plotting the resulting series. Compare the waveform with the sketch in Fig. 2-23.

2-6 Using the Fourier series representation of a sawtooth wave, determine the result of differentiating a sawtooth wave. Do this by differentiating Eq. (2-84) term by term and plotting the result.

2-7 Show how an RC low pass filter integrates a square wave by computing the amplitude and phase angle of the first three terms in the Fourier series representation of a square wave, Eq. (2-83), at the output of the filter if the fundamental frequency is equal to 10 times the half-power frequency of the filter. Determine the output waveform graphically as in Fig. 2-21.

2-8 Repeat Exercise 2-7 if the fundamental frequency of the square wave is equal to the half-power frequency of the filter.

2-9 By a procedure analogous to that used in preparing Fig. 2-32, determine the observed waveform on an oscilloscope screen if the frequency of the horizontal sawtooth is one-half of the vertical sine-wave frequency.

2-10 By a procedure analogous to that used in preparing Fig. 2-33, determine the observed waveform on an oscilloscope screen if a sine wave is applied to both the horizontal and vertical deflection plates and the phase angle between them is $30°$. Use the resulting pattern to confirm Eq. (2-97).

CHAPTER THREE

AC-CIRCUIT ANALYSIS

The principles of ac circuits treated in Chap. 2 can be used to find the currents in any network. However, solving the differential equation pertaining to each network of most practical circuits is cumbersome. More powerful techniques of ac-circuit analysis permit solutions for the circuit currents with much less labor. These techniques are, of course, based on the same differential equation of the circuit. The procedures are only slightly more complicated than for dc-circuit analysis, and both Ohm's law and Kirchhoff's rules are used in modified form. In fact, all the techniques of dc-network analysis treated in Chap. 1 are applicable.

IMPEDANCE

Ohm's Law for ac

The steady-state solutions to ac circuit differential equations developed in the previous chapter invariably contain several terms in sin ωt and cos ωt. These expressions can usually be simplified by introducing appropriate trigonometric relations, but the procedure proves to be laborious for any but the simplest of circuits. It turns out that ac-circuit analysis is greatly facilitated if the sinusoidal currents and voltages are represented by means of complex numbers such that

$$i = \mathbf{I} = I_p(\cos \omega t + j \sin \omega t) = I_p e^{j\omega t} \tag{3-1}$$

and

$$v = \mathbf{V} = V_p(\cos \omega t + j \sin \omega t) = V_p e^{j\omega t} \tag{3-2}$$

where $j = \sqrt{-1}$. The symbol j, rather than i, is commonly used to signify $\sqrt{-1}$ in electronic-circuit analysis so as to avoid confusion with the conventional symbol for current. In effect, the complex number representation brings in both sin ωt and cos ωt terms simultaneously, as Eqs. (3-1) and (3-2) illustrate.

The circuit differential equations may be solved quite directly when sinusoidal signals are represented by complex numbers. Consider, for example, the circuit differential equation for a simple RL series circuit, Eq. (2-18),

$$v = R_i + L\frac{di}{dt} \tag{3-3}$$

Insert Eqs. (3-1) and (3-2) into Eq. (3-3), allowing for a phase angle between the current and the voltage,

$$V_p e^{j(\omega t + \phi)} = RI_p e^{j\omega t} + j\omega LI_p e^{j\omega t}$$

$$= (R + j\omega L)e^{j\omega t} \tag{3-4}$$

$$\mathbf{V} = (R + j\omega L)\mathbf{I} \tag{3-5}$$

The quantity in parenthesis is called the *complex impedance*, $\mathbf{Z}$, of the circuit. Thus the circuit differential equation reduces to

$$\mathbf{V} = \mathbf{ZI} \tag{3-6}$$

which is the *ac form* of Ohm's law. It relates the current and the voltage in terms of the complex impedance of the circuit. Thus the solution of ac networks is reduced simply to determining the complex impedance.

Note that in the absence of reactance, Eq. (3-6) reduces to the standard dc form of Ohm's law. This means that series and parallel impedances combine in the same way as do series and parallel resistances. Thus the equivalent impedance of a series of individual impedances is

$$\mathbf{Z}_{eq} = \mathbf{Z}_1 + \mathbf{Z}_2 + \mathbf{Z}_3 + \cdots \tag{3-7}$$

Similarly, the equivalent of parallel impedances is

$$\frac{1}{Z_{eq}} = \frac{1}{Z_1} + \frac{1}{Z_2} + \frac{1}{Z_3} + \cdots \tag{3-8}$$

In applying Eqs. (3-7) and (3-8) due consideration must be given to the complex nature of impedances and the rules pertaining to how complex numbers combine. Several illustrative examples are considered in subsequent sections.

Complex Impedance

Equation (3-5) shows that the real part of the complex impedance is associated with resistance while the imaginary part stands for reactance. According to the standard graphical representation of a complex number, Fig. 3-1, the real part is plotted on the horizontal axis and the imaginary part on the vertical axis. Therefore, the impedance angle is given by

$$\phi = \arctan \frac{X_L}{R} \tag{3-9}$$

where X_L is the reactance. The term impedance comes from the fact that the effect of both reactances and resistances is to impede the current in a circuit. According to Eq. (3-6) the unit of impedance is the ohm.

It is useful to recall some of the properties of complex numbers in order to facilitate their use in circuit analysis. Two complex numbers are equal only if their real and imaginary parts are equal. Therefore, if it is true that

$$R_1 + jX_1 = R_2 + jX_2$$

then it must be that

$$R_1 = R_2 \quad \text{and} \quad X_1 = X_2 \tag{3-10}$$

From this definition of equality it follows that the addition of two complex numbers is accomplished by separately adding the real and imaginary parts. That is,

$$(R_1 + jX_1) + (R_2 + jX_2) = (R_1 + R_2) + j(X_1 + X_2) \tag{3-11}$$

Clearly, the same rule applies to the subtraction of two complex numbers.

Complex numbers are multiplied in conventional fashion,

$$Z_1 Z_2 = Z_1 e^{j\phi_1} Z_2 e^{j\phi_2} = Z_1 Z_2 e^{j(\phi_1 + \phi_2)} \tag{3-12}$$

Figure 3-1 Complex impedance is represented graphically by resistive component and reactive component.

According to (3-12), the product of two complex numbers is a complex number of magnitude equal to the product of the individual magnitudes and with an angle which is the sum of the individual angles. Division of complex numbers also follows immediately,

$$\frac{\mathbf{Z}_1}{\mathbf{Z}_2} = \frac{Z_1 e^{j\phi_1}}{Z_2 e^{j\phi_2}} = \frac{Z_1}{Z_2} e^{j(\phi_1 - \phi_2)} \tag{3-13}$$

This relation states that the result of dividing two complex numbers is a complex number whose magnitude is the ratio of the individual magnitudes and whose angle is the difference between the individual angles.

It is often useful to *rationalize* the reciprocal of a complex number. Thus

$$\frac{1}{\mathbf{Z}} = \frac{1}{Z(\cos \phi + j \sin \phi)} \tag{3-14}$$

This fraction is rationalized by multiplying numerator and denominator by the *complex conjugate*, obtained by replacing the imaginary part of the complex number by its negative. Therefore, Eq. (3-14) becomes

$$\frac{1}{\mathbf{Z}} = \frac{1}{Z(\cos \phi + j \sin \phi)} \frac{\cos \phi - j \sin \phi}{\cos \phi - j \sin \phi} = \frac{1}{Z} \frac{\cos \phi - j \sin \phi}{\cos^2 \phi + \sin^2 \phi}$$

$$= \frac{1}{Z}(\cos \phi - j \sin \phi) \tag{3-15}$$

Evidently the reciprocal of a complex number is simply a complex number with a magnitude given by the reciprocal of the magnitude of the complex number and an angle which is its negative. Using the exponential form

$$\frac{1}{Z e^{j\phi}} = \frac{1}{Z} e^{-j\phi} \tag{3-16}$$

which could have been written immediately.

The equivalent forms of representing impedance,

$$\mathbf{Z} = Z e^{j\phi} = Z(\cos \phi + j \sin \phi) = R + jX \tag{3-17}$$

should be noted. Evidently addition and subtraction are easiest in the component form, $R + jX$, while multiplication and division are easiest in the *polar* form, $Z e^{j\phi}$. Equation (3-17) is used to pass directly from one representation to another as needed. These operations, together with that of rationalization, are useful in computing the equivalent complex impedance of circuits comprising a number of individual impedances, as is illustrated in the next section. Note also that a complex number may be written in terms of its real and imaginary parts as

$$\mathbf{Z} = Z e^{j\phi} = \sqrt{R^2 + X^2} \, e^{j \arctan (X/R)} \tag{3-18}$$

In particular, the magnitude is equal to the square root of the sum of the squares of the real and imaginary components.

RLC CIRCUITS

Series Resonance

As the first example of ac-circuit analysis using the complex-impedance method, consider the series RLC circuit of Fig. 3-2. According to Kirchhoff's law

$$v = Ri + \frac{q}{C} + L\frac{di}{dt} \tag{3-19}$$

The differential equation of the circuit is, after differentiating with respect to t,

$$L\frac{d^2i}{dt^2} + R\frac{di}{dt} + \frac{1}{C}i = \frac{dv}{dt} \tag{3-20}$$

Note that this is a second-order differential equation. It can be solved for the steady-state current i by the procedure used in Chap. 2. The result is

$$i = \frac{V_p}{\sqrt{R^2 + (1/\omega C - \omega L)^2}} \sin(\omega t - \phi) \tag{3-21}$$

where

$$\phi = \arctan \frac{\omega L - 1/\omega C}{R} \tag{3-22}$$

The complex-impedance technique for the solution of the same circuit proceeds as follows. The total impedance of the series combination is, from Eq. (3-7),

$$Z = R + j\omega L - j\frac{1}{\omega C} = R + j\left(\omega L - \frac{1}{\omega C}\right) \tag{3-23}$$

Using Ohm's law to calculate the current,

$$I = \frac{V}{Z} = \frac{V}{R + j(\omega L - 1/\omega C)} \tag{3-24}$$

Equation (3-24) is the final solution, obtained directly in only two simple steps. The voltage V usually is given in terms of its rms value, so that the rms value of I is obtained.

Figure 3-2 Series RLC circuit.

To illustrate more clearly the correspondence between this solution and that given by (3-21) and (3-22), (3-24) is rationalized,

$$I = \frac{V}{R + j(\omega L - 1/\omega C)} \frac{R - j(\omega L - 1/\omega C)}{R - j(\omega L - 1/\omega C)}$$

$$= V \frac{R - j(\omega L - 1/\omega C)}{R^2 + (\omega L - 1/\omega C)^2}$$

$$= \frac{V}{\sqrt{R^2 + (\omega L - 1/\omega C)^2}} \left[\frac{R}{\sqrt{R^2 + (\omega L - 1/\omega C)^2}} \right.$$

$$\left. - j \frac{\omega L - 1/\omega C}{\sqrt{R^2 + (\omega L - 1/\omega C)^2}} \right] \tag{3-25}$$

Referring to the complex-impedance diagram of the circuit, Fig. 3-3, Eq. (3-25) may be written

$$I = \frac{V}{\sqrt{R^2 + (\omega L - 1/\omega C)^2}} (\cos \phi - j \sin \phi)$$

$$= \frac{V_p e^{j\omega t}}{\sqrt{R^2 + (\omega L - 1/\omega C)^2}} e^{-j\phi} = \frac{V_p}{\sqrt{R^2 + (\omega L - 1/\omega C)^2}} e^{j(\omega t - \phi)} \tag{3-26}$$

where

$$\phi = \arctan \frac{\omega L - 1/\omega C}{R} \tag{3-27}$$

The exact correspondence between the differential-equation solution, Eq. (3-21), and the result of the complex-impedance approach, Eq. (3-26), is evident. In subsequent sections the latter technique is used exclusively because of its simplicity.

Consider the amplitude of the current in the RLC circuit as a function of the frequency of the applied voltage. According to Eq. (3-26), the current is very small at low frequencies ($\omega \to 0$), because the capacitive reactance is great. Similarly, the current is small at high frequencies ($\omega \to \infty$), because the inductive reactance becomes large. Between these two extremes, the phase angle is zero when

$$\omega L - \frac{1}{\omega C} = 0 \tag{3-28}$$

Figure 3-3 Complex impedance of an RLC circuit.

At this frequency the circuit is said to be in *resonance*, and the current is a maximum,

$$I = \frac{V}{R} \tag{3-29}$$

The circuit appears as a pure resistance, and the current is in phase with the applied voltage. This is so because the capacitive reactance cancels the inductive reactance at resonance. The *resonant frequency* is, from Eq. (3-28),

$$\omega_0 = 2\pi f_0 = \frac{1}{\sqrt{LC}} \tag{3-30}$$

To illustrate the properties of a series resonant circuit, consider the current variation in the specific circuit of Fig. 3-4. After substituting the component values on the diagram into Eq. (3-26), it is found that the current in the circuit changes with frequency as illustrated in Fig. 3-5. The current maximum is $I = \frac{10}{100} = 0.1$ A at the resonant frequency,

$$f_0 = \frac{1}{2\pi\sqrt{LC}} = \frac{1}{6.28\sqrt{250 \times 10^{-3} \times 0.1 \times 10^{-6}}} = 1000 \text{ Hz} \tag{3-31}$$

and the decrease on either side of resonance is clearly evident.

Figure 3-4

Figure 3-5 Resonance curve of circuit in Fig. 3-4.

The voltage drops around the circuit at resonance further illustrate an important feature of alternating currents. The drop across the resistor is

$$V_R = RI = 100 \times 0.1 = 10 \text{ V} \tag{3-32}$$

The corresponding voltage drop across the capacitor is the current times the capacitive reactance

$$V_C = I \frac{1}{\omega C} = \frac{0.1}{6.28 \times 10^3 \times 10^{-7}} = 158 \text{ V} \tag{3-33}$$

and, similarly, the voltage across the inductor is

$$V_L = I\omega L = 0.1 \times 6.28 \times 10^3 \times 0.25 = 158 \text{ V} \tag{3-34}$$

It is evident that the rms voltage drops around the circuit do not sum to zero. In fact, the voltages across both reactances rise to large values at resonance. Note, however, that the phase angle of the capacitor voltage is $+90°$ with respect to the current, while the phase angle of the voltage across the inductor is $-90°$. This means that the instantaneous voltages across the two reactances cancel each other and that the drop across the resistance equals the applied emf, as confirmed by Eq. (3-32). Kirchhoff's voltage rule is valid when the phases of the currents and voltages in ac circuits are taken into account. This matter is considered further in the exercises.

A voltmeter connected across, say, the capacitor in Fig. 3-4 indicates a voltage corresponding to Eq. (3-33) at resonance. Therefore, the capacitor (and the inductor) must be capable of withstanding high voltages without breakdown, even though the applied voltage is well below the ratings of the components. Resonance is widely used in electronic circuits to increase the current at the resonant frequency.

Parallel Resonance

Resonance also occurs in a parallel circuit, such as the one shown in Fig. 3-6. The current in the circuit is obtained by first computing the total complex impedance. Since the inductance and capacitance are connected in parallel, their equivalent impedance is found with the aid of Eq. (3-8),

$$\frac{1}{Z_1} = \frac{1}{-j(1/\omega C)} + \frac{1}{j\omega L} = j\omega C + \frac{1}{j\omega L}$$

$$= \frac{-\omega^2 LC + 1}{j\omega L} \tag{3-35}$$

Figure 3-6 Parallel resonant circuit.

Therefore, the impedance of the LC combination is

$$\mathbf{Z}_1 = j\frac{\omega L}{1 - \omega^2 LC} \tag{3-36}$$

According to Eq. (3-36) the impedance is very large, actually infinite, when

$$\omega_0^2 LC = 1$$

or

$$\omega_0 = \frac{1}{\sqrt{LC}} \tag{3-37}$$

This is the same relation found in the case of series resonance, Eq. (3-30). Note, however, that in series resonance the impedance is a minimum, according to Eq. (3-23), while in parallel resonance Eq. (3-36) shows that the impedance is a maximum at the resonant frequency.

The total impedance of the circuit includes the series combination of R with $\mathbf{Z}_1$

$$\mathbf{Z} = R + j\frac{\omega L}{1 - \omega^2 LC} \tag{3-38}$$

so that the current in the circuit is

$$\mathbf{I} = \frac{\mathbf{V}}{R + j\omega L/(1 - \omega^2 LC)} \tag{3-39}$$

According to Eq. (3-39) the current is zero at resonance, when the impedance becomes infinite. The voltage $\mathbf{V}$ then appears across the LC combination, independent of the value of R. This feature of the parallel resonant circuit is extensively used in practical circuits. The output voltage $\mathbf{V}_o$ is simply the current times the impedance $\mathbf{Z}_1$ or

$$\mathbf{V}_o = \mathbf{I}\mathbf{Z}_1 = \frac{\mathbf{V}}{R + j\omega L/(1 - \omega^2 LC)} \frac{j\omega L}{1 - \omega^2 LC}$$

$$= \mathbf{V}\frac{j\omega L}{R(1 - \omega^2 LC) + j\omega L} \tag{3-40}$$

Rationalizing

$$\mathbf{V}_o = \mathbf{V}\frac{j\omega L}{R(1 - \omega^s LC) + j\omega L}\frac{R(1 - \omega^2 LC) - j\omega L}{R(1 - \omega^2 LC) - j\omega L}$$

$$= \mathbf{V}\frac{(\omega L)^2 + j\omega LR(1 - \omega^2 LC)}{(\omega L)^2 + R^2(1 - \omega^2 LC)^2}$$

$$= \mathbf{V}\frac{1 + j(R/\omega L)(1 - \omega^2 LC)}{1 + (R/\omega L)^2(1 - \omega^2 LC)^2} \tag{3-41}$$

The magnitude of the output voltage is found by taking the square root of the sum of the squares of the real and imaginary parts. Therefore, the ratio of the output voltage to the input voltage becomes

$$
\frac{V_o}{V} = \left\{ \frac{1 + (R/\omega L)^2(1 - \omega^2 LC)^2}{[1 + (R/\omega L)^2(1 - \omega^2 LC)^2]^2} \right\}^{1/2}
$$

$$
= \frac{1}{[1 + (R/\omega L)^2(1 - \omega^2 LC)^2]^{1/2}} \tag{3-42}
$$

The behavior of Eq. (3-42) is illustrated with the specific component values given on the parallel-circuit diagram of Fig. 3-7. The output voltage rises to equal the input voltage at the resonant frequency, as shown in Fig. 3-8. Parallel resonance is commonly used in electronic circuits to achieve a high impedance which develops an appreciable signal voltage at resonance. Resonance is also important in circuits designed to emphasize one single frequency over all others. By adjusting the value of, say, the capacitance, the circuit may be *tuned* to different frequencies. This is the principle used to select different channels in radio and TV receivers.

It is interesting to compare the currents in various components of a parallel resonant circuit at resonance. According to Eq. (3-42) the full input voltage is applied across both the capacitor and inductance at resonance, so that the currents are, respectively

$$
I_C = \frac{V}{1/\omega C} \tag{3-43}
$$

Figure 3-7

Figure 3-8 Resonance curve of circuit in Fig. 3-7.

and

$$I_L = \frac{V}{\omega L} \tag{3-44}$$

Yet the total current in the circuit is zero, according to Eq. (3-39). This is another situation where the phase angle of the currents must be considered. Since the capacitive reactance equals the inductive reactance at resonance, the magnitudes of the currents in the inductance and capacitance are equal. Their phase angles are such that at, say, the upper branch point in Fig. 3-6 these two currents cancel at every instant and no current is present in the resistor. In this sense a circulating current exists in the parallel LC combination at resonance.

Q Factor

The resistance in a resonant circuit is significant in determining the current at frequencies removed from the resonant frequency, as well as at resonance. This is an important consideration since, for example, the resistance of the turns of wire is present in all inductors. The effect is similar in both series and parallel resonance cases, but the former is simpler mathematically and can be used to illustrate all the important features.

Consider the magnitude of the current in a series circuit, Fig. 3-2, given by Eq. (3-26)

$$I = \frac{V}{[R^2 + (\omega L - 1/\omega C)^2]^{1/2}} \tag{3-45}$$

This can be rearranged as

$$I = \frac{V}{R} \frac{1}{\sqrt{1 + (\omega L/R)^2 (1 - 1/\omega^2 LC)^2}} \tag{3-46}$$

The ratio of inductive reactance to resistance is called the Q of the circuit. For present purposes, it is convenient to consider the Q at resonance.

$$Q_0 = \frac{\omega_0 L}{R} \tag{3-47}$$

Introducing (3-47) and the resonant frequency from (3-30) into (3-46),

$$\frac{I}{I_M} = \frac{1}{\{1 + Q_0^2 (\omega/\omega_0)^2 [1 - (\omega_0/\omega)^2]^2\}^{1/2}} \tag{3-48}$$

where I_M is the maximum current at resonance. Finally, an alternative form to Eq. (3-48) is, after multiplying the two quantities in ω/ω_0,

$$\frac{I}{I_M} = \left[1 + Q_0^2 \left(\frac{\omega}{\omega_0} - \frac{\omega_0}{\omega} \right)^2 \right]^{-1/2} \tag{3-49}$$

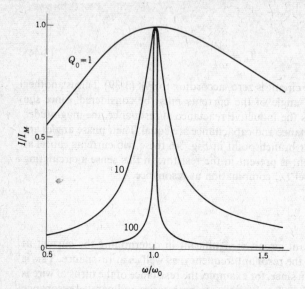

Figure 3-9 Sharpness of resonance curve is greatest for high-Q circuits.

This expression, plotted in Fig. 3-9 for representative values of Q_0, shows that a large Q_0 (that is, a small value of R) leads to a very sharply resonant circuit. Under this condition the bandwidth of the resonant circuit, that is, the frequency interval between half-power points, is quite small.

Values of Q_0 in the range from 10 to 100 are common in electronic circuits and are useful in very selective tuned circuits, as illustrated by the two lower curves of Fig. 3-9. Special resonant components can have Q's as high as several thousand, and these result in very high frequency selectivity indeed. On the other hand, it is sometimes desirable to broaden the frequency response of a resonant circuit so that it is useful over a wide band of frequencies on either side of the resonant frequency. In this case a resistor is purposely included to yield a lower Q than that of the inductor alone.

BRIDGE CIRCUITS

Figure 3-10 shows the ac analogy of the dc Wheatstone bridge. It has a complex impedance in each arm and uses a sine-wave generator and an ac detector, such as an oscilloscope. Analysis of this circuit proceeds as in the dc case considered in Chap. 1, except that complex impedances and currents are used. Only the balance condition is of interest here, and this is obtained in the following way. At balance, the voltage across the detector is zero, which means that the current in this branch is zero and therefore that the current in Z_1 is equal to that in Z_3; also, the current

Figure 3-10 The ac Wheatstone bridge uses sine-wave generator and ac detector such as an oscilloscope.

in Z_2 is equal to that in Z_4. Furthermore, since the voltage across the detector is zero, the voltage drops across Z_1 and Z_2 are equal and so are the voltage drops across Z_3 and Z_4.

Equating the voltage drops across corresponding arms of the bridge

$$Z_1 I_1 = Z_2 I_2 \tag{3-50}$$

and

$$Z_3 I_1 = Z_4 I_2 \tag{3-51}$$

Dividing Eq. (3-50) by Eq. (3-51), the condition for balance is found to be

$$\frac{Z_1}{Z_3} = \frac{Z_2}{Z_4} \tag{3-52}$$

which should be compared with the condition for balance of the dc Wheatstone bridge, Eq. (1-46). According to Eq. (3-52) the balance condition involves the equality of two complex numbers. This means that both the real and imaginary parts must be equal and implies that two independent balance adjustments are necessary in ac bridge circuits. Specific illustrations of this situation are taken up in the following sections.

Inductance and Capacitance Bridge

Bridge circuits in Figs. 3-11 and 3-12 can be used to measure inductance and capacitance in the same way a Wheatstone bridge is used to measure resistance. Consider the inductance bridge, Fig. 3-11, and note the comparison with Fig. 3-10. The impedance $R_u + j\omega L_u$ represents the inductance and resistance of

Figure 3-11 Inductance bridge used to measure resistance and inductance of unknown coil.

Figure 3-12 Capacitance bridge.

an unknown inductor. The balance condition, Eq. (3-52), means

$$\frac{R_1}{R_2} = \frac{R_3 + j\omega L_3}{R_u + j\omega L_u} \qquad (3\text{-}53)$$

Cross-multiplying,

$$R_1 R_u + j\omega R_1 L_u = R_2 R_3 + j\omega R_2 L_3 \qquad (3\text{-}54)$$

After equating real and imaginary parts, the two balance conditions are written as

$$R_1 R_u = R_2 R_3 \qquad \text{and} \qquad R_1 L_u = R_2 L_3 \qquad (3\text{-}55)$$

The first condition is the same as the Wheatstone bridge condition and may be carried out with dc instruments. The second equation is obtained using ac excitation of the bridge subsequent to obtaining the dc balance.

Rewriting Eqs. (3-55),

$$R_u = \frac{R_2}{R_1} R_3 \quad \text{and} \quad L_u = \frac{R_2}{R_1} L_3 \tag{3-56}$$

which illustrates how the ratio R_2/R_1 acts as a multiplying factor on the variable components R_3 and L_3 to yield the unknown resistance and inductance. This suggests that both R_3 and L_3 should be variable, but since variable inductors are difficult to make, the following procedure is more satisfactory. Either R_2 or R_1 is adjusted until balance is obtained with a given value of L_3. Then R_3 is adjusted to satisfy the dc balance condition. In this way the unknown inductance is compared with the standard inductance L_3, and the only variable components needed are resistances.

An interesting feature of the balance conditions, Eqs. (3-56), is that they are independent of frequency, so that the generator frequency may have any convenient value. It is usually best to choose a frequency such that the inductive reactance is approximately equal to the resistances. A similar situation exists in the case of the capacitance bridge, Fig. 3-12, which is analyzed in Exercise 3-8. Since there is no dc path in the lower arms of the capacitance bridge, both balance adjustments are made with ac excitation (this can be done in the inductance bridge case as well). Although calibrated variable capacitors are usually used in this circuit, it is equally possible to employ a fixed standard capacitor together with variable resistors. This is particularly convenient when the same instrument is used as an inductance bridge as well as a capacitance bridge by replacing the standard capacitor C_3 with a standard inductance L_3. Since it is also possible to employ a standard resistor in this position, such an instrument is quite versatile in that it can measure resistance, capacitance, and inductance.

Wien Bridge

A bridge which has a parallel combination in one arm and a series combination in an adjacent arm is known as a *Wien* bridge. A useful example employing only resistors and capacitors, shown in Fig. 3-13, is analyzed by first calculating the impedances Z_3 and Z_4. Considering the parallel combination first,

$$\frac{1}{Z_3} = j\omega C_3 + \frac{1}{R_3} = \frac{1 + j\omega R_3 C_3}{R_3} \tag{3-57}$$

which is rationalized to give

$$Z_3 = \frac{R_3}{1 + (\omega R_3 C_3)^2} (1 - j\omega R_3 C_3) \tag{3-58}$$

The series combination is

$$Z_4 = R_4 - j \frac{1}{\omega C_4} \tag{3-59}$$

Figure 3-13 Wien bridge.

Inserting Eqs. (3-58) and (3-59) into the balance equation (3-52) results in

$$\frac{R_1}{R_2} = \frac{R_3/(1 + j\omega R_3 C_3)}{R_4 - j(1/\omega C_4)} \tag{3-60}$$

Cross-multiplying,

$$(1 + j\omega R_3 C_3)\left(R_4 - \frac{j}{\omega C_4}\right) = \frac{R_2}{R_1} R_3 \tag{3-61}$$

$$R_4 + \frac{R_3 C_3}{C_4} + j\left(\omega R_3 R_4 C_3 - \frac{1}{\omega C_4}\right) = \frac{R_2}{R_1} R_3 \tag{3-62}$$

Upon equating the real and imaginary parts, the balance conditions are found to be

$$\frac{C_3}{C_4} + \frac{R_4}{R_3} = \frac{R_2}{R_1}$$

and

$$\omega^2 R_3 C_3 R_4 C_4 = 1 \tag{3-63}$$

This result differs from those for the inductance and capacitance bridges in that the frequency ω appears in the balance equations. Therefore, it is possible to achieve balance by adjusting the frequency and only one component, say R_1, rather than using two variable impedances. Alternatively, by adjusting two of the components for balance, the bridge is capable of determining the frequency of a sine-wave source.

The Wien bridge is also useful as a frequency-selective network which has properties similar to those of a resonant circuit. Since inductors suitable for use over a wide frequency interval are expensive and difficult to construct, the Wien bridge has considerable advantages in many applications. This is particularly true for low-frequency circuits where inconveniently large inductance values are necessary. Often, the resistors and capacitors in the series and parallel branches are equal, as in Fig. 3-14. This means that the characteristic frequency of the network is, from Eq. (3-63),

$$\omega_0 = \frac{1}{RC} \tag{3-64}$$

The frequency-selective properties of the Wien bridge are illustrated by calculating the output voltage of the network as a function of the frequency of the input voltage. The output voltage is obtained by subtracting the voltages across Z_3 and R_1. The drop across Z_3 is simply the current in this arm times the impedance,

$$v_1 = \frac{v_i}{Z_3 + Z_4} Z_3 = \frac{v_i}{1 + Z_4/Z_3} \tag{3-65}$$

where it has been assumed that the current in the output circuit is negligible, even when the bridge is not balanced. This condition is satisfied in practice by connecting a high-impedance load to the output terminals. The impedance ratio in Eq. (3-65) is evaluated from Eqs. (3-58) and (3-59), after making all the R's and C's equal. Also, introducing ω_0,

$$\frac{Z_4}{Z_3} = \frac{R(1 - j\omega_0/\omega)}{R(1 - j\omega/\omega_0)/[1 + (\omega/\omega_0)^2]} = 2 + j\left(\frac{\omega}{\omega_0} - \frac{\omega_0}{\omega}\right) \tag{3-66}$$

Note that according to Eq. (3-66) the reactive term vanishes at the characteristic frequency when $\omega = \omega_0$. This means that the network is resistive at this frequency (and at this frequency only) and that the output voltage is in phase with the input voltage, just as for a resonant circuit.

The drop across R_1 is

$$v_2 = \frac{v_i}{R_1 + R_2} R_1 = \frac{v_i}{1 + R_2/R_1} \tag{3-67}$$

Finally, the output voltage is

$$v_o = v_1 - v_2 = v_i\left[\frac{1}{3 + j(\omega/\omega_0 - \omega_0/\omega)} - \frac{1}{1 + R_2/R_1}\right] \tag{3-68}$$

Figure 3-14 Wien bridge used as frequency-selective network.

Figure 3-15 Frequency characteristic of Wien bridge network.

The ratio of the output to the input v_o/v_i can be found after rationalizing. Two plots of Eq. (3-68) are presented in Fig. 3-15 for different values of the ratio R_2/R_1. The output voltage clearly goes through a minimum at $\omega = \omega_0$, very analogous to the situation in series resonance. The minimum is quite sharp and mathematically discontinuous in the case of true balance, $R_2 = 2R_1$. Therefore, the frequency selectivity of the circuit is quite good.

Bridged-T and Twin-T Networks

Although the so-called *bridged-T* filter is not a true bridge circuit, it has properties similar to those of the Wien bridge. In contrast to the Wien bridge, the bridged-T, shown in Fig. 3-16, generally does not employ equal values of resistance and capacitance in the two positions. To solve for the output voltage, the current in the upper resistor is determined and the output is obtained by subtracting the voltage drop across R_2 from the input voltage.

Figure 3-16 Bridged-T filter.

Kirchhoff's law applied to the i_1 loop in Fig. 3-16 is

$$v_i = i_1 \frac{-j}{\omega C_1} - i_2 \frac{-j}{\omega C_1} + i_1 R_1 \qquad (3\text{-}69)$$

Adding up the voltages around the second loop,

$$0 = -i_1 \frac{-j}{\omega C_1} + i_2 \frac{-j}{\omega C_1} + i_2 R_2 + i_2 \frac{-j}{\omega C_2} \qquad (3\text{-}70)$$

After simplification, i_1 is determined from (3-70),

$$i_1 = \left(1 + \frac{C_1}{C_2} + j\omega R_2 C_1\right) i_2 \qquad (3\text{-}71)$$

and substituted into (3-69). The result is solved for i_2 and the output voltage is then given by

$$v_o = v_i - R_2 i_2 = v_i - \frac{v_i \omega R_2 C_2}{\omega C_2 [R_1(1 + C_1/C_2) + R_2] + j(\omega^2 R_1 C_1 R_2 C_2 - 1)} \qquad (3\text{-}72)$$

The imaginary part vanishes at the critical frequency

$$\omega_0 = \frac{1}{\sqrt{R_1 C_1 R_2 C_2}} \qquad (3\text{-}73)$$

which should be compared with the corresponding expression for the Wien bridge, Eq. (3-64). After inserting (3-73) into (3-72), the complex ratio of output to input voltage may be written

$$\frac{v_o}{v_i} = 1 - \left\{ \left[1 + \frac{R_1}{R_2}\left(1 + \frac{C_1}{C_2}\right)\right] + j\sqrt{\frac{R_1 C_1}{R_2 C_2}}\left(\frac{\omega}{\omega_0} - \frac{\omega_0}{\omega}\right) \right\}^{-1} \qquad (3\text{-}74)$$

The frequency-response characteristic of the bridged-T filter, Eq. (3-74), is similar to that of the Wien bridge, except that the minimum is not as sharp at the characteristic frequency. Nevertheless, its relative simplicity, and the fact that both input and output terminals have a common connection, make the bridged-T filter a useful frequency-selective network. Specific applications of this circuit are considered in a later chapter.

Somewhat more complicated than the bridged-T filter is the *twin-T* filter shown in Fig. 3-17, which may be analyzed by the technique used in the previous section. The twin-T filter is equivalent in response characteristic to the Wien bridge, but the minimum is much sharper. It also has the advantage of a common input-output terminal. The choice between the Wien bridge, bridged-T network, or twin-T network for any given application depends on factors such as frequency range, desired performance, and complexity; all three circuits are commonly used.

Figure 3-17 Twin-T filter.

TRANSFORMERS

Mutual Inductance

Suppose a changing magnetic flux resulting from current in one circuit is intercepted by another circuit. Then an emf is induced in the second circuit, and the *mutual inductance* between circuit 1 and circuit 2 is defined by analogy with Eq. (2-15) as

$$v_2 = M \frac{di_1}{dt} \tag{3-75}$$

where v_2 is the induced emf in circuit 2 produced by the current i_1 in circuit 1. A very important application of mutual inductance is the *transformer*, which has two multiturn coils wound on the same iron core. This makes the mutual inductance between the two coils as large as possible. Schematically, a transformer appears as in Fig. 3-18a, with a *primary* winding which is part of one circuit and a *secondary* winding which is part of the second circuit. The symbol for a transformer is shown in Fig. 3-18b.

Consider an ideal transformer, in which all of the magnetic flux from the primary winding is intercepted by the secondary winding. Suppose the secondary winding is open-circuited and the primary is connected to a sinusoidal voltage source. Current in the primary winding is determined by the inductance of the primary. The voltage induced in the primary winding v_1 is proportional to the primary inductance according to Faraday's law, and the inductance depends

(a) (b)

Figure 3-18 (a) Sketch of a transformer and (b) circuit symbol.

upon the number of turns on the primary winding. Since all of the flux is also intercepted by the secondary winding, the voltage v_2 induced in the secondary depends upon the number of turns in the same way. That is,

$$\frac{v_1}{v_2} = \frac{n_1}{n_2} \qquad (3\text{-}76)$$

Note that it is possible to achieve either a *step-up* transformer or a *step-down* transformer, depending upon whether $n_1 < n_2$ or $n_1 > n_2$, so that the secondary voltage is respectively greater or less than the primary voltage.

Suppose now the secondary is connected to a load such as a resistor. Current in the secondary circuit results in $I^2 R$ losses in the resistor, and this power must come from the primary winding. The way this comes about is as follows. Both the primary current and the secondary current set up magnetic flux in the core. The magnetic flux caused by the secondary current acts to oppose the magnetic flux set up by the primary current. The weaker magnetic flux means that the induced primary voltage is correspondingly smaller. Therefore the voltage source connected to the primary winding increases the primary current until the voltages around the primary circuit are zero again as required by Kirchhoff's rules. Thus current in the secondary winding requires current in the primary winding and the peak magnetic flux remains constant at its no-load value. The near cancellation of fluxes also means that the induced voltages are essentially independent of frequency. Since the rate of change of current increases with frequency, Eq. (3-75) suggests that the secondary voltage should increase with frequency. The induced voltage in the primary winding increases with frequency correspondingly, however, so that the magnitude of the primary current is reduced, and the secondary voltage remains independent of frequency.

Note that the total magnetic flux in the core does not change with current, because the magnetic flux from the primary and secondary currents are equal and opposite. Since both inductances depend upon the number of turns, it follows that

$$n_1 i_1 = n_2 i_2 \qquad (3\text{-}77)$$

The primary current in Eq. (3-77) really refers only to the additional current accompanying a load on the secondary. The current under no-load conditions is usually so small, however, that it may be neglected.

Transformer Ratio

The ratio of the number of turns on the secondary to that on the primary, called the *transformer ratio*, is particularly significant when the impedance connected to the secondary is compared with the apparent impedance in the primary circuit. Consider the situation depicted in Fig. 3-19, where the secondary load is a simple resistance R.

Figure 3-19 Apparent resistance at primary terminals of transformer differs from resistance connected to secondary.

The magnitude of R determines the secondary current

$$i_2 = \frac{v_2}{R} \tag{3-78}$$

Substituting for i_2 from Eq. (3-76) and for v_2 from Eq. (3-75),

$$\frac{n_1}{n_2} i_1 = \frac{n_2 v_1}{n_1 R}$$

or

$$i_1 = \left(\frac{n_2}{n_1}\right)^2 \frac{v_1}{R} \tag{3-79}$$

Viewed from the primary side, the current i_1 is in an equivalent resistance R_{eq} such that

$$i_1 = \frac{v_1}{R_{eq}} \tag{3-80}$$

Comparing (3-79) and (3-80), the secondary load resistance R appears to be a resistance on the primary side given by

$$R_{eq} = \left(\frac{n_1}{n_2}\right)^2 R \tag{3-81}$$

This means that a transformer may be used to match the resistances in a circuit to obtain maximum power transfer between a given power source and a fixed-load resistance. This very useful property is applied, for example, in coupling the large internal resistance of an amplifier to the low resistance of a loudspeaker load. Complex impedances are transformed by the turns ratio in exactly the same way as the resistance in Eq. (3-81).

According to Eq. (3-81) the transformer appears from the primary side to be a pure resistance. In particular, the self-inductance of the primary winding is not evident. The reason for this is that the fluxes caused by current in the primary and secondary windings cancel each other. This means that a transformer acts as a device which changes the effective resistance from one circuit to another but which has no inductance of itself. Note also that the primary and secondary circuits in a transformer are not connected electrically and that the two circuits are isolated for direct currents.

Practical Transformers

Most transformers are constructed in a fashion similar to iron-core chokes described in an earlier section, except, of course, that more than one winding is present on the core. For large transformer ratios it is common practice to make the low-voltage high-current winding of heavy-gauge wire in order to reduce I^2R losses in this winding. The other winding is a large number of turns of fine wire since only a small current is present. Laminated cores are ordinarily used to reduce eddy-current losses, but transformers for frequencies in excess of 100,000 Hz generally employ high-resistivity ferrite cores.

Transformers with one primary winding and several separate secondary windings are used to supply different voltages to vacuum-tube and transistor devices. Such a *power* transformer may, for example, have a primary winding suitable for connection to a 115-V 60-Hz source with secondaries providing 700, 6.3, and 5.0 V. Applications of such power transformers are studied in a later chapter.

Although for many purposes the inherent inductive effects of the transformer windings may be neglected, in careful work it is necessary to account for the properties of the transformer more exactly. An equivalent circuit of a practical transformer is illustrated in Fig. 3-20. The inductances in the primary and secondary circuits are caused by leakage magnetic flux which does not link both windings, so that the opposing fluxes do not quite cancel. The resistances are included to account for the resistance of the wire in the windings. The inductance L_m accounts for the small magnetizing current corresponding to the no-load primary current. The capacitors on the primary and secondary sides result from the layer-to-layer capacitances between windings.

According to this equivalent circuit, a transformer is ineffective at low frequencies where the reactance of L_m becomes so small that current is shunted from the primary winding of the ideal transformer. High-frequency performance is impaired by the leakage flux inductances and winding capacitances. In spite of these limitations, transformers can be designed for effective performance over useful frequency intervals. It is common practice to indicate the approximate impedance level at which the primary and secondary windings are designed to be used, even though Eq. (3-81) suggests that only the transformer ratio is significant.

Figure 3-20 Equivalent circuit of practical transformer.

The specification indicates the conditions under which the transformer inductances and resistances in Fig. 3-20 are negligible in comparison with primary and secondary load impedances, and, correspondingly, the transformer operates very nearly like an ideal transformer.

SUGGESTIONS FOR FURTHER READING

A. M. P. Brookes: "Basic Electric Circuits," The Macmillan Company, New York, 1963.

S. Fich and J. L. Potter: "Theory of AC Circuits," Prentice-Hall, Inc., Englewood Cliffs, N.J., 1959.

"The Radio Amateur's Handbook" (published annually by the American Radio Relay League, West Hartford, Conn.).

EXERCISES

3-1 Determine the current in a series RC circuit using the ac form of Ohm's law and show that the result is identical with the differential-equation solution in Chap. 2.

3-2 Calculate the equivalent impedance of the network shown in Fig. 3-21. Is the reactive term capacitive or inductive?

    Ans.: Capacitive

Figure 3-21

3-3 Determine the rms current in the 1000-Ω resistor of the circuit in Fig. 3-22. Is the current inductive or capacitive?

    Ans.: 6.5 mA; capacitive

Figure 3-22

3-4 Calculate the equivalent impedance of the circuit in Fig. 3-23 at a frequency of 100 Hz. Repeat for 1000 Hz.

    Ans.: $0.198 + j2.63 \times 10^6 \ \Omega$; $384 + j2.47 \times 10^9 \ \Omega$

Figure 3-23

3-5 Determine the current and the rms voltages across each component in the circuit of Fig. 3-4 at a frequency of 900 Hz. Show that, considering the phase angles of the voltages, Kirchhoff's rule concerning the sum of the voltages around a loop is valid.

Ans: $2.75 \times 10^{-2} e^{j1.29}$ A; $2.75 e^{j1.29}$ V; $48.6 e^{-j0.27}$ V; $38.8 e^{j2.87}$ V.

3-6 The impedance at resonance of a parallel resonant circuit is limited by the resistance of the windings of the inductor. Derive an expression for the impedance of the circuit of Fig. 3-24 and calculate the value at resonance appropriate to the components given on the circuit diagram.

Ans.: 4×10^5 Ω

Figure 3-24

3-7 Compute the currents in the components of the circuit of Fig. 3-7 at resonance. Repeat for a frequency $\omega = \omega_0/2$. In both cases show that, considering the phase angles of the currents, Kirchhoff's current rule is valid.

Ans.: 0 A; $3.16 \times 10^{-2} e^{-j1.57}$ A; $3.16 \times 10^{-2} e^{j1.57}$ A; $2 \times 10^{-3} e^{-j0.042}$ A;

$$2.67 \times 10^{-3} e^{-j0.042} \text{ A}; \ 6.67 \times 10^{-4} e^{j3.10} \text{ A}.$$

3-8 Determine the balance conditions for the capacitance bridge, Fig. 3-12.

Ans.: $R_2 R_3 = R_1 R_u$, $R_1/R_2 = C_u/C_3$.

3-9 Plot the frequency and phase characteristics of a bridged-T filter using Eq. (3-74). Select component values such that the characteristic frequency is 1 kHz while $R_2 = 1000 R_1$, $C_1 = C_2$, and $R_1 = 10^4$ Ω.

3-10 Plot the frequency and phase characteristics of the twin-T filter, Fig. 3-17, for which the response characteristic is

$$\frac{v_o}{v_i} = \left(1 + j\omega_0 R_1 \frac{C_1 + 2C_2}{\omega/\omega_0 - \omega_0/\omega}\right)^{-1} \tag{3-82}$$

where the resonance frequency is given by

$$\omega_0^2 = \frac{2}{R_1^2 C_1 C_2} \tag{3-83}$$

when $R_1 C_1 = 4R_2 C_2$. Take $C_1 = 2C_2$ and choose appropriate component values to make the characteristic frequency equal to 1 kHz.

FOUR

DIODE CIRCUITS

Resistors, capacitors, and inductors are called linear components, because the current increases in direct proportion to the applied voltage, in accordance with Ohm's law. Components for which this proportionality does not hold are termed nonlinear devices, and they are the basis for all practical electronic circuits. This chapter examines the properties of an important nonlinear device, the diode rectifier. The term "diode" comes from the fact that rectifiers have two active terminals, or electrodes.

A rectifier is nonlinear in that it passes a greater current for one polarity of applied voltage than for the other. If a rectifier is included in an ac circuit, the current is negligible whenever the polarity of the voltage across the rectifier is in the reverse direction. Therefore, only a unidirectional current exists and the alternating current is said to have been rectified.

A major application of rectifiers is in power-supply circuits which convert conventional 115-V, 60-Hz ac line voltages to dc potentials suitable for use with vacuum tubes and transistors. Two types of rectifiers, the vacuum diode and the semiconductor junction diode, are commonly used in power supplies, but, because of its superior properties, the junction diode has largely supplanted the vacuum diode.

NONLINEAR COMPONENTS

Current-Voltage Characteristics

The most useful description of the electrical properties of a nonlinear component is the relationship between the current in the unit and the applied voltage. Indeed, this is analogous to the importance of Ohm's law for linear components since Ohm's law forms the basis for all linear-circuit operation and analysis. The relationship between current and voltage is conventionally displayed graphically in the *current-voltage* (or *IV*) *characteristic* of the device. The performance of a nonlinear circuit element in any circuit may be analyzed with the aid of the appropriate current-voltage characteristic.

There exists a rich variety of current-voltage characteristics associated with various practical electrical components, rectifiers included, which are discussed in this and subsequent chapters. In many cases it is possible to understand the origin of these nonlinear relationships in terms of basic physical principles. For example, the motions of electrons in the device may lead directly to the variation of current with applied voltage. This is the case with Ohm's law discussed in Chap. 1, where the drift of electrons in a conductor results in a linear *IV* characteristic. While an understanding of the physical principles involved is often important in the development of new device designs having favorable current-voltage properties, such an understanding is not necessary for the analysis of device performance and circuit operation.

Actually, the *IV* characteristics of practical nonlinear devices are universally determined by empirical measurement anyway, rather than by theoretical analysis. This so because of the complexities inherent in practical units. Characteristic current-voltage curves of commercial devices are presented in manufacturer's compilations and are widely available. All the principles of circuit analysis, in particular Kirchhoff's rules, are applicable to circuits with nonlinear components. Most often it is convenient to employ graphical techniques in the analysis because the *IV* characteristic is not expressible in a manageable mathematical form.

The Ideal Rectifier

An elementary but very important nonlinear current-voltage characteristic is that of the *ideal rectifier*, Fig. 4-1. This shows that an ideal rectifier has zero resistance for one polarity of applied voltage and infinite resistance for the reversed polarity. It is, in effect, a voltage-sensitive switch which is closed when the voltage polarity is in the *forward* direction and open for *reverse* polarity. This ideal characteristic is approximated by several practical types of diodes and is the basis for many important electronic circuits. The vacuum diode, which is discussed in Appendix A, and the semiconductor junction diode are both widely used, although the junction diode has nearly replaced the earlier vacuum tube version except for certain special applications.

Figure 4-1 Current-voltage characteristic of an ideal rectifier.

The Junction Diode

As will be shown in Chap. 5, the addition of certain foreign atoms to otherwise pure semiconductor materials such as germanium or silicon produces free electrons which carry electric current. A semiconductor crystal containing such foreign atoms is called an *n-type* crystal because of the negative charge of the current carrying electrons. Similarly, by incorporating certain other kinds of foreign atoms, the semiconductor appears to conduct current by positive current carriers. Such a crystal is called a *p-type* conductor because the current carriers have a positive charge. It is possible to change from *n*-type to *p*-type conductivity in the same crystal by introducing an abrupt change from one impurity type to the other across a given region of the semiconductor. The junction between an *n*-type region and a *p*-type region in a semiconductor crystal, called a *pn junction*, exhibits distinctive nonlinear electrical characteristics.

Figure 4-2 Current-voltage characteristics of germanium and silicon junction diodes.

In fact the *pn junction diode* is very nearly an ideal rectifier for the voltages commonly encountered in practical applications. This is illustrated by the current-voltage characteristics of typical germanium and silicon *pn* junctions shown in Fig. 4-2. In both units the forward current increases very rapidly for applied potentials greater than a few tenths of a volt, and even at appreciable reverse voltages the reverse currents are negligible by comparison. The forward voltage required for conduction is about 0.2 V for germanium diodes at room temperature compared to approximately 0.6 V in the case of silicon junctions. This difference can be related to the difference in electrical properties of the two semiconductor materials and is examined further in the next chapter.

The reverse current is several microamperes at room temperature in germanium *pn* junctions and only a few picoamperes in silicon diodes, a difference also related to the basic properties of the two semiconductors. These values are sufficiently minute to approximate the current-voltage characteristic of an ideal diode. Of greater concern is the fact that the reverse current increases exponentially with temperature. This rapid increase means that only a modest temperature rise can be tolerated and silicon diodes, for example, become nearly inoperative above about 200°C. Temperature changes also influence the forward characteristic, as can be seen in Fig. 4-3 where the potential for forward conduction in a silicon diode is reduced more than 0.1 V by only a 20°C temperature rise.

Junction diodes used in circuits at power levels above a few watts or so are cooled in order to dissipate joule heat caused by the current in the device. It is common practice to attach junction diodes firmly to metal heat sinks having radiator fins to conduct the heat away from the diode.

In spite of their temperature sensitivity, a characteristic common to most semiconductor devices, junction diodes are extremely good rectifiers. Commercial

Figure 4-3 Effect of temperature on current-voltage characteristics of silicon junction diode.

Figure 4-4 Circuit symbol for semiconductor diode.

junction diodes are made of either silicon or germanium by processes similar to those described in Chap. 5, and practical devices are completely encapsulated to protect the semiconductor surface from contamination. This is accomplished by placing the diode in a small metal can filled with an inert atmosphere or by encasing the diode in plastic. Because of their superior electrical characteristics, silicon and germanium junction diodes have supplanted older forms of semiconductor rectifiers made of selenium or copper oxide. It is interesting to note that the latter rectifiers were the first commercially useful semiconductor devices. The circuit symbol for all semiconductor diodes, given in Fig. 4-4, has an arrow to indicate the direction of conventional forward current in the device.

The excellent forward conductivity of a *pn* junction means that practical diodes can be quite small. Therefore the stray capacitive reactances are correspondingly small and junction diodes are useful at high frequencies. This feature is enhanced in *point-contact* diodes, in which a metal probe is placed in contact with a semi-conductor crystal. During processing a minute *pn* junction is formed immediately under the point. Such tiny devices can operate at frequencies corresponding to millimeter wavelengths and are used, for example, in radar and high-speed computer circuits. An early type of point-contact diode is shown in Fig. 4-5.

Figure 4-5 Point-contact diode. (*Ohmite Manufacturing Company.*)

RECTIFIER CIRCUITS

Half-wave Rectifier

An elementary rectifier circuit, Fig. 4-6, has a junction diode in series with an ac source and a resistive load. When the polarity of the source is in the forward direction, the diode conducts and produces a current in the load. On the reverse half-cycle the diode does not conduct and the current is zero. The output current is therefore a succession of half sine waves, as indicated in Fig. 4-6, and the circuit is called a *half-wave* rectifier. The average value of the half sine waves is clearly not zero, so that the output current has a dc component. That is, the input sine wave has been rectified.

Actually, of course, the peak output voltage is smaller than the peak value of the source voltage because of the voltage drop across the diode. Similarly, the current during the reverse half-cycle is not truly zero but is equal to the reverse current of the diode. In most rectifier applications both the forward voltage drop and the diode reverse current are small enough to be neglected, however.

Full-wave Rectifier

The half-wave rectifier is inactive during one-half of the input cycle and is therefore less efficient than is possible. By arranging two diodes as in Fig. 4-7 so that each diode conducts on alternate half-cycles, *full-wave* rectification results. This is accomplished by using a center-tapped transformer winding. Then, diode D_1 conducts on one half-cycle while diode D_2 is reversed-biased. On the alternate half-cycle, the conditions are interchanged so that the output-current waveform, Fig. 4-8, has only momentary zero values, in contrast to the half-wave circuit.

Note that each diode in the full-wave rectifier must withstand the full end-to-end voltage of the transformer winding. Therefore the *peak inverse voltage* rating of the diodes must be at least twice the peak output voltage. This is usually not a serious drawback except for circuits designed to operate at the highest voltages. Comparing the output waveforms of half-wave and full-wave rectifiers, Figs. 4-6 and 4-8, reveals that the fundamental frequency equals the supply voltage in the half-wave rectifier but is twice the supply frequency in the case of the full-wave circuit. This is an important consideration in power-supply circuits, as will be demonstrated in a later section. The center-tapped transformer in the full-wave

Figure 4-6 Elementary half-wave rectifier circuit.

Figure 4-7 Full-wave rectifier.

Figure 4-8 Waveforms in full-wave rectifier.

circuit supplies current on both half-cycles of the input voltage, which permits a more efficient transformer design than is possible in the case of the half-wave circuit. Note, however, that the output voltage is only one-half the total voltage of the transformer secondary.

Bridge Rectifier

Full-wave rectification without a center-tapped transformer is possible with the *bridge rectifier*, Fig. 4-9. The operation of this circuit may be described by tracing the current on alternate half-cycles of the input voltage. Suppose, for example, the upper terminal of the transformer is positive. This means that diode D_2 conducts, as does diode D_3, and current is present in the load resistor. On the alternate

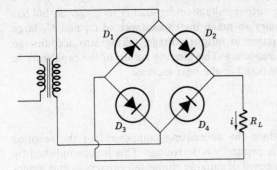

Figure 4-9 Bridge rectifier is full-wave circuit without center-tapped transformer.

half-cycle diodes D_4 and D_1 conduct and the current direction in the load resistance is the same as before. The voltage across R_L corresponds to full-wave rectification and the peak voltage is equal to the transformer voltage less the potential drops across the diodes.

Since two diodes are in series with the load, the output voltage is reduced by twice the diode drop. On the other hand, the peak inverse voltage rating of the diodes need only be equal to the transformer voltage, in contrast to the previous full-wave circuit.

Voltage Doubler

Consider the circuit of Fig. 4-10, in which two diodes are connected to the same voltage source but in the opposite sense. On the half-cycle during which the upper terminal of the source is positive, diode D_1 conducts and capacitor C_1 charges to the peak value of the input voltage. On the reverse half-cycle D_2 conducts and C_2 also charges to the full input voltage. Meanwhile, the charge on C_1 is retained, since the potential across D_1 is in the reverse direction. Thus both C_1 and C_2 charge to the peak supply voltage and the dc output voltage is equal to twice the peak input voltage. Accordingly, this circuit is called a *voltage doubler*.

The above analysis applies to the case when no current is delivered to the load. When a load resistance is connected, current is supplied by the discharge of the capacitors. On alternate half-cycles the capacitors are subsequently recharged.

Figure 4-10 Voltage-doubler rectifier yields dc output voltage equal to twice peak input voltage.

This implies, however, that the output voltage under load is no longer dc, but has an ac component. It is necessary to make the capacitance of C_1 and C_2 large enough to minimize this variation in output voltage, taking into account the current drain and the supply frequency. The filtering action of the capacitors in this circuit is treated in greater detail in the next section.

FILTERS

It is usually desirable to reduce the alternating component of the rectified waveform so that the output is primarily a dc voltage. This is accomplished by means of *filters* which are composed of suitably connected capacitors and inductances. A power-supply filter is a low-pass filter which reduces the amplitudes of all alternating components in the rectified waveform and passes the dc component. A measure of the effectiveness of a filter is given by the *ripple factor r*, which is defined as the ratio of the rms value of the ac component to the dc or average value. That is,

$$r = \frac{V_{\text{rms}}}{V_{\text{dc}}} \tag{4-1}$$

According to this definition it is desirable to make the ripple factor as small as possible. Although rectified waveforms contain many harmonics of the input voltage frequency, it is generally satisfactory to determine the ripple factor at the fundamental frequency only. This is so because the fundamental frequency is predominant and also because higher harmonics are attenuated more than the fundamental by the low-pass filter characteristic.

Capacitor Filter

The simplest filter circuit consists of a capacitor connected in parallel with the load resistance, as in the typical low-voltage bridge-rectifier power supply shown in Fig. 4-11. If the reactance of the capacitor at the power-line frequency is small

Figure 4-11 Practical low-voltage power supply using capacitor filter.

compared with the load resistance R_L, the ac component is shorted out and only dc current remains in the load resistor.

Characteristics of the *capacitor filter* are determined by examining waveforms in the circuit, Fig. 4-12. The capacitor is charged to the peak value of the rectified voltage V_p and begins to discharge through R_L after the rectified voltage decreases from the peak value. The decrease in capacitor voltage between charging pulses depends upon the relative values of the RC time constant and the period of the input voltage. A small time constant means the decrease is large and the ripple voltage is also large. On the other hand, a large time constant results in a small ripple component. The diodes conduct only during the portion of the cycle that the capacitor is charging, because only during this interval is the sum of the supply voltage and the capacitor voltage such that the potential across the diodes is in the forward direction.

The ripple voltage is approximately a triangular wave if the RC time constant is long compared with the period, $R_L C \gg T$. In this case the exponential decrease of capacitor voltage during one period is given approximately by $V_p \times T/R_L C$. The rms value of a triangular wave is calculated in Chap. 2, Eq. (2-87), so that the ripple factor of the capacitor filter is

$$r = \frac{1}{V_p} \frac{1}{\sqrt{3}} \frac{V_p T}{2 R_L C} = \frac{1}{2\sqrt{3}} \frac{1}{f R_L C} \qquad (4\text{-}2)$$

where the dc output voltage has been taken to be equal to V_p for simplicity.

This result shows that ripple is reduced by increasing the value of the filter capacitor. When the load current is equal to zero $(R_L \to \infty)$, the ripple factor becomes zero, which means that the output voltage is pure dc. As the load current is increased (smaller values of R_L), the ripple factor increases. Inserting component values given on the circuit diagram of Fig. 4-11, the ripple factor is found to be 0.05. Thus the ripple voltage is about 5 percent of the dc output. Note that the ripple voltage of a full-wave rectifier is approximately one-half that of the half-wave circuit, because the frequency of the rectified component is twice as great.

Figure 4-12 Output voltage of capacitor filter is dc voltage and small triangular ripple voltage.

The actual dc output voltage from the capacitor filter is equal to the peak capacitor voltage less the average ripple component,

$$V_{dc} = V_p - \frac{V_p T}{2R_L C} = V_p - \frac{1}{2fC} I_{dc} \tag{4-3}$$

where, again, the approximation $I_{dc} \approx V_p/R_L$ is used for simplicity. According to Eq. (4-3), the dc output voltage decreases linearly as the dc current drawn by the load increases. The constancy in output voltage with current is called the *regulation* of the power supply; Eq. (4-3) shows that a large value of filter capacitance improves the regulation. It should be noted that the decrease in output voltage given by this equation refers only to the change accompanying the increase in the ac ripple component. The IR drops associated with diode resistances and the resistance of the transformer winding further reduce the output voltage as the load current increases.

The simple capacitor filter provides very good filtering action at low currents and is often used in high-voltage low-current power supplies. Because of its simplicity, the circuit also is found in those higher current supplies where ripple is relatively less important. The dc output voltage is high, equal to the peak value of the supply voltage. The disadvantages of the capacitor filter are poor regulation and increased ripple at large loads.

L-section Filter

It is useful to add a series inductor to the capacitance filter, as in the circuit of Fig. 4-13. The series inductance in this *L-section* or *choke-input* filter opposes rapid variations in the current and so contributes to the filtering action. The ripple factor is calculated by noting that the ac voltage components of the rectified waveform divide between the inductance and the impedance Z of the resistor-capacitor combination, while the dc output voltage is simply equal to the dc component of the rectified waveform. Therefore, the ripple factor of the L-section filter is written as

$$r = \left(\frac{V_{rms}}{V_{dc}}\right)_{rect} \frac{Z}{Z + X_L} \tag{4-4}$$

Inserting the rms and dc values appropriate for a full-wave rectified waveform, Eqs. (2-92) and (2-88),

$$r = \frac{\sqrt{2}}{3} \frac{1}{1 + \omega^2 LC\sqrt{1 + (1/\omega R_L C)^2}} \approx \frac{\sqrt{2}}{3} \frac{1}{\omega^2 LC} \tag{4-5}$$

Figure 4-13 L-section filter.

where the approximation is true for $R_L C \gg 1$, as is always carried out in practice. Note that the ripple factor is independent of the load, in contrast to the case of the capacitor filter, and that large values of L and C improve the filtering action.

The inductance chokes off alternating components of the rectified waveform and the dc output voltage is simply the average, or dc, value of the rectified wave. For a series of half-sinusoids this means that the dc output voltage is $2V_p/\pi$, or approximately $0.6V_p$. Therefore, the output voltage of the choke-input filter is considerably less than that of the capacitor filter.

According to Eq..(4-5) the ripple is independent of the load current. This means that there is no decrease in output voltage due to a decrease in filtering action at high currents, such as in Eq. (4-3). Therefore, the output voltage is independent of the load, except for IR drops in the diodes and transformer windings. For this reason the L-section filter is used in applications where wide variations in the load current are expected. The advantage of good regulation must be balanced against the comparatively low output voltage in considering any given application.

When the load current is zero the capacitor charges to the peak value of the input waveform as in the capacitor filter. The output voltage is then equal to V_p and regulation is poor. As discussed in connection with the capacitor filter, in this circumstance the diodes conduct for only a part of the applied voltage cycle. The transition between this performance and that of the true L-section filter occurs at the load current for which the diodes just conduct over the entire cycle, and the voltage across the capacitor remains at the dc component of the rectified waveform.

A comparison of the voltage-regulation curves of a capacitor filter and a choke-input filter, Fig. 4-14, shows the poor regulation but high output of the former and the good regulation but lower voltage of the latter. The minimum current necessary to ensure good regulation in the case of the L-section filter is clearly evident.

Figure 4-14 Voltage-regulation characteristics of capacitor-input filter and L-section filter.

π-section Filter

The combination of a capacitor-input filter with an L-section filter, as shown in the power-supply diagram of Fig. 4-15, is a very popular circuit. The output voltage of this *π-section filter* is nearly equal to that of the capacitor-input filter, and the regulation characteristics are about the same. The ripple is very much reduced by the double filtering action, however. In fact, the overall ripple factor is essentially the product of the ripple factor of the capacitor filter times the impedance ratio of the L-section filter. In spite of the inferior regulation properties of the π-section filter, it is widely used because of its excellent filtering action.

Other combinations of filter types are also used. Two L-section filters, Fig. 4-16a for example, provide the good regulation of the choke-input filter together with very low ripple. The ripple factor of any filter design is calculated by considering each simple filter separately, as in the case of the π-section filter described above.

A useful variant of the π-section filter particularly suitable for low-current circuits replaces the inductance with a resistor, Fig. 4-16b. This circuit is useful only if the current is low enough so that the *IR* drop across the resistor is not excessive and if regulation is of secondary concern. Within these restrictions, the circuit provides better filtering action than a simple capacitance filter.

Figure 4-15 A practical 100-V power supply.

Figure 4-16 (a) Double L-section filter and (b) π-section RC filter.

VOLTAGE REGULATORS

It is often desirable that the voltage of a power supply remain fixed independent of the load current. Although the regulation of the choke-input filter is good, the output voltage decreases with increasing current even in this circuit because of the IR drops across the transformer and filter choke winding resistances. Other variables, such as changes in the line voltage and component aging, may also contribute to variations in the power-supply voltage. In order to maintain the output voltage constant, *voltage regulator* devices are made part of the power supply. These are fairly elaborate electronic circuits if very precise regulation is necessary, as discussed in a later chapter. In other cases, special components that maintain modest voltage stability are used.

Zener Diodes

At some particular reverse-bias voltage the reverse current in a *pn* junction increases very rapidly. This happens when electrons are accelerated to high velocities by the electric field at the junction and produce other free electrons by ionization collisions with atoms. These electrons are similarly accelerated by the field and in turn cause other ionizations. This avalanche process leads to very large current and the junction is said to have suffered breakdown. The breakdown is not destructive, however, unless the power dissipation is allowed to increase the temperature to the point where local melting destroys the semiconductor. The voltage across the junction remains quite constant over a wide current range in the breakdown region. This effect can be used to maintain the output of a power supply at the breakdown voltage.

These *pn* junctions are called *zener diodes*, because Clarence Zener first suggested an explanation for the rapid current increase at breakdown. The current-voltage curve of a zener diode, Fig. 4-17, has a sharp transition potential and a flat current plateau above breakdown. Zener diodes may be obtained with breakdown voltages ranging from about two to several hundred volts and with current ratings of a few milliamperes to many amperes.

The way in which a zener diode is used to regulate the output voltage of a power supply is illustrated in Fig. 4-18. The unregulated output voltage of the supply V_s must be greater than the zener breakdown voltage of the diode. Then the voltage drop across R_S caused by the diode current plus the diode breakdown potential adds up to the power-supply potential. As the load current increases, the diode current decreases, so the drop across R_s always is the difference between the breakdown potential and the power-supply voltage. Even if the power-supply voltage varies under load, the regulated output voltage remains constant at the diode breakdown potential.

The zener diode voltage regulator is analyzed by plotting the current-voltage characteristic pertaining to R_s on the same axes as the current-voltage characteristic of the diode. Kirchhoff's rule applied to the circuit of Fig. 4-18 is

$$V_s - IR_s - V = 0 \tag{4-6}$$

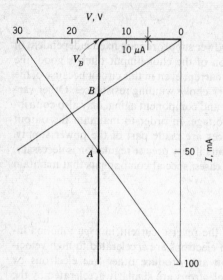

Figure 4-17 Current-voltage characteristic of zener diode.

Figure 4-18 Zener diode used as voltage regulator.

Solving for I,

$$I = \frac{V_s}{R_s} - \frac{1}{R_s} V \qquad (4-7)$$

which is the equation of a straight line of slope $-1/R_s$ and intercept $I = V_s/R_s$. Using numerical values given on the circuit diagram, Eq. (4-7) intersects the current-voltage characteristic of the diode at point A on Fig. 4-17. This intersection gives the current in R_s. The load current is just V_B/R_L, where V_B is the breakdown voltage of the diode.

It is useful to plot the IV characteristic of the load,

$$I_L = \frac{1}{R_L} V \qquad (4-8)$$

on Fig. 4-17. The load current is given by the intersection of Eq. (4-8) with the diode characteristic, point B. Note that the load current must be less than the current in R_s, for otherwise the voltage drop across R_s is too large to maintain V_B across the diode and regulating action ceases. So long as intercept B is at a smaller current than point A, the circuit performs properly.

Zener diodes also provide some filtering action since they maintain the output voltage constant against changes in the power supply voltage, including ripple. For this reason it is usually possible to employ only rudimentary capacitance filtering in conjunction with voltage regulators.

Controlled Rectifiers

It is often necessary to control the power delivered to some load, such as an electric motor or the heating element of a furnace. Series resistances or potentiometers waste power, a serious drawback in high-power circuits. *Controlled rectifiers* have been developed which are capable of adjusting the transmitted power with little waste.

The most satisfactory unit of this kind is the *silicon controlled rectifier*, or *SCR*. This semiconductor device contains four parallel *pn* junctions; it is described in detail in Chap. 5. For present purposes it is sufficient to note that it is similar to a junction rectifier in which forward conduction is controlled by the current in a control electrode, called the *gate*. The inclusion of the gate electrode is shown on the SCR symbol (Fig. 4-19).

The current-voltage characteristic of a typical SCR, Fig. 4-20, is identical to a junction rectifier in the reverse direction. The forward direction has both an "on"

Figure 4-19 Circuit symbol for silicon controlled rectifier.

Figure 4-20 Current-voltage characteristic of SCR. Unit can be triggered into on state by gate current. Here $I_{g2} > I_{g1} > 0$.

state, which is equivalent to normal forward conduction in a junction rectifier, and a low-current "off" state. So long as the forward anode-cathode potential remains below a certain critical value (actually it is the corresponding current which is critical), the forward current is small. Above this critical value the SCR is in the high-current low-voltage on condition. The critical current may be suplied by the gate electrode, which means that the SCR may be triggered into the on position by a small current (as low as 100 μA) supplied to the gate terminal. It is not necessary to maintain the gate current, for once the SCR is triggered into the high-conduction state it remains in this condition until the anode potential is reduced to zero.

Consider the simple SCR circuit of Fig. 4-21, in which the motor speed is controlled by the power delivered to it. The peak value of the transformer voltage V_p is less than the critical value, so that, unless the SCR is triggered by a gate current, the load current is zero. Suppose the gate is supplied with a current pulse each cycle that lags the transformer voltage by a phase angle α, as illustrated in Fig. 4-22. Forward conduction is delayed until this point in each cycle and the rectified output waveform is similar to that of a half-wave rectifier, except that the first part of each cycle is missing. The average, or dc, value of the current in the motor is found by integrating the output current over a complete cycle, or

$$I_{dc} = \frac{I_p}{2\pi} \int_\alpha^\pi \sin \omega t \, d\omega t \tag{4-9}$$

where I_p is the peak value of the current. Integrating,

$$I_{dc} = \frac{I_p}{2\pi} (1 + \cos \alpha) \tag{4-10}$$

According to Eq. (4-10) the motor current can be adjusted from a maximum $(\alpha = 0)$ to zero $(\alpha = \pi)$ simply by changing the phase angle of the gate pulses. Simple circuits are available to generate such pulses, as described in a later chapter.

It is not necessary to use pulses in the gate circuit, so long as care is taken to limit the power dissipated by the control electrode below that which might damage the SCR. Consider the *phase-shift control* circuit of Fig. 4-23, in which the gate current is shifted in phase with respect to the anode voltage. Since the gate voltage is the sum of the secondary voltage v_1 and the drop across R, the gate voltage is

$$v_g e^{j\phi_g} = v_1 - Rie^{j\phi} \tag{4-11}$$

This is easily solved for the phase angle between the gate voltage and the anode voltage, which has the same phase as v_1. The result

$$\phi_g = \arctan \frac{2R\omega C}{(R\omega C)^2 - 1} = -2 \arctan R\omega C \tag{4-12}$$

shows that the gate voltage may be put in phase with the anode potential $(R = 0)$ or nearly 180° out of phase $(R = \infty)$.

Figure 4-21 Simple SCR circuit to control speed of dc motor.

Figure 4-22 Output current waveform depends upon phase angle between transformer voltage and gate-current pulse.

Figure 4-23 Phase-shift control of SCR.

Figure 4-24 Waveforms in phase-shift control SCR circuit.

The waveforms of Fig. 4-24 show how variation of ϕ_g alters the time at which the diode conducts on each forward cycle as the gate voltage becomes larger than the critical value. Note that the gate signal remains on during part of the reverse half-cycle as well, but this has no effect upon the reverse characteristic. The purpose of resistor R_1 and diode D_1 in the circuit (Fig. 4-23) is to limit the gate current to a safe value and to prevent reverse gate current, both of which are deleterious to the SCR.

DIODE CIRCUITS

Diodes prove to be useful in circuits other than power supplies. A major application is in circuits designed to operate with square-wave pulses, as will be discussed in a later chapter. The rectification characteristics of diodes are also put to work in circuits dealing with sine-wave signals. In these applications the nonlinear diode modifies the sine-wave signals in a specified manner.

Clippers

Consider the diode *clipper* circuit, Fig. 4-25, in which two diodes are connected in parallel with an ac voltage source. Batteries V_1 and V_2 bias each diode in the reverse direction. Whenever the input voltage signal is greater than V_1, diode D_1 conducts and causes a voltage drop across the series resistor R. Similarly, every time the input voltage becomes more negative than V_2, diode D_2 conducts. Thus, the output waveform is *clipped* or *limited* to the voltages set by the reverse biases V_1 and V_2. The clipping action is most efficient when the series resistance is much greater than the load impedance.

Figure 4-25 Diode clipper.

Figure 4-26 Maximum amplitudes in output waveform of diode clipper are limited to values of bias voltages.

Clipper circuits produce square waves from a sine-wave generator, as illustrated by the waveforms in Fig. 4-26. If $V_1 = V_2$ and the amplitude of the input signal is considerably greater than the bias voltages, the output waveform is very nearly a square wave. Note that if, say, $V_2 = 0$, the output can never swing negative, so the waveform is a train of positive-going square pulses. The clipping action operates on any input waveform; it can be adjusted by altering the reverse-bias voltages V_1 and V_2.

The clipper circuit is also a useful protective device which limits input voltages to a safe value set by the bias voltage. In radio-receiver circuits limiters are often used to reduce the effect of strong noise pulses by limiting their amplitude to that of the desired signal. The signal waveform is transmitted undistorted so long as the instantaneous amplitude remains smaller than the bias voltages.

Clamps

The diode *clamp* circuit is shown in Fig. 4-27. Consider first the situation in which the bias voltage V is set equal to zero. The diode conducts on each negative cycle of input voltage and charges the capacitor to a voltage equal to the negative peak

Figure 4-27 Diode clamp.

Figure 4-28 Negative peak of output waveform is clamped at zero when $V = 0$ in diode clamp circuit of Fig. 4-27.

value of the input signal. If the load current is zero, the capacitor retains its charge on the positive half-cycle, since the diode voltage is in the reverse direction. Thus, the output voltage is

$$v_o = v_i + V_p \tag{4-13}$$

where V_p is the negative peak value of the input voltage. According to Eq. (4-13) the output voltage waveform replicates the input signal except that it is shifted by an amount equal to the dc voltage across the capacitor.

The output waveform corresponding to a sinusoidal input, Fig. 4-28, has the negative peaks of the sine wave *clamped* at zero voltage. This is always the case, independent of the amplitude of the input voltage. Furthermore, the negative peaks are always clamped at zero volts no matter what form the input waveshape takes. When the terminals of the diode are interchanged, the same circuit analysis applies; the positive peaks of the output wave are clamped at zero.

If the bias voltage V in Fig. 4-27 is set at a potential other than zero, the capacitor charges to a voltage equal to $V_p + V$. Therefore, the negative peaks are clamped at the voltage V. Similarly, when the bias is $-V$, the negative peaks are clamped at this potential. Reversing the diode polarity makes it possible to clamp the positive peaks of the input wave at the voltage equal to the bias potential.

The diode clamp is used in circuits which require that the voltages at certain points have fixed peak values. This is important, for example, if the waveform is subsequently clipped at a given voltage level. In most diode clamp circuits it is useful to connect a large resistance across the output terminals so that the charge on the capacitor can eventually drain away. This makes it possible for the circuit to adjust to changes in amplitude of the input voltage.

Ac Voltmeters

The rectifying action of diodes makes it possible to measure ac voltages with a dc voltmeter, such as the d'Arsonval meter discussed in the first chapter. Two ways in which this is accomplished in VOM meters are illustrated in Fig. 4-29. In Fig. 4-29a diode D_1 is a half-wave rectifier and the dc component of the rectified waveform registers on the dc voltmeter. Diode D_2 is included to rectify the negative cycle of the input waveform. Although no meter current results from the action of diode D_2, it is included so that both halves of the ac voltage are rectified. This ensures that the voltmeter loads the circuit equally on both half-cycles and avoids possible waveform distortion. The bridge circuit, Fig. 4-29b, has greater sensitivity than the half-wave circuit because it is a full-wave rectifier.

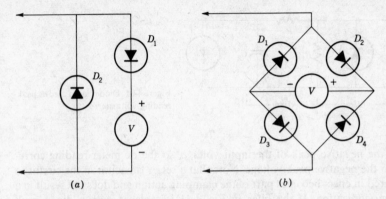

Fig. 4-29 (*a*) Half-wave and (*b*) full-wave ac voltmeters.

The deflection of a d'Arsonval meter is proportional to the average current, so that the voltmeters of Fig. 4-29 measure the average value of the ac voltage. The meter scale is commonly calibrated in terms of rms readings, however, in order to facilitate comparison with dc readings. This calibration assumes that the ac waveform is sinusoidal, and, if this is not the case, the meter readings must be interpreted in terms of the actual waveform measured. Series resistor multipliers are used with ac voltmeters to increase the range, just as in the case of dc meters.

The half-wave rectifier, Fig. 4-30, makes a useful *peak-reading* voltmeter if the current drain through the meter circuit is small. In this case the capacitor charges to the positive peak value of the input waveform on each half-cycle, and the dc meter deflection corresponds to this value. Here again, it is common practice to calibrate the meter scale in terms of rms readings, assuming that the unknown voltage is sinusoidal. This is accomplished by designing the scale to indicate the peak value divided by $\sqrt{2}$. In case the unknown voltage is not sinusoidal, the peak value is obtained by multiplying the scale reading by $\sqrt{2}$. A difficulty with this circuit is that the system being measured must contain a dc path. If this is not so and a series capacitor is present, the voltage across C depends upon the relative values of the two capacitors, since they are effectively in series during the forward cycle. This results in erroneous voltage readings.

The diode clamp is a peak-reading circuit which does not have this difficulty. It is customary to include an RC filter, Fig. 4-31, to remove the ac component of the clamped wave. Referring to the waveform in Fig. 4-28, the average value is

Figure 4-30 Peak-reading ac voltmeter.

|← Clamp →|← Filter →|

Figure 4-31 Diode clamp used as peak-reading voltmeter.

equal to the negative peak of the input voltage, so the dc meter reading corresponds to the negative peak voltage. Note that a series capacitor in the circuit to be measured in effect becomes part of the clamping action and does not result in a false meter indication. If the filter in Fig. 4-31 is replaced with a diode peak rectifier, Fig. 4-30, the combination reads the peak-to-peak value of the input wave (Exercise 4-9). Although usually calibrated in terms of rms readings for a sine wave, a peak-to-peak voltmeter is very useful in measuring complex waveforms.

Detectors

Nonlinear diode properties are useful in other ways than those corresponding to actual rectification. Suppose that the amplitude of the ac voltages applied is very small, so that the diode current-voltage characteristic can be represented by the expression

$$i = a_1 v + a_2 v^2 \qquad (4\text{-}14)$$

where a_1 and a_2 are constants. An expression of this form is a satisfactory approximation of the diode characteristic for sufficiently small values of applied voltage. According to Eq. (4-14), one part of the current is proportional to the square of the input voltage.

Figure 4-32 Diode mixer uses nonlinear properties of diode.

Consider the circuit shown in Fig. 4-32, in which two sinusoidal signals of somewhat different frequencies, ω_1 and ω_2, are supplied to a diode. The total diode voltage is the sum of the individual waves,

$$v = V_1 \sin \omega_1 t + V_2 \sin \omega_2 t \tag{4-15}$$

where V_1 and V_2 are the peak amplitudes. The current is found by substituting Eq. (4-15) into Eq. (4-14), assuming for the moment that the diode impedance is greater than other impedances in the circuit.

$$i = a_1 V_1 \sin \omega_1 t + a_1 V_2 \sin \omega_2 t + a_2 V_1^2 \sin^2 \omega_1 t$$
$$+ a_2 V_2^2 \sin^2 \omega_2 t + 2a_2 V_1 V_2 \sin \omega_1 t \sin \omega_2 t \tag{4-16}$$

It is instructive to express Eq. (4-16) in the form of a Fourier series type expression in which each term represents a given frequency in the total waveform. This is accomplished by introducing trigonometric substitutions for $\sin^2$ and $\sin \omega_1 t \sin \omega_2 t$ and rearranging,

$$i = \frac{a_2}{2}(V_1^2 + V_2^2) + a_1(V_1 \sin \omega_1 t + V_2 \sin \omega_2 t)$$

$$- \frac{a_2}{2}(V_1^2 \cos 2\omega_1 t + V_2^2 \cos 2\omega_2 t)$$

$$+ a_2 V_1 V_2 \cos (\omega_1 - \omega_2)t - a_2 V_1 V_2 \cos (\omega_1 + \omega_2)t \tag{4-17}$$

The first term in Eq. (4-17) is a direct current, while the second corresponds to the input voltages. The last two terms are at frequencies not present in the original inputs. Suppose ω_1 and ω_2 are not too different; then the term involving $\omega_1 - \omega_2$ is a comparatively low frequency.

In the *diode mixer* circuit of Fig. 4-32 the output circuit $L_2 C_2$ is tuned to the frequency $\omega_1 - \omega_2$, so only this component has an appreciable amplitude at the output terminals. In effect, an input signal of frequency ω_1 is mixed with a constant-amplitude sine wave at ω_2 and is converted to an output frequency $\omega_1 - \omega_2$. This is called *heterodyning* and is of considerable importance in, for example, radio receivers. The incoming high-frequency signal is converted into a lower-frequency signal which is easier to amplify. Furthermore, changing the frequency ω_2 means that the receiver is tuned to a different ω_1, such that $\omega_1 - \omega_2$ is a constant. This is important because the amplification of the $\omega_1 - \omega_2$ signal is accomplished by fixed tuned circuits that are inexpensive and easy to keep in adjustment, yet the receiver may be tuned to a different input frequencies. In this application Fig. 4-32 is called a *first detector* because the signal frequency is modified in the first stages of the receiver.

This circuit has another important use, which may be illustrated by an alternate rearrangement of Eq. (4-17), using the trigonometric substitution $2 \sin a \sin b = \cos (a - b) - \cos (a + b)$

$$i = \frac{a_2}{2}(V_1^2 + V_2^2) + a_1 V_2 \sin \omega_2 t - \frac{1}{2}(V_1^2 \cos 2\omega_1 t + V_2^2 \cos 2\omega_2 t)$$

$$+ a_1 V_1 \sin \omega_1 t + 2a_2 V_1(V_2 \sin \omega_2 t) \sin \omega_1 t \tag{4-18}$$

Suppose ω_2 is much smaller than ω_1 and that the output circuit $L_2 C_2$ is tuned to the frequency ω_1. Then only the last two terms in Eq. (4-18) have an appreciable output amplitude. The first of these is simply the current resulting from the input voltage at the frequency ω_1. Careful inspection of the last term in Eq. (4-18) shows that it is a sinusoidal waveform at a frequency ω_1 but with an amplitude that varies sinusoidally at a frequency corresponding to ω_2.

The waveform of the last term in Eq. (4-18), illustrated in Fig. 4-33, shows how the amplitude varies. The frequency ω_1 is said to be *modulated* at the frequency of ω_2. This is the basis for radio and television communication wherein a high-frequency *carrier* ω_1 is modulated in accordance with a low-frequency signal ω_2 corresponding to the sound or picture to be transmitted. The high-frequency voltage is radiated from the sending station to a receiver where the waveform of the signal is recovered.

Figure 4-33 Modulated waveform.

Figure 4-34 Diode peak rectifier used to demodulate modulated waveform.

Figure 4-35 Modulating waveform recovered at output of second detector.

The signal is recovered by *demodulating* the modulated waveform with a diode peak rectifier, Fig. 4-34. Since the output of the peak rectifier is equal to the peak value of the input voltage, the waveform across the output capacitor corresponds to that of modulating signal, as in Fig. 4-35. In this application the peak rectifier is called a second detector, because the waveform is changed for the second time in the receiver circuit. The second detector circuit, Fig. 4-34, is widely used in radio and television receivers.

SUGGESTIONS FOR FURTHER READING

Paul M. Chirlian and Armen H. Zemanian: "Electronics," McGraw-Hill Book Company, New York, 1961.
W. Ryland Hill: "Electronics in Engineering," McGraw-Hill Book Company, New York, 1961.
Jacob Millman and Samuel Seely: "Electronics," McGraw-Hill Book Company, New York, 1951.

EXERCISES

4-1 Determine the dc output current from the half-wave rectifier circuit in Fig. 4-6, if $v = 10$ V, $R_L = 100 \, \Omega$, and the rectifier is a silicon diode. Repeat for the full-wave rectifier in Fig. 4-7.
    Ans: 0.45 A, 0.90 A

4-2 In the voltage-doubler circuit, Fig. 4-10, the input voltage is 115 V rms at a frequency of 60 Hz. If the load resistance is 10,000 Ω, what is the minimum value of the capacitors necessary to be sure the voltage drop between charging pulses is less than 10 percent of the output voltage?
    Ans.: 4.2×10^{-6} F

4-3 Calculate the output voltage and ripple factor of the circuit of Fig. 4-11 after diodes D_1 and D_4 are removed.
    Ans.: 12.9 V; 0.048

4-4 Plot the dc output and the ripple voltage of the full-wave rectifier circuit given in Fig. 4-15 up to a current of 200 mA.

4-5 Determine the ripple factor of an L-section filter comprising a 10-H choke and a 8-μF capacitor used with a full-wave rectifier. Compare with a simple 8-μF capacitor filter at a load current of 50 mA and also 150 mA, assuming an output of 50 V.
    Ans.: 0.0104; 0.300; 0.900

4-6 In the zener diode voltage regulator of Fig. 4-18 determine the range of load resistances over which the circuit is useful if $R_s = 1500 \, \Omega$ and the supply voltage $V = 150$ V; the diode breakdown voltage is 100 V and maximum rated current is 100 mA. For a fixed $R_L = 10,000 \, \Omega$, over what range of input voltages does the circuit regulate?
    Ans.: Greater than 3 kΩ; 115 to 265 V

4-7 Plot the dc output voltage of the zener diode voltage regulator in Fig. 4-18 as a function of load current from 0 to 75 mA. Over what current range is regulation effective?
    Ans.: 0 to 50 mA

4-8 Using the phase-shift controlled rectifier circuit, Fig. 4-23, plot the dc load current as a function of control resistance R if the transformer secondary voltage is 100 V rms, the load resistance is 10 Ω, and $C = 0.1 \, \mu$F. Assume the SCR turns on whenever the gate voltage is positive with respect to the cathode.

4-9 Analyze the peak-to-peak voltmeter circuit, Fig. 4-36, by sketching the voltage waveforms at points A, B, and C, assuming the input voltage A is sinusoidal. Note that the circuit is a diode clamp followed by a peak rectifier.

Figure 4-36

4-10 The output reading of the peak-to-peak voltmeter in Fig. 4-36 measuring a sawtooth waveform is 1.4 V. What is the voltage of the waveform?

    Ans.: 1.15 V (4 V peak-to-peak)

SEMICONDUCTOR DEVICES

The versatility inherent in nonlinear electronic components is enhanced immeasurably by the capability to influence current in the device in accordance with signals introduced at a control electrode. Both the older vacuum triode (so named because of the third electrode) and the modern transistor exhibit this feature. These are considered to be active devices, rather than passive components, because the control electrode permits active interaction with signal currents in the device.

Electrical properties of active devices are described by current-voltage characteristics as for other nonlinear components. It is often useful to develop an appreciation for the origin of these current-voltage characteristics in terms of basic physical processes. In this way it is possible to optimize performance of a nonlinear element in any application, and also to design improved components having desirable electrical features. It turns out that the current-voltage characteristics of electronic devices depend primarily upon the motions of free electrons in them.

The electronic functions of transistors take place within solid materials called semiconductors. Accordingly, the properties of transistors and other semiconductor devices, such as the junction diode, stem directly from the behavior of electrons in semiconductor crystals. The great variety of semiconductor devices that have been developed since the discovery of the transistor are possible because of the versatility of semiconductor materials.

SEMICONDUCTORS

Energy Bands

The properties of any solid material, including semiconductors, depend upon the nature of the constituent atoms and upon the way in which the atoms are grouped together. That is, the properties are a function of both the atomic structure of the atoms and the crystal structure of the solid. Experiments have shown that an atom consists of a positively charged nucleus surrounded by electrons located in discrete orbits. Actually electrons can exist in stable orbits near the nucleus only for certain discrete values of energy called *energy levels* of the atom.

The allowed energies of electrons in an atom are depicted by horizontal lines on an *energy-level diagram*, Fig. 5-1*a*. The curved lines in the diagram represent the potential energy of an electron near the nucleus. As a consequence of the *Pauli exclusion principle*, only a certain maximum number of electrons can occupy a given energy level and the result is that in any atom, electrons fill up the lowest possible levels first.

When atoms come close together to form a solid crystal, electrons in the upper levels of adjacent atoms interact to bind the atoms together. Because of the strong interaction between these outer, or *valence*, electrons, the upper energy levels are drastically altered. This can be illustrated by an energy-level diagram for the entire crystal. Consider first two isolated atoms, each with an energy-level diagram pertaining to the outer electrons as in Fig. 5-1*a*. When these are brought close together, Fig. 5-1*b*, the valence electrons in both atoms are attracted by both nuclei. The result is that the energy required to remove an electron from one nucleus and place it on the other is reduced. This means that an outer electron is equally likely to be located near either nucleus. The appropriate energy-level diagram for the combination of two atoms has two energy levels near each atom core. The higher, unoccupied, levels are similarly split, indicating that these levels, too, can each contain two electrons. When three atoms are brought together, Fig. 5-1*b*, the outer electrons of all three atoms can be associated with any of the three nuclei. Consequently, three energy levels are available.

Even the tiniest crystal contains many hundreds of millions of atoms, so that very many energy levels are associated with each nucleus and the energy-level diagram appropriate for the entire crystal is a band of levels. The lowest energy band, called the *valence band* (Fig. 5-1*c*), is completely filled with electrons for there is one electron for each of the available energy levels. Conversely, the upper energy band is empty of electrons because it corresponds to the unoccupied higher levels in the isolated atom. It is called the *conduction band*, for reasons explained below. The energy region between the valence band and the conduction band is called the *forbidden energy gap* since no electrons with such energies exist in the crystal. The forbidden energy gap corresponds to the energy region between energy levels in the isolated atom, as can be seen by comparing the energy-level diagrams in Fig. 5-1.

(a)

(b)

Empty conduction band
Forbidden energy gap
Filled valence band

(c)

Figure 5-1 Energy-level diagram for (a) isolated atom, (b) two and three atoms close together, and (c) solid crystal. In the crystal, energy levels are broadened into bands.

This picture of the electronic energy levels in a crystal is known as the *energy-band model* of a crystal. It is very useful in determining the electrical properties of any solid, since it shows how electrons can move in the crystal. While the general features of the band model for any solid are as described, many important details depend upon the specific atomic and crystal structure. In particular, the differences between metals, semiconductors, and insulators are reflected in their energy-band models.

The atomic and crystal structures of metals are such that the valence and conduction bands overlap, as indicated in the conventional energy-band model for a metal, Fig. 5-2a. Since there is no forbidden energy gap in a metal crystal, any of the many valence electrons are free to roam throughout the solid and to move in response to an electric field. Therefore, metals are excellent electric conductors.

(a)   (b)   (c)

Figure 5-2 Energy-band models for (a) metal, (b) semiconductor, and (c) insulator.

Table 5-1. Forbidden energy gaps of typical semiconductors

Name	Chemical symbol	Forbidden energy gap, eV
Silicon	Si	1.1
Germanium	Ge	0.72
Gallium arsenide	GaAs	1.34
Indium antimonide	InSb	0.18
Cadmium sulfide	CdS	2.45
Zinc oxide	ZnO	3.3

An insulating crystal has a wide forbidden energy gap, Fig. 5-2c. The valence band is completely filled with electrons and the conduction band is completely empty. Obviously the upper band cannot contribute to electric conductivity since no electrons are present to act as carriers. It may seem paradoxical at first, but electrons in the completely filled valence band also cannot conduct electricity. When an electron moves in response to an electric field, it must move slightly faster than before. Consequently, it has greater energy and must find an empty level at a slightly higher energy. Every nearby level is filled, however, so that it is impossible for any electron in the filled valence band to be accelerated by the electric field and the crystal is therefore an insulator.

The energy-band model of a semiconductor, Fig. 5-2b, is similar to that of an insulator except that the forbidden energy gap is comparatively narrow. A few electrons can be promoted from the valence band to the conduction band across the forbidden energy gap by virtue of the thermal energy of the crystal at room temperature. Electrons promoted to the conduction band can conduct electricity. The corresponding electron vacancies in the valence band make it possible for electrons in this band to contribute to conductivity as well. Since the number of carriers is much fewer than in the case of a metal, semiconductors are poorer conductors than metals but better than insulators.

The width of the forbidden energy gap of semiconductors is of the order of 1 electronvolt, as shown in Table 5-1 for several typical semiconductor crystals. The electronvolt, abbreviated eV, is equal to the kinetic energy gained by an electron in traversing a potential difference of 1 V. It is a convenient energy unit in semi-conductor studies. In general, materials with a wide forbidden energy gap are desirable for semiconductor devices. The number of electrons promoted to the conduction band at high temperatures is small, and the change in device charac-teristics with temperature is less severe when the forbidden energy gap is wide. For this reason, silicon crystals are more widely used than germanium crystals.

Electrons and Holes

According to the preceding discussion, the net current resulting from electrons in a filled valence band is zero. Electrons in the valence band of a semiconductor at room temperature can conduct current, however, because of the few vacant levels left behind by electrons excited to the conduction band. The vacant levels in the valence band are called *holes*. It turns out that holes in the valence band can be treated as positively charged carriers fully analogous to the negatively charged electrons in the conduction band.

The energy-band model, Fig. 5-2b, refers to a perfect crystal structure which contains no chemical impurities and in which no atoms are displaced from their proper sites. The properties of the solid are therefore characteristic of an ideal structure, and the crystal is called an *intrinsic semiconductor*. Although it is not possible to achieve perfect structures in real crystals, this ideal may be approached and intrinsic behavior observed experimentally.

Extrinsic Semiconductors

The electrical properties of a semiconductor are drastically altered when foreign, or *impurity*, atoms are incorporated into the crystal. Since the properties now depend strongly upon the impurity content, the solid is called an *extrinsic* semiconductor. Consider, for example, a single crystal of the important semiconductor material silicon. Each silicon atom has four valence electrons which are part of the filled valence band of a silicon crystal. Suppose now a pentavalent atom, such as phosphorus, arsenic, or antimony, substitutes for a silicon atom in the crystal. Four of the impurity atom's electrons play the same role as the four valence electrons of the replaced silicon atom and become part of the valence band. The fifth valence electron is easily detached by thermal energy and moves freely in the conduction band.

Phosphorus, arsenic, or antimony impurity atoms in silicon donate electrons to the conduction band and are called *donor* impurities. One electron is present in the conduction band for each donor atom in the crystal; note that there is not an equivalent number of holes in the valence band. The crystal conducts electricity mainly by virtue of electrons in the conduction band, and such a crystal is called an *n-type* semiconductor because of the negative charge on the current carriers.

By comparison, a trivalent atom like boron, aluminum, gallium, or indium substituted for a silicon atom produces a hole in the valence band. Since only three electrons are available, an electron from an adjacent silicon atom transfers to the impurity atom. The foreign atom is said to have accepted a valence electron, and such impurities are termed *acceptors*. Acceptors in a crystal create holes in the valence band and produce a *p-type* semiconductor because of the effective positive charge on each hole.

A crystal containing both donor and acceptor impurities is either *n*-type or *p*-type, depending upon which impurity concentration is greater, because electrons from donor atoms fill up all available acceptor levels. Intrinsic crystals can be produced by including equal concentrations of donor and acceptor impurities, and such crystals are said to be *compensated*.

Note that a few holes are present in the valence band of an *n*-type crystal since some electrons are excited across the forbidden energy gap. These holes are called *minority* carriers and the electrons are called *majority* carriers because of their relative concentration. Conversely, holes are majority carriers in a *p*-type semiconductor, while electrons in the conduction band are minority carriers. Intentionally introducing impurity atoms in semiconductors to obtain a desired concentration of majority carriers is called *doping*.

SEMICONDUCTOR DIODES

The *pn* Junction

The junction between a *p*-type region and an *n*-type region in the same semiconductor single crystal, a basic structure in many semiconductor devices, is called a *pn junction*. Several different techniques for producing *pn* junctions are described in a later section. All result in a transition from acceptor impurities to donor impurities at a given place in the crystal. Electrons in the *n* region tend to diffuse into the *p* region at the junction. At equilibrium, this is compensated by an equal flow of electrons in the reverse direction. The concentration of electrons is much larger in the *n* material, however, and the electron current from the *n* region would be greater except for a potential rise at the junction which reduces the electron current from the *n* side. An identical argument applies to the hole currents diffusing across the junction. The magnitude and polarity of the internal potential rise at the junction make the two currents equal, and the built-in potential rise leads to excellent rectification in the *pn* junction, as already noted in Chap. 4.

Notice that the polarity of the potential rise shown in the energy-band model for a *pn* junction, Fig. 5-3, tends to keep electrons in the *n* region and holes in the *p* region, as discussed above. The magnitude of the potential step, V_0, depends upon

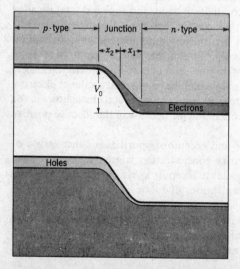

Figure 5-3 Energy-band model of *pn* junction.

the number of donor electrons at the junction that transfer into nearby acceptor levels, that is, upon the impurity concentrations, and is approximately equal to the size of the forbidden energy gap. The width of the junction region, d, is simply related to the impurity concentration and the junction voltage,

$$d^2 = A \frac{V_0}{N_a} \tag{5-1}$$

where A is a constant and N_a is the acceptor impurity concentration. In arriving at Eq. (5-1) it has been assumed that the donor impurity concentration on the n-type side is much larger than the acceptor concentration on the p-type side.

According to Eq. (5-11), the junction is narrow for high impurity concentrations. This principle is used to set the breakdown voltage of zener diodes at the desired potential. The avalanche breakdown process discussed in Chap. 4 depends upon the electric field in the junction. Therefore, the voltage at which breakdown occurs is small in narrow junctions and large in wide junctions.

When the pn junction is in equilibrium the current I_1 resulting from electrons diffusing from the n side is equal to the current I_2 which arises from electrons leaving the p side, Fig. 5-4a. Suppose now an external potential is applied to the

(a) Equilibrium

(b) Reverse bias

(c) Forward bias

Figure 5-4 Current across (a) pn junction depends upon height of potential rise at junction, which is (b) increased under reverse bias and (c) decreased by forward bias.

junction so as to increase the internal barrier, as in Fig. 5-4*b*. The number of electrons diffusing across the junction from the *n* region is much reduced since very few electrons have sufficient energy to surmount the larger potential barrier. On the other hand, the number moving from the *p* to the *n* side is not affected because these electrons encounter no barrier. Thus a net current exists, but it is limited by the small number of electrons in the *p* region. If the polarity of the external potential is reversed, Fig. 5-4*c*, the internal barrier is reduced and I_1 is large because the number of electrons in the *n* region is so great. Again, the electron current I_2 from *p*-type to *n*-type remains unaffected. The net current in this case is large and corresponds to the forward direction.

In determining the current-voltage characteristic of the junction diode, the same considerations apply to the current carried by the holes as to that carried by electrons, and the total current is the sum of the two. Focusing attention on the electrons first, the electron current from the *n* region to the *p* region is controlled by the potential rise at the junction in accordance with Boltzmann's law,

$$I_1 = C_n e^{-e(V_0 - V)/kT} \tag{5-2}$$

where k is Boltzmann's constant, T is the absolute temperature, and C_n is a constant determined by the junction area and properties of the semiconductor. Note that in equilibrium when the applied voltage V is equal to zero, then $I_1 = I_2$, so that the net electron current can be written as

$$I_n = I_1 - I_2 = C_n e^{-eV_0/kT}(e^{eV/kT} - 1) \tag{5-3}$$

A similar expression applies to the hole currents, which means that the total junction current is just

$$I = I_0(e^{eV/kT} - 1) \tag{5-4}$$

where I_0 is called the *saturation current*,

$$I_0 = C e^{-eV_0/kT} \tag{5-5}$$

Equation (5-4) is known as the *rectifier equation*.

Figure 5-5 Current-voltage characteristic of *pn* junction according to rectifier equation.

The polarity of the applied potential is such that the p region is positive for forward bias. According to Eq. (5-4) the current increases exponentially in the forward direction. In contrast, the reverse current is essentially equal to I_0, independent of reverse potentials greater than a few tenths of a volt. A plot of the rectifier equation for small values of applied voltage is shown in Fig. 5-5. Experimental current-voltage characteristics of practical junction diodes, Fig. 4-2, are in good agreement with the rectifier equation.

The Tunnel Diode

A useful effect occurs in pn junctions in which the impurity concentrations on both the n and p sides are very great. The energy-band model of such a junction is shown in Fig. 5-6a for the case of zero applied voltage.

Very large impurity concentrations mean that the junction is very narrow, according to Eq. (5-1). Widths of the order of 10 to 100 atomic diameters are easily achieved in germanium diodes and similar devices. In this situation it is

Equilibrium

Small forward bias

Increased forward bias

Figure 5-6 Energy-band model of tunnel diode (a) in equilibrium, (b) at small forward bias, and (c) at increased forward bias. Note current is greater at (b) than at (c) because of electron tunneling.

possible for an electron in the conduction band on the n side to jump to the valence band on the p side by a process called electron *tunneling*. The tunneling transition takes place with no change in the energy of the electron. The ability of an electron to tunnel through the potential barrier of the junction is a result of the wavelike nature of the electron and the probability that an electron will penetrate the potential barrier can be calculated from the principles of quantum mechanics. The tunneling current depends upon the junction width, upon the number of electrons capable of tunneling, and upon the number of empty energy levels into which they can transfer. In equilibrium, Fig. 5-6a, the tunneling currents across the junction in the two directions are equal and the net current is zero.

At a small forward-bias potential, electrons tunnel from the conduction band on the n side to empty levels in the valence band on the p side and a forward current results. The current increases with voltage until the electrons on the n side are in line with the holes on the p side, Fig. 5-6b. As the bias is increased beyond this point, electrons in the conduction band are raised above the valence-band states and the tunnel current is reduced. In this range of forward bias each increment in voltage causes a decrease in current. Finally, Fig. 5-6c, the electrons and holes are completely out of line and the junction current corresponds to the normal forward pn-junction current.

The current-voltage characteristic of a *tunnel diode* (also referred to as an *Esaki diode* after its discoverer), Fig. 5-7, shows how the current rises to a maximum, corresponding to the condition of Fig. 5-6b, and then decreases as the forward bias is increased. The interval between the peak and valley of the characteristic curve exhibits a negative resistance effect since each voltage increase reduces the current. The usefulness of this effect stems from the fact that conventional electronic circuit components have positive resistance and therefore dissipate power. If a tunnel diode is placed in a resonant circuit so that the net resistance vanishes, no power loss results. The circuit therefore oscillates at its resonant frequency, as analyzed in Chap. 9.

Figure 5-7 Experimental current-voltage characteristic of tunnel diode. Negative resistance region exists between peak and valley currents.

Figure 5-8 Hole concentration near *pn* junction under foward bias.

Minority Carrier Injection

The internal potential barrier in a *pn* junction is reduced when the junction is biased in the forward direction. The forward current results from holes diffusing across the junction from the *p*-type side and electrons diffusing across from the *n*-type side. The result is that holes are injected into the *n* region and electrons are injected into the *p* region, where in each case they are minority carriers. This *minority-carrier injection* at a *pn* junction is the basis for junction transistors, as described in the next section. It is usually desirable that the forward current be carried predominantly by either holes or electrons in order to enhance the injection effect.

If the *n* region is lightly doped and the *p* region is heavily doped, the forward current is carried mostly by holes and, consequently, a large excess hole concentration is injected into the *n* region. If the doping ratio is interchanged, the reverse situation is true and electrons are injected into the *p* region. Injected holes diffuse away from the junction because of their concentration gradient. As they diffuse, they combine with the majority carriers, electrons, so that far from the junction the hole concentration is characteristic of the *n*-type semiconductor. The concentration of holes in the various regions is indicated schematically in Fig. 5-8.

JUNCTION TRANSISTORS

Collector Characteristics

A junction transistor consists of two parallel *pn* junctions juxtaposed in the same single crystal. Two distinct types are possible, the *pnp* transistor and the *npn* transistor, depending upon the conductivity type of the common region. The operation of the two is conceptually identical except for the interchange of minority and majority carrier types and the polarity of the bias potentials, so that it suffices to discuss the *pnp* transistor. Examples of structures having this dual junction arrangement are described in a following section.

(a)

(b)

(c)

Figure 5-9 (a) Energy-band diagram of *pnp* transistor in equilibrium, and (b) under operating bias; (c) hole concentration for case (b).

The energy-band model for a *pnp* structure in the absence of applied bias voltages, Fig. 5-9a, is simply that of two *pn* junctions placed back to back. In operation, one junction, called the *emitter*, is biased in the forward direction and the other, the *collector*, is biased in the reverse direction, as in Fig. 5-9b. Holes injected into the *n*-type *base* region at the emitter junction diffuse across to the collector junction where they are collected by the electric field at the junction. The hole concentration in the various regions is sketched in Fig. 5-9c.

Variations in the emitter-base bias voltage change the injected current correspondingly, and this signal is observed at the collector junction. The forward-biased emitter represents a small resistance and the reverse-biased collector a large resistance. Since nearly the same current is in both, a large power gain results. For maximum amplification, it is desirable that the collector current be as large a fraction of the emitter current as possible. The current through the emitter junction should be carried primarily by holes, since electrons injected from the base to the emitter cannot influence the collector current. This can be accomplished if the n-type base region is lightly doped. Secondly, the base region must be thin so that few holes are lost before reaching the collector junction. It is convenient to dope the collector region lightly in order to increase the junction width. This reduces the capacitance of the collector junction and also increases the reverse breakdown voltage.

A useful figure of merit for a transistor is the *current-gain factor* α, which is the ratio of the change in collector current to the change in emitter current for constant collector voltage. Carrier currents in the various regions of a transistor are given in Fig. 5-10, using the emitter current I_e as a starting point. The current I_e in the emitter junction results in a collector current αI_e. The difference between the emitter and collector currents $(1 - \alpha)I_e$ appears at the external connection to the base. Obviously, when the current gain is unity, all the emitter current appears in the collector circuit and the base current is zero.

Actually, α is very nearly unity for most devices so that the base-collector current gain β is a more sensitive measure of transistor quality. The relation between the base-collector current gain and α is, from Fig. 5-10,

$$\beta = \frac{I_c}{I_b} = \frac{\alpha}{1 - \alpha} \tag{5-6}$$

The base-collector current gain is large when α approaches unity. Typical values of β range from 20 to 10^3 in practical transistors.

Minority carriers injected into the base move to the collector by diffusion, and the finite time taken by the carriers to cross the base limits the high-frequency

Figure 5-10 Currents in *pnp* transistor expressed in terms of emitter current and current-gain factor.

usefulness of transistors. By ingenious fabrication techniques it is possible to achieve very narrow base widths ($\sim 5 \times 10^{-7}$ m) and obtain useful amplification at frequencies of 10^9 Hz. It is also necessary to reduce the area of the junctions in high-frequency transistors in order to minimize adverse effects of the junction capacitances.

A convenient way to represent the current-voltage characteristics of a transistor is the collector characteristic, Fig. 5-11, for different values of emitter current. When $I_e = 0$ the characteristic is simply the reverse saturation curve of the collector junction. Emitter current translates the curve along the current axis. These characteristics follow directly from the preceding discussion and pertain to the *grounded-base* circuit configuration already considered in Fig. 5-10. In the grounded-base connection the base terminal is common to both input and output circuits. As is considered in greater detail in the following chapter, the *grounded-emitter* configuration in which the emitter electrode is common to both input and output circuits, exhibits more favorable properties.

The quality of practical transistors is so satisfactory that the grounded-base current gain α is very nearly equal to unity and the reverse leakage current at the collector junction is negligible for all practical purposes. Consequently, the grounded-base collector characteristics are very flat, straight, and uniformly spaced. As a matter of fact, so little new information is presented in such curves that they are almost never measured. The grounded-emitter collector characteristics, Fig. 5-12, are much more illustrative. In particular, a value for the grounded-emitter current gain β may be estimated directly from the curves by noting the collector current for a given base current.

The operation of a junction transistor results from the properties of the emitter junction under forward bias and the collector junction under reverse bias. The

Figure 5-11 Experimental grounded-base collector characteristics of *pnp* transistor.

Figure 5-12 Grounded-emitter collector characteristics for type 2N3521 *npn* silicon transistor.

Figure 5-13 Grounded-emitter collector characteristics for type 2N3521 transistor at a temperature of 125°C.

properties of both the emitter juction and the collector junction depend sensitively upon temperature as shown by Eqs. (5-4) and (5-5). Comparing collector characteristics at an elevated temperature, Fig. 5-13, with those for the same transistor at room temperature, Fig. 5-12, shows that the currents are increased considerably. The general shape of the curves is preserved, however, so that successful transistor action is still possible.

Silicon Controlled Rectifier

It is now pertinent to describe in greater detail the *silicon controlled rectifier* (SCR) originally discussed in Chap. 4. The useful performance of this device is obtained by adding a third *pn* junction to the junction transistor structure. The resulting *pnpn* four-layer device, Fig. 5-14a, is also called a *four-layer diode* when connections are made only to the two outer layers. A positive potential applied to the *p*-type terminal puts the center *pn* junction under reverse bias while the two outer junctions are forward-biased. The device may be looked upon as the back-to-back combination of a *pnp* transistor together with an *npn* transistor, Fig. 5-14b, the two transistors having a common collector junction.

Note that I_1 is the emitter current in the *pnp* transistor and that $\alpha_1 I_1$ is the collector current. Similarly, I_2 is the emitter current of the *npn* unit and $\alpha_2 I_2$ is the collector current. Using Kirchhoff's current rule at the collector junction,

$$I_2 = \alpha_1 I_1 + \alpha_2 I_2 \tag{5-7}$$

Considering the overall current input to the device

$$I_2 = I_g + I_1 \tag{5-8}$$

where I_g is the current into the gate terminal. Substituting for I_1 in Eq. (5-7) and solving the result for I_2 yields

$$I_2 = \frac{-\alpha_1}{1 - (\alpha_1 + \alpha_2)} I_g \tag{5-9}$$

Figure 5-14 (a) Sketch of SCR and (b) its interpretation in terms of *pnp* transistor coupled to *npn* transistor.

According to Eq. (5-9), if the sum of the current gains $\alpha_1 + \alpha_2$ is near unity, the current I_2 can be very large even though the gate current is small. In fact, if $\alpha_1 + \alpha_2 = 1$, the current is large and limited only by the ohmic resistance of the semiconductor, even if there is no gate current.

The current-voltage characteristics of a SCR, Fig. 4-20, are interpreted as follows. When the gate current I_g is zero, the *pnp* transistor in Fig. 5-14*b* has no base-emitter bias and consequently does not conduct. At a critical applied voltage, avalanche breakdown at the collector junction increases the current and this biases both transistors into conduction. The conducting state is attained at lower values of applied voltage by increasing I_g to bias the *pnp* transistor into conduction. This in turn biases the *npn* transistor into conduction. Once triggered into the conducting state, the gate current may be reduced to zero, since the device current itself maintains bias current in both transistors.

The SCR current-voltage characteristic, Fig. 4-20, indicates negative resistance properties since a voltage decrease results in a current increase when a transition from the off state to the on state occurs. This negative resistance effect is inherently different from that of the tunnel diode, Fig. 5-7. The difference between the two may be identified by noting that a line through the negative-resistance region parallel to the voltage axis in Fig. 5-7 intercepts the characteristic curve at another point. By contrast, the line must be parallel to the current axis if it is to intercept the curve again in Fig. 4-20. The shape of the tunnel-diode curve is often called *N-type* negative-resistance characteristic and the SCR curve called an *S-type* negative-resistance characteristic because of the relative shapes of the two types.

The Unijunction Transistor

An S-type negative-resistance characteristic somewhat similar to that of the SCR is produced by the *unijunction* transistor. The circuit symbol for a unijunction transistor, Fig. 5-15, shows schematically that the device consists of a semiconductor bar with two ohmic contacts and a single, small-area, emitter *pn* junction positioned between them. If no emitter current is present, a fraction of the voltage V_b applied between the base contacts appears at the emitter junction because of the voltage-divider action of the semiconductor bar. Only reverse current is present in the emitter circuit so long as the emitter voltage is less than this fraction

Figure 5-15 Circuit symbol of unijunction transistor.

Figure 5-16 Current-voltage characteristics of type 2N492 unijunction transistor.

and the emitter junction remains under reverse bias. If, however, the junction voltage V_e is increased such that the emitter becomes biased in the forward direction, emitter current increases and carriers are injected into the semiconductor in the region between the emitter and B_1. The resistance of this portion decreases correspondingly and the voltage divider action reduces the fraction of V_b at the junction. Thus the emitter current increases even if the emitter voltage decreases. This results in a very stable negative-resistance characteristic, illustrated in Fig. 5-16. Circuit applications of the unijunction transistor are considered in a subsequent chapter.

Fabrication of Transistors

Of the several ways to produce a semiconductor single crystal containing two *pn* junctions, four types have been most popular: the grown-junction, alloy-junction, diffused-mesa, and planar transistors, Fig. 5-17. Many modifications of these four types have been devised, but they differ only in minor details from the ones shown. The grown-junction transistor was historically the first junction type to be fabricated and has now been supplanted. It is prepared during crystal growth from the melt by altering the impurity content of the melt as the solidified crystal is slowly withdrawn. A typical crystal may be 1 to 2 cm in diameter, while a single transistor is only about 1 mm square and many transistors can be obtained by cutting up the original crystal. Connections are soldered to the emitter, base, and collector regions. Solder containing a small amount of impurity corresponding to the conductivity type being soldered is used in order to provide good electric contacts. Thus, indium is introduced into the solder for the contacts attached to the *p* regions and antimony is added to the solder for the base. This is particularly important in making the connection to the very narrow base region because the *n*-type impurity produces a good electric connection to the *n*-type base and at the same time forms a *pn* junction with the *p* regions. This means that the soldered base contact is effectively electrically isolated from the emitter and collector regions, even though it may physically extend beyond the thin base region.

(a) Grown junction

(b) Alloy junction

(c) Diffused mesa

(d) Planar

Figure 5-17 Structure of junction transistors.

The *pnp* alloy-junction transistor, now also largely supplanted by other processes, is produced from a thin wafer of *n*-type single-crystal material by placing indium pellets on opposite surfaces and heating. As the indium melts, it dissolves some of the germanium beneath it. During subsequent cooling the dissolved germanium recrystallizes upon the base crystal and incorporates many indium atoms into its structure. The recrystallized material is therefore *p*-type, and a transistor structure results. It is convenient to make the collector pellet larger than the emitter pellet, for in this way carriers injected at the emitter junction are collected more efficiently. Thus, the alloy junction is inherently a more satisfactory transistor geometry than the simple grown-junction shape.

As discussed earlier, it is necessary to make the dimensions of the transistor region as small as possible in order to achieve satisfactory high-frequency performance. This is difficult in the case of the alloy-junction transistor, although many ingenious variations of the above process have been developed to accomplish this goal. The diffused-mesa transistor, Fig. 5-17c, however, is much more adaptable to high-frequency transistors because the junction fabrication process can be precisely controlled. The collector junction is formed by placing a p-type single-crystal wafer in a hot gas of, say, antimony atoms. As the wafer is heated, antimony atoms diffuse into the crystal to a depth of approximately 10^{-3} mm. The crystal is subsequently masked and chemically etched to produce a small elevated region, or mesa, about 2 mm in diameter. The mesa top is then suitably masked so that two regions 0.3 mm in diameter are exposed and appropriate metals are deposited by evaporation in high vacuum. Next, the slab is maintained at an elevated temperature while the two metal deposits diffuse into the n region. This results in the formation of a p-type emitter region by one of the metal deposits while the other one forms a contact to the base region.

The area of the collector junction is defined exactly by the size of the mesa while the emitter junction is set by the area of the vapor-deposited metal. Both processes can be precisely controlled and result in an extremely tiny volume. The width of the base region is determined by the gaseous diffusion process and subsequent diffusion of the deposited metal. Both processes can be adjusted easily by the temperature of the wafer to yield a very thin, yet accurately defined, base region. Therefore, the entire fabrication technique lends itself to precise control of transistor geometry. Very tiny active regions are produced on a wafer which is large enough to handle conveniently.

The mesa transistor, like the grown-junction and alloy-junction transistors, is so constructed that the emitter and collector junctions are exposed at the surface of the semiconductor. The electrical properties of the junction are extremely sensitive to chemical impurities from the surrounding atmosphere because of the high electric fields at the junction. Minute traces of foreign atoms on the surface in the junction region greatly degrade the performance of actual transistors compared with theoretical expectations. For this reason, transistors are often hermetically sealed in carefully cleansed metal enclosures after fabrication.

A more satisfactory solution to the surface-contamination problem is to grow a thin oxide film on the surface of the semiconductor by heating in an oxidizing atmosphere. This technique, applicable primarily to silicon because of the favorable properties of silicon oxide, results in a semiconductor device completely protected from atmospheric contamination. In addition, the oxide film is a barrier to gaseous diffusion of impurities so the film itself acts as a mask during fabrication. This technique makes possible the planar transistor, Fig. 5-17d, so called because the entire transistor appears to be a simple plane wafer of silicon.

Construction of a planar npn transistor begins with a single-crystal wafer of n-type silicon. A thin oxide layer is grown on the top surface by heating in an oxygen atmosphere. The oxide is chemically etched away in a circular region of diameter a. The wafer is then exposed to a hot boron gas atmosphere. Boron atoms diffuse into the exposed silicon, forming a p-type base region and the collector junction. Boron atoms also diffuse laterally under the oxide film so that

Figure 5-18 Photomicrograph of a planar transistor. (*Raytheon Semiconductor.*)

the base region is somewhat larger than the diameter a and the collector junction is protected at the surface by the oxide layer. Next, the wafer is reoxidized to produce a film covering the entire area once again. After the oxide is etched away in an area of diameter b, the wafer is again exposed to a hot impurity gas, this time phosphorus. These n-type impurity atoms diffuse into the silicon, forming the emitter region and the emitter junction. Lateral diffusion of phosphorus atoms under the oxide film again ensures that the emitter junction is protected by the oxide layer.

Finally, electric contacts are provided by vapor deposition and alloying as in the case of the mesa transistor. The result is an *npn* transistor entirely contained in one plane wafer and completely encased in an impervious oxide film. Geometric definition of the junction positions and the base width is extremely accurate because of the easily controlled gaseous diffusion doping technique. Several hundred transistors can be fabricated simultaneously from one silicon wafer a centimeter or so in diameter. These are subsequently cut into individual units by sawing up the wafer. A micrograph of a typical planar transistor is shown in Fig. 5-18. Connections to the emitter and base regions are clearly visible.

FIELD-EFFECT TRANSISTORS

Drain Characteristics

Very effective control action is obtained in the *field-effect transistor*, or *FET*, in which the flow of majority carriers is controlled by signal voltages applied to a reverse-biased *pn* junction. Consider the *n*-type semiconductor bar (most often of silicon) in Fig. 5-19 which has an ohmic contact, the *source*, at one end and a

Figure 5-19 *n*-channel junction FET.

similar contact, the *drain*, at the other. Electrons moving from the source to the drain in response to the drain voltage V_d pass through a channel between two *p* regions. This *pn* junction is called the *gate* because the width of the reverse-biased gate junction determines the width of the channel and consequently the magnitude of the current between source and drain. Signal voltages applied to the gate result in corresponding variations in drain current.

In effect the FET is a variable resistor which is controllable by the electric fields associated with the *pn* junction, whence the quite descriptive name. Very little power is expended by the applied signal because of the small reverse current in a *pn* junction. Note that the unit described is an *n*-channel FET. Clearly a *p*-channel FET is equally possible. In this case the potentials applied to the drain and gate have the opposite polarity so that the gate junction is again under reverse bias. The circuit symbol for an *n*-channel FET is shown in Fig. 5-20. The corresponding symbol for a *p*-channel unit has the arrow representing the gate junction pointing in the other direction. Note that the direction of the arrow is consistent with the direction of conventional current flow in a junction diode.

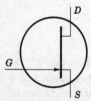

Figure 5-20 Circuit symbol for *n*-channel FET.

The current-voltage characteristic of a FET is derived from an analysis of current flow in the channel. As sketched in Fig. 5-19, the width of the junction is greatest near the drain end of the gate because the reverse bias here is the sum of the gate potential plus the drain potential. The junction is narrowest at the source end because the voltage drop along the channel means that the reverse bias is essentially due to the gate potential alone. The sum of V_g and V_d may be large enough for the junction to completely pinch off the channel. It turns out, in fact, that this is the normal operating condition and the reverse bias required for pinchoff is labelled V_p.

The nonuniform voltage drop along the channel and the nonlinear variation in channel width combine to make the exact analysis complicated. A simplified study can be carried out by noting that the current in the channel resistance is proportional to V_d in accordance with Ohm's law, and to $V_g - V_p$ because of pinchoff, as discussed above. A finite drain voltage also results in an additional reverse bias varying from zero at the source end to $-V_d$ at the drain end. As an approximation, the added reverse bias is taken to be the simple average, $-V_d/2$. The drain current is then,

$$I_d = K\left(V_g - V_p - \frac{V_d}{2}\right)V_d \tag{5-10}$$

where K is a constant.

Figure 5-21 Current-voltage characteristics of type 2N2499 p-channel FET.

At pinchoff the sum of the gate voltage and the drain voltage is equal to the pinchoff voltage, or $V_g + V_d = V_p$. Therefore, Eq. (5-10) becomes

$$I_d = K\left(V_g - V_p - \frac{V_g - V_p}{2}\right)(V_g - V_p) = \frac{K}{2}(V_g - V_p)^2$$

$$= I_{dss}\left(1 - \frac{V_g}{V_p}\right)^2 \tag{5-11}$$

where I_{dss} is a constant that depends primarily upon the geometrical design of the FET. Evidently the drain current does not depend upon the drain voltage above pinchoff. Equations (5-10) and (5-11) are a good representation of the *drain characteristics* of practical FETs, as illustrated in Fig. 5-21, for the case of a *p*-channel device. The maximum source-drain voltage is limited by reverse breakdown at the gate junction.

Figure 5-22 Transfer characteristic of *n*-channel FET.

The Transfer Characteristic

The variations in drain current corresponding to voltage changes at the gate electrode described by Eq. (5-11) are most clearly displayed by a plot of the drain current as a function of gate voltage, which is called the *transfer characteristic* because it relates changes in the output circuit to signals at the input. The typical transfer characteristic in Fig. 5-22 shows the square-law dependence given by Eq. (5-11).

The transfer characteristic is an alternative way of describing the nonlinear electrical properties of the FET. It is complementary to the drain characteristics and shows directly the change in drain current per unit change in gate voltage. This ratio is a measure of the device's control feature, and a large value implies favorable amplification properties. It should be noted that the FET control characteristics are fundamentally different from those of the junction transistor in that the input signal is a voltage rather than a current. The consequences of this difference are explored in various ways in succeeding chapters. Actually, both types of control lead to quite effective amplifying action, but the circuit configurations and operating properties of FETs do differ from those of junction transistors.

Since current flow in a FET is by majority carriers, in contrast to the case of the junction transistor, its electrical properties tend to be less sensitive to temperature. Nevertheless, the change in the transfer characteristic with temperature illustrated in Fig. 5-23 shows that the effect is not negligible. Suitable circuit

Figure 5-23 Transfer characteristics of type 2N5267 *p*-channel FET at three different temperatures.

configurations have been devised to minimize changes in circuit performance with temperature so that both types of transistors are operable over practical temperature ranges.

IGFETs and MOSFETs

Quite analogous control over current carriers in a semiconductor is possible without a *pn* junction by using a suitable metal gate electrode insulated from the semiconductor. Such insulated-gate field-effect transistors are known as *IGFETs*. The most successful practical units are made of silicon and employ a thin silicon dioxide layer as the insulator. The descriptive terminology for a metal-oxide-semiconductor field-effect transistor is *MOSFET*.

The MOSFET illustrated in Fig. 5-24*a* has *n*-type source and drain regions imbedded in a *p*-type crystal. The metal gate electrode is deposited on top of a thin oxide layer on the surface. Usually the *p* substrate is connected to the source and a negative potential is applied to the gate.

The negative gate potential repels electrons in the *n*-type channel and a *depletion* layer devoid of carriers forms. This depletion layer is quite analogous to the region between the *n*-type and *p*-type sides of a *pn* junction. Creation of the depletion layer by the gate voltage decreases the conductivity of the *n*-channel and the drain current is reduced. Pinchoff occurs when the depletion layer extends entirely across the *n*-layer and the drain characteristics of the *depletion* MOSFET are quite similar to those of the junction FET.

If the *n*-channel between source and drain is eliminated, Fig. 5-24*b*, successful operation is still obtained when the gate voltage is made positive with respect to the source and the *p*-type substrate. The drain current is zero without gate bias because of the *pn* junctions at the source and drain ends. As the gate potential is increased positively from zero, holes in the *p* substrate immediately under the gate are repelled and the surface layer tends to become *n*-type. At the threshold gate voltage, V_T, an *n*-channel forms and current passes between source and drain. Greater gate voltage increases the drain current further. Since the gate voltage increases the *n*-channel conductivity, this design is called an *enhancement* MOSFET.

(a) Depletion MOSFET   (b) Enhancement MOSFET

Figure 5-24 *n*-channel MOSFETS. (*a*) Depletion and (*b*) enhancement versions.

Actually, a depletion MOSFET can operate in the enhancement mode as well, since the channel can be widened as well as narrowed by the gate potential. A convenient comparison between the several possible FET types is provided by their transfer characteristics, Fig. 5-25. The junction FET employs the depletion mode, Fig. 5-25*a*, since the gate *pn* junction cannot have forward bias. The depletion MOSFET can be used with either gate polarity, Fig. 5-25*b*, while the enhancement MOSFET requires positive gate bias, Fig. 5-25*c*. Note, however, that except for lateral displacement, all three transfer characteristics are quite similar. The drain characteristics of a practical MOSFET are similar to those of a junction FET, as illustrated in Fig. 5-26, except that both positive and negative gate voltages are operational.

The circuit symbol for *n*-channel MOSFETs, Fig. 5-27, suggests the insulated-gate structure. As might be anticipated, *p*-channel MOSFETs are indicated by reversing the direction of the arrowhead that suggests the direction of conventional current in the channel. Further evidence for the versatility of the MOSFET design is the *dual-gate* MOSFET in which two separate gate electrodes are

(*a*) Depletion

(*b*) Depletion-enhancement

(*c*) Enhancement

Figure 5-25 Transfer characteristics of (*a*) depletion, (*b*) depletion-enhancement, and (*c*) enhancement FETs.

Figure 5-26 Drain characteristics of type 2N3797 n-channel MOSFET.

Figure 5-27 Circuit symbol for n-channel MOSFET.

disposed along the channel. A large n-type region similar to the source is placed between the two gates but is not externally connected. The gate closest to the drain isolates the control gate from the drain by reducing the gate-drain capacitance. Much improved high-frequency performance is thereby obtained.

INTEGRATED CIRCUITS

Principles of Integrated Circuits

The properties of semiconductors, most particularly silicon, make it possible to produce an entire electronic circuit within one single crystal. Such an *integrated circuit* miniaturizes electronic networks and also reduces the number of individual components in complex electronic circuits. This is so because an entire integrated circuit represents, in effect, only one component. The basis for the integrated-circuit development is that many electronic components, such as transistors, diodes, and resistors, can be made of silicon having suitably disposed n-type and p-type regions. It remains only to position the various components in a judicious geometric arrangement in order to combine them all into one single-crystal slab.

The close juxtaposition of components within a single slab leads to electrical coupling between them. It proves possible, however, to isolate components by suitable arrangement of the p-type and n-type regions. Consider the case of a resistor of n-type silicon in a p-type crystal, for example. Clearly the resistor is isolated from the crystal by the pn junction between resistor and the slab. This isolation is most effective only at dc because of pn-junction capacitance. The stray coupling introduced by all the junction capacitances is taken into account in the design of the circuit.

Actually, the capacitance of a pn junction may also serve as a capacitor in an integrated circuit. Inductors of small inductance in the form of a spiral conductor are also possible. Most often, however, the integrated circuit per se is composed of transistors, diodes, and resistors, and the necessary capacitors and inductors are connected as separate, discrete components.

Fabrication Processes

Silicon integrated circuits are fabricated by processes similar to those used in the case of planar transistors. Appropriate patterns in the oxide layer prior to impurity diffusions are produced by first coating the oxide layer with a special photopolymeric material. A photopolymer becomes resistant to chemical etching when exposed to strong light. Thus, when the photopolymer is exposed to light through a suitable mask, the unexposed portions can be subsequently dissolved away. This exposes a corresponding pattern in the oxide layer which is etched away with a different chemical, in turn exposing the underlying silicon. After gaseous diffusion of n-type or p-type impurities, the wafer may be reoxidized and the process repeated to develop an additional array of doped regions.

The photographic process produces integrated circuits of extremely small size. Accurate, large-scale masks are optically demagnified to microscopic dimensions. This also means that many integrated circuits may be fabricated simultaneously from the same silicon wafer simply by constructing a mask having the desired circuit repeated many times. Subsequent to the last impurity-diffusion step, a thin metal layer is deposited (also through a photopolymeric layer) to interconnect the various components and to provide terminals for external connections.

This process is best illustrated by following actual steps in the fabrication of a simple circuit, Fig. 5-28. To begin, a p-type single-crystal slab is exposed to silicon vapor containing n-type impurity atoms. The result, Fig. 5-28a, is an n-type *epitaxial* layer. Epitaxial means that the n-type layer has the same crystal structure as the p-type substrate. The surface is oxidized, masked, and etched, and a p-type diffusion isolates several n regions, Fig. 5-28b. These steps are repeated in Fig. 5-28c and Fig. 5-28d. Finally, Fig. 5-28e, the metallic interconnections are added.

Oxide film

n-epitaxial layer

p-substrate

(a) Epitaxial growth

(b) Isolation diffusion

(c) Base diffusion

(d) Emitter diffusion

Resistor   Diodes   Transistor   Resistor

(e) Metalization

(f) Circuit diagram

Figure 5-28 Processing steps (a) through (e) result in complete integrated circuit (f).

This procedure results in the circuit diagrammed in Fig. 5-28*f*, as can be appreciated from a detailed comparison with Fig. 5-28*e*. Note, in particular, how the resistors and diodes are isolated from the *p*-type substrate in each instance by a *pn* junction. Note also that the conventional circuit symbol for an *npn* transistor is used in the circuit diagram. The circuit, incidentally, is termed a DTL NAND gate, which is considered in greater detail in Chap. 11.

Practical Circuits

It is possible to recognize the various components in the micrograph of a practical integrated circuit, Fig. 5-29. This so-called *difference-amplifier* circuit is diagrammed in Fig. 5-30*a* and is analyzed in the next chapter. Note that the metallic interconnections appear lighter in the micrograph and that it is often necessary to zigzag the resistors in order to achieve the desired resistance. Although this is a relatively simple circuit, it is noteworthy that all 12 components are contained in one small case about the size of a single transistor, Fig. 5-30*b*.

Much more complicated circuits are available as integrated circuits and these are discussed in subsequent sections. Each such circuit usually can be employed in several ways and is often associated with other integrated circuits to perform an entire complex task. In certain instances, most notably in digital computers, the

Figure 5-29 Integrated circuit of difference amplifier in Fig. 5-30. (*Motorola Semiconductor Products, Inc.*)

(a) Circuit diagram

(b)

Figure 5-30 (a) Difference amplifier circuit and (b) practical unit.

combination of several integrated circuits can itself be made as an integrated circuit. Such *large-scale integration*, or *LSI*, further miniaturizes electronic circuits and yields the further advantage that signal propagation throughout the circuit is very rapid. LSI circuits containing 1000 or more transistors are feasible.

The advantages of LSI are most readily achieved in MOSFET integrated circuits because fabrication processes are easier than in the case of integrated circuits based on junction transistor design. This is so because of the simpler structure of MOSFET transistors compared to junction transistors and because of the versatility inherent in MOSFET structures. In particular, complementary symmetry MOSFET integrated circuits prove to be very advantageous in digital electronic LSI applications, as discussed in subsequent chapters.

SUGGESTIONS FOR FURTHER READING

Leonid V. Azaroff and James J. Brophy: "Electronic Processes in Materials," McGraw-Hill Book Company, New York, 1963.

James J. Brophy: "Semiconductor Devices," McGraw-Hill Book Company, New York, 1964.

Edwin R. Jones: "Solid-State Electronics," Intext Educational Publishers, Scranton, Penn., 1974.

Rodney Bruce Sorkin: "Electronics," McGraw-Hill Book Company, New York, 1970.

EXERCISES

5-1 Plot the forward characteristics of a germanium and a silicon *pn* junction up to a current of 10 mA. Use the rectifier equation and take $I_0 = 10^{-6}$ A for germanium and $I_0 = 10^{-12}$ A for silicon. How much more forward bias is required in the case of the silicon diode to make the current 10 mA? Repeat for 1 mA.

    Ans.: 0.4 V; 0.4 V

5-2 By differentiating the rectifier equation for a junction diode, Eq. (5-4), determine an expression for the junction resistance $R = (dI/dV)^{-1}$. Given that $e/kT = 38$ V^{-1} at room temperature, calculate the rectification ratio at a potential of 1 V.

    Ans.: 10^{33}

5-3 Sketch the energy-band model for an *npn* transistor in equilibrium and biased for transistor operation.

5-4 Calculate and plot the drain characteristics of a type 2N2499 FET using Eqs. (5-10) and (5-11). Do this by evaluating I_{dss} and V_p from the $V_g = 0$ curve in Fig. 5-21. Compare with Fig. 5-21.

    Ans.: 9 mA; 6.6 V

5-5 Calculate and plot the grounded-base collector characteristics of a silicon transistor at room temperature and for collector potentials up to a maximum of 0.5 V. Assume the current gain is unity and that $I_0 = 1$ μA.

5-6 Repeat Exercise 5-5 for a temperature of 150°C. Assume that I_0 increases according to Eq. (5-5) and take $V_0 = 0.7$ V.

5-7 Calculate the transfer characteristic of a type 2N2499 FET using Eq. (5-11). Do this by using values of I_{dss} and V_p from Exercise 5-4. Compare with an experimental curve derived from Fig. 5-21 at a drain potential of 15 V.

5-8 Plot experimental transfer characteristics for all transistors for which current-voltage characteristics are given in Figs. 5-11, 5-12, 5-13, 5-21, 5-26, 6-14, 6-28, 7-12, and 7-13. Note the range of currents and voltages exhibited in this data.

5-9 For each device in Exercise 5-8 calculate the change in output current per change in input signal near the middle of the range over which the data is given. Note any trend in the magnitude of this ratio with power handling capability.

    Ans.: Smaller at high power

5-10 Compare the ratio of the change in output current divided by the change in input voltage for the junction transistor of Fig. 5-12 and the FET of Fig. 5-21. Do this by dividing the current gain ratio of the junction transistor by the emitter junction resistance to convert the input signal current to an input signal voltage.

    Ans.: 41 mho, 2.6×10^{-3} mho

SIX

TRANSISTOR AMPLIFIERS

The advent of electronics is reckoned from the discovery that the current in a vacuum diode can be controlled by introducing a third electrode, which makes the triode an exceedingly effective amplifier. The possibilities inherent in electronic circuits were greatly expanded by the invention of the transistor in 1948. Since then a wide variety of semiconductor devices has to a large extent supplanted vacuum tubes in most applications.

Both vacuum tubes, which are discussed in Appendix A, and transistors are nonlinear devices, and their operation in any circuit is determined by graphical analysis using the description of their electrical properties given by current-voltage characteristics. The analysis differs in detail for voltage-controlled devices such as the triode and field-effect transistor compared to current-controlled devices such as the junction transistor, but is not different in principle. Furthermore, in either case it proves possible to develop useful equivalent-circuit representations which are most satisfactory for circuit analysis.

THE OPERATING POINT

Load Lines

The operation of a FET as an amplifier is examined most easily with the aid of the elementary amplifier circuit employing a p-channel FET shown in Fig. 6-1. In this circuit the drain is maintained at a negative potential with respect to the source by the battery V_{dd} while the gate is *biased* positively by the gate bias battery V_{gg}. Variations in gate voltage resulting from an input signal ΔV_{gs} produce changes in the drain current which are observed as a voltage signal ΔV_{ds} across the load resistor R_L.

The circuit is analyzed using Kirchhoff's rules in the usual way except that, since the current-voltage properties of the FET are nonlinear, the solution must be carried out graphically. The voltage equation around the output circuit is

$$V_{dd} - I_d R_L + V_{ds} = 0 \qquad (6\text{-}1)$$

which may be put in the form

$$I_d = \frac{V_{dd}}{R_L} + \frac{1}{R_L} V_{ds} \qquad (6\text{-}2)$$

Equation (6-2) plots as a straight line on the drain characteristics of the FET and has intercepts at $I_d = 0$, $V_{ds} = V_{dd}$ and $V_{ds} = 0$, $I_d = V_{dd}/R_L$. It is, in fact, the current-voltage characteristic of R_L and is called the *load-line*.

The load line together with the drain characteristics of the FET represent two relations in the two unknowns V_{ds} and I_d. By plotting the load line on the drain characteristics as in Fig. 6-2, the drain current corresponding to each gate voltage is given by the intersection of the load line with the characteristic curve for that gate voltage.

Figure 6-1 Simple p-channel FET amplifier.

Figure 6-2 Operating point is the intersection of load line with gate bias curve.

The *operating point* is the intersection of the load line with the characteristic curve corresponding to the gate bias voltage. As the gate voltage varies in accordance with an applied input signal, the drain-current excursions move back and forth along the load line so that Eq. (6-2) is satisfied at every instant. In Fig. 6-2, for example, an input signal of 0.4 V causes the drain current to change by approximately 0.6 mA and results in an output signal of about 6 V. That is, the circuit amplifies the input signal by a factor of $\frac{6}{0.4} = 15$.

Note that the input power is very small because the gate current is negligible. In contrast, the output power, which is equal to the product of the change in drain current times the change in drain voltage, may be quite appreciable. This power is derived from the drain supply battery, V_{dd}, and is controlled by the valve-like action of the gate.

Source Bias

A separate gate bias battery is economically impractical in most circuits. Instead, gate bias can be obtained by inserting a resistor in series with the source, Fig. 6-3. The gate potential is set by a voltage divider connected to the drain voltage source. Then the current through the source resistor R_s results in a gate bias of the proper polarity and magnitude to maintain the gate junction under reverse bias. A useful feature of this circuit is that if, for example, a change in ambient temperature increases the drain current, the voltage drop across the source resistor increases the gate bias, which tends to return the drain current to the original value. That is, the drain current and the operating point are stabilized against external influences.

Figure 6-3 Practical *n*-channel FET amplifier using source bias.

Note that the source resistor is shunted by a large capacitor to prevent ac signals caused by ac drain currents in the source resistor from appearing in the gate circuit. Also, capacitors are included at the input and output to isolate the amplifier from external dc voltages that could change the operating point.

The operating point is determined by writing the equation for the load line from the dc voltage drops in the output circuit,

$$V_{dd} - I_d R_L - V_{ds} - I_d R_s = 0 \qquad (6\text{-}3)$$

Since, however, the gate bias voltage is not known, in contrast to the simpler circuit previously analyzed, another relation is required. This is found from the dc voltage drops around the input circuit,

$$\frac{V_{dd} R_2}{R_1 + R_2} - V_{gs} - I_d R_s = 0 \qquad (6\text{-}4)$$

Both equations may be put in the same form. The load line is, from Eq. (6-3),

$$I_d = \frac{V_{dd}}{R_s + R_L} - \frac{1}{R_s + R_L} V_{ds} \qquad (6\text{-}5)$$

and the *bias line*, from Eq. (6-4), is

$$I_d = \frac{V_{dd} R_2}{(R_1 + R_2) R_s} - \frac{1}{R_s} V_{gs} \qquad (6\text{-}6)$$

Note that Eq. (6-5) is similar to Eq. (6-2), and that Eq. (6-6) has the same form. These expressions, together with the current-voltage characteristics of the FET, represent three relations in the three unknowns, V_{ds}, I_d, and V_{gs}, and are solved graphically.

The bias line, Eq. (6-6), is a straight line on the transfer characteristic, Fig. 6-4, and the intersection gives the drain current and gate voltage immediately. The drain voltage is then found from the intersection of the gate bias curve with the load line plotted on the drain characteristics. Thus the operating point is completely determined.

An input signal applied to the gate results in corresponding changes in the drain current, as shown by the transfer characteristic in Fig. 6-4. As in the previous circuit, drain signal currents through the load resistor produce an output signal. This is illustrated graphically in Fig. 6-4 by plotting the *ac load line* on the drain characteristics. The ac load line is simply the dc load line, Eq. (6-5), with the source resistor set equal to zero to account for the shunting action of the source bypass capacitor at signal frequencies.

Note that a sinusoidal input signal results in a sinusoidal output signal of larger amplitude. The output waveform is not an exact amplified replica of the input signal, however, because of curvature in the transfer characteristic. This *distortion* is minimized by proper circuit design and suitable choice of the operating point. Note also in Fig. 6-4 that an increase in gate voltage produces a

Figure 6-4 Intersection of bias line with transfer characteristic yields operating point of source-bias amplifier.

decrease in the output voltage. This means that the circuit introduces a 180° phase shift between input and output signals.

A simpler version of the circuit in Fig. 6-3 eliminates resistor R_1. Quite satisfactory operation is achieved, but because the source resistor must then usually be made small to obtain the correct gate bias, the circuit is not quite so stabilized against shifts in operating point accompanying changes in ambient temperature or supply potential. In either circuit it often proves more convenient to find the operating point by a "cut-and-try" process rather than by resorting to the graphical technique. This is done by first choosing a reasonable value for the gate bias, V_{gs}, and estimating I_d from the drain characteristics. This value is used to calculate V_{gs} from the bias characteristic, Eq. (6-6), to compare with the original estimate. If the two agree, the operating point is determined; if not, a revised estimate for V_{gs} is taken and the process repeated. Usually two or three trys are sufficient to attain satisfactory agreement.

It is inconvenient to measure electrode voltages with respect to the source in circuits employing source bias. This is particularly true when more than one transistor is used in a circuit. Rather, usual practice is to refer all potentials to a common point called *ground*. The ground point is considered to be electrically neutral so that, for example, the ground points of two separate circuits may be connected with no influence upon the operation of either circuit. A typical use of the circuit symbol for the ground point in Fig. 6-3 is at the junction of the lower input and output terminals and the bottom end of the source resistor.

Small-Signal Parameters

Very frequently the signal amplitudes applied to an FET amplifier are small compared with the full range of voltages covered by the drain characteristics. In this situation graphical analysis of FET performance is inaccurate because the drain characteristics are not given with sufficient precision. A more satisfactory procedure is to replace the FET with an equivalent circuit which can be studied by conventional circuit analysis. After the operating point is determined graphically, small departures about the operating point caused by small signals are treated by assuming that the FET is a linear device. According to the drain characteristics, drain current depends upon both the gate voltage and the drain voltage. Therefore, a small change in drain current, ΔI_d, away from the operating point may be written

$$\Delta I_d = k_1 \ \Delta V_{gs} + k_2 \ \Delta V_{ds} \tag{6-7}$$

where ΔV_{gs} and ΔV_{ds} are small changes in the electrode potentials and k_1 and k_2 are constants.

The ratio $\Delta I_d / \Delta V_{ds}$ can be identified as the reciprocal of an equivalent resistance called the *drain resistance*,

$$\frac{1}{r_d} = \frac{\Delta I_d}{\Delta V_{ds}} \tag{6-8}$$

Similarly, the ratio $\Delta I_d / \Delta V_{gs}$ gives the change in drain current resulting from a change in voltage in the gate circuit. It has the dimensions of conductance and is called the *mutual transconductance*,

$$g_m = \frac{\Delta I_d}{\Delta V_{gs}} \tag{6-9}$$

The total change in drain current is, from Eqs. (6-7), (6-8), and (6-9),

$$\Delta I_d = g_m \, \Delta V_{gs} + \frac{1}{r_d} \Delta V_{ds} \tag{6-10}$$

Consider now that the small departures about the operating point are ac signals, so that Eq. (6-10) becomes

$$i_d = g_m v_{gs} + \frac{1}{r_d} v_{ds} \tag{6-11}$$

This expression may be interpreted as the circuit equation for a current generator in parallel with a resistor, as in Fig. 6-5; summing the currents at the upper node yields Eq. (6-11) directly. Figure 6-5 is the *ac equivalent circuit* of the FET. Note that it has the configuration of the Norton equivalent circuit discussed in Chap. 1, which is consistant with the constant-current properties of the FET drain characteristics.

The mutual transconductance and the drain resistance are called the *small signal parameters* of the FET. The mutual transconductance is just the slope of the transfer characteristic at the operating point, and the drain resistance is the reciprocal of the drain characteristic at the operating point. Typical values of small-signal parameters given in Table 6-1 show that the several different types of FETs are comparable. A rather wide range of parameters may be achieved by suitable choice of the FET design, however. Actually, since the FET is a nonlinear device, the magnitudes of the small signal parameters depend upon the dc drain current as well, Fig. 6-6. For this reason, it is usually necessary to evaluate g_m and r_d graphically at the operating point. According to Fig. 6-6 quite appreciable variations in the small signal parameters can be obtained by selecting the operating point. This is important in the design of FET circuits having specific performance requirements.

It is illustrative to analyze the simple FET amplifier in Fig. 6-3 by means of the ac equivalent circuit technique. According to Fig. 6-5, the equivalent circuit of the amplifier is obtained by replacing the FET with a constant-current generator

Figure 6-5 FET ac equivalent circuit.

Table 6-1. FET small signal parameters

Type		g_m, 10^{-6} mho	r_d, Ω
2N5484	n-channel FET	2000	50,000
2N5268	p-channel FET	1700	15,000
2N3797	n-channel MOSFET (depletion)	2300	40,000
3N157	p-channel MOSFET (enhancement)	2000	17,000

in parallel with a resistor, Fig. 6-7a. Note that the gate terminal is not connected to the remainder of the equivalent circuit. This is a consequence of the very high input resistance of the reverse-biased gate junction. This circuit ignores the capacitance of the gate junction, however, and is only accurate at frequencies low enough that its capacitive reactance can be neglected. This matter is treated more extensively in Chap. 7.

Similarly, it is usually assumed that the capacitive reactances of the coupling capacitors and the source bypass capacitor are negligible at signal frequencies of

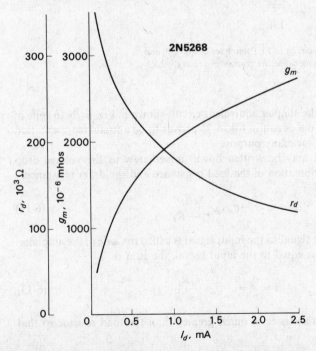

Figure 6-6 Variation of small-signal parameters of type 2N5268 p-channel FET with drain current.

Figure 6-7 (a) ac equivalent circuit of FET amplifier in Fig. 6-3, and (b) simplified version assuming capacitive reactances are negligible.

interest. This leads to the simpler equivalent circuit shown in Fig. 6-7b. In spite of these approximations, the circuit in Fig. 6-7b proves to be a useful representation of the actual amplifier for many purposes.

The output signal may be written down immediately as the voltage drop across the parallel combination of the load resistance and the drain resistance,

$$v_o = -g_m v_{gs} \frac{r_d R_L}{r_d + R_L} \tag{6-12}$$

The ratio of the output signal to the input signal is called the *gain* of the amplifier. Since the gate voltage is equal to the input signal, the gain is

$$a = \frac{v_o}{v_i} = -g_m \frac{r_d R_L}{r_d + R_L} \tag{6-13}$$

Most often the drain resistance is much greater than the load resistor so that Eq. (6-13) reduces to

$$a = -g_m R_L \tag{6-14}$$

According to Table 6-1, a gain of about $7000 \times 2000 \times 10^{-6} = 14$ can be anticipated. A more precise value is calculated after determining the actual mutual transconductance at the operating point. Also, the minus sign in Eq. (6-14) represents the 180° phase shift between input and output signals, as discussed previously. An important feature of the FET amplifier is the large input impedance which is essentially equal to R_{eq} in Fig. 6-7b. Also of interest is the output impedance, which is equal to the load resistor so long as the drain resistance is large, as assumed in arriving at Eq. (6-14).

FET AMPLIFIERS

Source Follower

The source is common to both input and output terminals in the simple FET amplifier circuit insofar as ac signals are concerned. This is the most common configuration and is sometimes referred to as the *grounded-source* connection. Another possible configuration is the *grounded-drain* amplifier, Fig. 6-8. The drain load resistor is omitted and the output signal is developed across the source resistor. Note that in this arrangement the drain is at ground potential insofar as ac signals are concerned and is common to both input and output circuits. The drain remains at a dc potential with respect to the source and to ground, however. Gate bias is developed across R_1, and the operating point is found in the same fashion as for the grounded-source amplifier. The circuit is commonly called a *source follower* because the output signal at the source follows the signal at the gate very closely.

The equivalent circuit of the source follower is shown in Fig. 6-9a for comparison with Fig. 6-8, and rearranged for greater clarity in Fig. 6-9b. The output

Figure 6-8 FET source-follower amplifier.

Figure 6-9 (a) Equivalent circuit of source-follower amplifier, and (b) the same circuit rearranged.

signal is given by the voltage drop across the combination of r_d in parallel with R_1 and R_2 in series.

$$v_o = g_m v_{gs} \frac{r_d(R_1 + R_2)}{r_d + R_1 + R_2} \cong g_m v_{gs} R_2 \tag{6-15}$$

where the approximation is valid since in practice the drain resistance is greater than $R_1 + R_2$ and R_2 is much greater than R_1. The gate signal is equal to the input signal less the voltage across the source resistor.

$$v_{gs} = v_i - g_m v_{gs} R_2 \tag{6-16}$$

where r_d is again taken to be large in comparison to $R_1 + R_2$. Equation (6-16) is solved for v_{gs} and inserted into Eq. (6-15). The result is

$$v_o = g_m R_2 \frac{v_i}{1 + g_m R_2} = \frac{g_m R_2}{1 + g_m R_2} v_i \cong v_i \tag{6-17}$$

the approximation applies for $g_m R_2 \gg 1$, as is usually the case.

According to Eq. (6-17), the output signal at the source is equal to the input signal at the gate, so that the name *source follower* is appropriate indeed. Although it is true that the voltage gain of the circuit is only equal to unity, the power gain is large. This is so because the source voltage replicates the input signal in a much lower resistance circuit. The ac signal power is proportional to the square of the signal voltage divided by resistance, so that the power amplification is great and the circuit can deliver appreciable power to a load.

Since the output voltage is included in the input circuit and the output signal opposes the input signal, the effective input impedance of a source follower is very large, much greater than, say, R_g. It turns out also that the effective output impedance of the circuit is very low, much less than $R_1 + R_2$. Thus, the source follower is an impedance-matching amplifier with a high input impedance and a low output impedance and for this reason proves useful at both the input and output of multiple-stage amplifier circuits. This aspect is explored more quantitatively in Chap. 8. In addition, Eq. (6-17) shows that circuit performance is relatively unaffected by changes in circuit parameters or by changes in the FET small signal parameters, so long as the drain resistance and the mutual transconductance are large enough for the approximations to remain valid.

FET Voltmeter

The amplification and large input impedance of a FET results in a sensitive electronic voltmeter when used in conjunction with a standard d'Arsonval milliammeter. A common circuit, Fig. 6-10, is a source-follower balanced amplifier with the meter connected between the source terminals. The unknown voltage is applied to one of the gates while the other gate is unused and returned to ground. Note that p-channel FETs are used in Fig. 6-10; a voltmeter employing n-channel units is equally possible. Since a source follower presents a high input impedance and a low output impedance, this circuit is particularly useful in the voltmeter application.

Figure 6-10 FET electronic voltmeter.

The balanced circuit provides a convenient connection for the d'Arsonval meter which results in zero deflection with zero input signal, even though the quiescent current in each FET is nonzero. Furthermore, the symmetry of the circuit maintains performance stability against component changes caused by aging. Actually the circuit may be looked upon as a bridge comprising a FET in each of two arms together with the source resistors in the other two arms. A dc voltage applied to the gate changes the FET resistance and unbalances the bridge. The circuit may be analyzed using equivalent circuits in the usual fashion.

The adjustable resistor R_a in series with the meter is a calibration resistor, and the potentiometer R_b is a balance adjustment which corrects for minor asymmetries in the FETs or other components. The circuit is balanced for zero current in the meter by adjusting R_b with zero voltage applied to the gate.

The input gate is connected to multiplier resistors in commercial instruments to achieve multiple-range performance, much as in the case of the VOM circuit. The resistor values are greater in the electronic instrument, however, because of the much greater sensitivity. An additional feature of interest is that the maximum meter current, which occurs when one of the FETs is completely cut off, does not damage the meter. Therefore, the meter movement is unharmed, even if the input is inadvertently connected to a voltage source much larger than the full-scale value.

Standard instruments provide for ac measurements by including a diode rectifier circuit, as discussed in the previous chapter. The high input impedance of the source-follower is ideally suited to measure the output of the rectifier circuit. In addition to ac and dc voltages, most instruments also measure resistance with an ohmmeter circuit similar to those discussed in Chap. 1. Again, the FET voltmeter is used as the indicator. This permits a very wide range of operation. For example, quite inexpensive instruments are capable of measuring unknown resistances from 1 Ω to 100 MΩ. An earlier version of this circuit employing a dual triode is called a *vacuum-tube voltmeter*, or *VTVM*.

MOSFET Amplifier

Circuit configurations for MOSFET amplifiers are quite similar to those for junction FETs. This is particularly so in the case of depletion MOSFETs since the gate bias polarity is most often the same as for the FET. Similar circuits apply to enhancement MOSFETs as well, so long as circuit parameters are such that the gate bias has the same polarity as the drain and that the magnitude exceeds the threshold gate voltage.

The properties of the enhancement MOSFET permit the particularly simple amplifier circuit shown in Fig. 6-11, which proves to be suitable for many applications. Gate bias is obtained simply by connecting the gate to the drain through a large resistor. The operating point is determined by noting the point on the load line for which the dc drain voltage and the dc gate voltage are equal. Subsequent circuit analysis using the ac equivalent circuit is straightforward.

The gate bias resistor is connected to the drain, rather than to the drain

V_{dd}
+20 V

R_d
10 kΩ

R_g
10 MΩ

Figure 6-11 Simple enhancement MOSFET amplifier.

supply voltage, in order to stabilize the operating point against external changes. If, for example, the drain current falls slightly, the voltage drop across the drain load resistor is reduced. This increases the gate bias, which in turn tends to return the drain current to the original value. This stabilization is similar to that achieved by the drain bias resistor in the conventional source bias configuration.

Signal voltages from the drain circuit are also returned to the input through the gate bias resistor. This unfavorable feature can usually be ignored because the impedance of signal sources is small compared to the very large gate bias resistor, and the drain signals appearing in the input circuit are reduced accordingly.

JUNCTION-TRANSISTOR AMPLIFIER

Bias Circuits

The emitter junction of a transistor requires forward bias and the collector junction reverse bias. Consider, for example, the simple grounded-emitter amplifier, Fig. 6-12, in which base bias current is supplied by the resistor R_B. Because the forward resistance of the emitter junction is very small, the base current is given by V_{cc}/R_B. An approximate value of collector current is then βI_b, and the operating point is completely determined. Since the current gain depends somewhat upon the operating point, it is more accurate to plot the load line corresponding to R_L on the appropriate collector characteristics, such as Fig. 5-12. The intersection of the load line with the base current curve for $I_b = V_{cc}/R_B$ is the operating point.

This simple bias circuit is not generally satisfactory because the operating point shifts drastically with temperature. Comparison of collector characteristics at an elevated temperature, Fig. 5-13, with those for room temperature, Fig. 5-12, reveals that a much greater collector current exists at the higher temperature. Since the base bias current is fixed by the circuit, it is possible for the operating point to move into an unusable region of the transistor characteristics.

Figure 6-12 Simple grounded-emitter amplifier.

The most satisfactory transistor bias circuit is obtained by including a resistor in the emitter circuit, Fig. 6-13a. The voltage drop across R_E tends to bias the emitter junction in the reverse direction and the voltage divider comprising R_1 and R_2 sets the base voltage so that the base-emitter potential is in the forward direction. This circuit has the advantage that an increase in transistor current changes the voltage drop across R_E such that the base bias current is reduced, much as in the case of source bias of a FET.

The quiescent operating point is found as follows. First, the appropriate load line is calculated from the voltage equation around the output circuit,

$$I_c = \frac{V_{cc}}{R_E + R_L} - \frac{1}{R_E + R_L} V_{ce} \qquad (6\text{-}18)$$

Figure 6-13 (a) Practical transistor bias circuit. Note type 2N338 *npn* transistor is used. Circuit applies equally to *pnp* type if polarity of V_{cc} is reversed. (b) Dc equivalent circuit used to determine operating point.

Next, the bias line is derived from voltage drops around the input circuit,

$$I_c = \frac{V_{eq}}{R_E} - \frac{R_{eq}}{R_E} I_b \qquad (6\text{-}19)$$

In arriving at Eqs. (6-18) and (6-19) it is assumed that $I_e = I_c$, for simplicity. The bias line may be plotted on the transfer characteristic and the operating point determined from the intersection, as in the case of the FET. Actually, however, the transfer characteristics of junction transistors are so linear that it is easier to use a simple analytical expression for the transfer characteristic,

$$I_c = h_{FE} I_b \qquad (6\text{-}20)$$

In Eq. (6-20), h_{FE}, called the *dc current gain*, is often specified by the manufacturer and in any event can be estimated directly from the collector characteristics. After inserting Eq. (6-20) into Eq. (6-19), the expression may be solved for the base current,

$$I_b = \frac{V_{eq} - V_{be}}{R_{eq} + h_{FE} R_E} \qquad (6\text{-}21)$$

Knowing I_b and I_c, a value for V_{ce} is then calculated from Eq. (6-18) and the operating point is completely determined. Note that V_{be} may be considered to be a constant equal to 0.2 V for germanium transistors and to 0.6 V in the case of silicon transistors, as can be seen from the forward characteristics of the *pn* junctions in Fig. 4-2. The operating point of the grounded-emitter amplifier in Fig. 6-13 found by this procedure is shown on the collector characteristics of Fig. 6-14.

Figure 6-14 Operating point for circuit of Fig. 6-13 using common-emitter collector characteristics.

To design a transistor amplifier circuit such as in Fig. 6-13, it is first necessary to choose the desired operating point. This is best done after consulting the transistor manufacturer's data sheets for a given transistor type. Next, a value for R_L is selected to suit gain requirements (see discussion in the next section) and output impedance considerations. In order that the potential divider made up of R_1 and R_2 may not seriously impair the amplification, R_{eq} should be larger than the transistor input impedance by a factor of 5 or 10. The emitter resistor should be chosen so that $R_E \approx R_{eq}/5$ to ensure satisfactory stability of the operating point. Then, the required dc supply voltage V_{cc} is $I_c(R_L + R_E) + V_{ce}$. The value of R_{eq} selected above, together with Eqs. (6-18) and (6-19), yields simultaneous equations for R_1 and R_2, which completes the design. It is apparent from this brief description that the design of a transistor circuit is not completely determined by the usual requirements specified in advance. For this reason many different circuits have been developed. Each of these optimizes some particular feature of interest such as gain, minimum current requirement, or stability of operating point.

T-equivalent Circuit

The electrical characteristics of a junction transistor can be represented by a *T-equivalent circuit*, Fig. 6-15, which corresponds to the grounded-base configuration already illustrated in Fig. 5-10. In the T-equivalent circuit the constant-current generator αi_e is in parallel with the collector-junction resistance r_c. The collector resistance is the reciprocal of the slope of the collector characteristic curves and in practice is of the order of 1 to 10 MΩ. The emitter resistance r_e is the forward resistance of the emitter junction, which turns out to be about 26 Ω at a quiescent bias current of 1 mA. The base resistance in the T-equivalent circuit arises from two sources; the ohmic resistance of the base region (which may be appreciable since the base is so thin) and a feedback effect between the collector and the base. The origin of this effect is the decrease in base width as the collector junction widens with collector voltage. A typical value for the base resistance arising from these combined effects is 500 Ω.

The T-equivalent circuit for transistors is used in circuit analysis analogously to the FET ac equivalent circuit. The T equivalent is particularly appropriate

Figure 6-15 T-equivalent circuit of transistor.

because its parameters are directly related to the basic physical structure of the transistor. Note that the T-equivalent circuit involves a direct connection between the input and output terminals, which differs from the FET case. This may be illustrated explicitly by solving the T-equivalent circuit, Fig. 6-15, for the various currents and voltages. Using Kirchhoff's rules,

$$i_e + i_b + i_c = 0$$

$$v_{eb} = i_e r_e - i_b r_b \qquad \qquad (6\text{-}21)$$

$$v_{cb} = (\alpha i_e + i_c)r_c - i_b r_b$$

Solving for i_c and v_{eb},

$$i_c = -\alpha i_e + \frac{1}{r_c} v_{cb} \qquad \qquad (6\text{-}23)$$

$$v_{eb} = [r_e + (1 - \alpha)r_b]i_e + \frac{r_b}{r_c} v_{cb} \qquad \qquad (6\text{-}24)$$

where the approximation $r_c \gg r_b$ has been introduced. Thus, two equations are necessary to describe the operation of a transistor, whereas a single expression is sufficient in the case of the FET. Actually, Eq. (6-23) is quite analogous to the corresponding expression for a FET and should be compared with Eq. (6-11). The other relation, Eq. (6-24), indicates that the input voltage depends upon the output voltage, v_{cb}, as well as upon the input current i_e. This is a direct result of the connection between input and output in a transistor.

Hybrid Parameters

Although the T-equivalent circuit is a satisfactory representation of transistor operation, it is fairly difficult to determine the various resistance parameters of the circuit by direct measurements on actual transistors. For this reason an equivalent circuit which involves so-called *hybrid parameters* is more commonly employed in circuit analysis. This representation may be illustrated by writing Eqs. (6-23) and (6-24) as

$$i_c = h_{fb} i_e + h_{ob} v_{cb} \qquad \qquad (6\text{-}25)$$

$$v_{eb} = h_{ib} i_e + h_{rb} v_{cb} \qquad \qquad (6\text{-}26)$$

where the h's are the hybrid parameters. The significance of the h-parameter subscripts is as follows: the subscripts $i, r, f,$ and o refer to input, reverse, forward, and output, respectively, which may be understood by examining the meaning of each coefficient in Eqs. (6-25) and (6-26). The subscript b signifies the common-base configuration for which these equations have been developed. This designation distinguishes these parameters from those appropriate for the common-emitter and common-collector connections discussed below. A summary of the subscript notations is presented in Table 6-2.

Table 6-2. Subscript notation for h parameters

Subscript	Meaning
i	Input parameter
r	Reverse parameter
f	Forward parameter
o	Output parameter
e	Common emitter
b	Common base
c	Common collector

According to Eq. (6-26) it is possible to determine h_{ib} from the ratio of the emitter-base voltage to the emitter current v_{eb}/i_e with the collector shorted to ground so that $v_{cb} = 0$. Similarly, the ratio of the collector current to emitter current i_c/i_e with the collector shorted yields h_{fb}, using Eq. (6-25). When the emitter is open-circuited, $i_e = 0$, so that $h_{rb} = v_{eb}/v_{cb}$ and $h_{ob} = i_c/v_{cb}$. Note that the conditions $v_{cb} = 0$ and $i_e = 0$ refer to ac signals; the dc potentials are maintained at the proper operating point for the transistor. Because of the small emitter resistance and large collector resistance of a transistor it is particularly easy to achieve an open-circuited emitter or a short-circuited collector for ac signals while maintaining the desired dc potentials. In this way the small-signal h parameters can be measured using ac bridge techniques. The great virtue of the hybrid parameters is the fact that they can be measured directly with relative ease.

The *hybrid equivalent circuit* representing the grounded-base configuration is given in Fig. 6-16, as can be determined by inspecting Eqs. (6-25) and (6-26). Kirchhoff's rule applied to the input loop yields Eq. (6-26) directly. The voltage equation around the output loop gives

$$v_{cb} = \frac{1}{h_{ob}}(i_c - h_{fb} i_e) \tag{6-27}$$

Solving for i_c yields Eq. (6-25). Therefore, this hybrid equivalent circuit represents the operation of the transistor as given by Eqs. (6-25) and (6-26). These equations, in turn, are based on the fundamental processes associated with pn junctions in the transistor structure.

Figure 6-16 Grounded-base hybrid equivalent circuit.

The direct connection between input and output is not so immediately apparent in the hybrid equivalent circuit as compared with the T-equivalent. Note, however, that the voltage generator in the input circuit $h_{rb}v_{cb}$ involves the collector-base voltage and that the current generator in the output circuit $h_{fb}i_e$ includes the emitter current. Thus, the input and output circuits are indeed coupled. According to Fig. 6-16, h_{ib} represents a resistance, h_{ob} is a conductance, while h_{rb} and h_{fb} are simple numerics. These quantities represent different physical entities and are therefore termed hybrid parameters. The relation between hybrid parameters and T-equivalent parameters may be determined directly by comparing Eq. (6-25) with (6-23) and Eq. (6-26) with (6-24). These relations are summarized in Table 6-3 together with typical quantitative values for the parameters.

The common-emitter configuration is more widely used than is the common-base circuit. While it is possible to rearrange the equivalent circuit appropriate for the common-base configuration to apply to the common-emitter case, it is much more convenient to employ the same form of equivalent circuit and adjust parameters of the circuit for the common-emitter configuration. The relations between grounded-emitter h parameters and those corresponding to the grounded-base circuit are developed in the next section.

The h parameters depend upon the dc operating point of the transistor as well as upon the particular transistor type. It is common practice for transistor data sheets to specify the h parameters for the grounded-emitter or grounded-base configurations. The relations in Table 6-3 can be used to determine other parameters as needed. In addition, the variation of the parameters with operating point is usually specified so that appropriate corrections can be applied to the given values.

Table 6-3. Transistor parameter relations

		Common emitter	Common base	Common collector	T-equivalent
Common emitter	$h_{ie} =$	1400 ohm	$h_{ib}/(1 + h_{fb})$	h_{ic}	$r_b + r_e/(1 - \alpha)$
	$h_{re} =$	3.4×10^{-4}	$h_{ib}h_{ob}/(1 + h_{fb}) - h_{rb}$	$1 - h_{rc}$	$r_e/(1 - \alpha)r_c$
	$h_{fe} =$	44	$-h_{fb}/(1 + h_{fb})$	$-(1 + h_{fc})$	$\alpha/(1 - \alpha)$
	$h_{oe} =$	27×10^{-6} mho	$h_{ob}/(1 + h_{fb})$	h_{oc}	$1/(1 - \alpha)r_c$
Common base	$h_{ib} =$	$h_{ie}/(1 + h_{fe})$	31 ohm	$-h_{ic}/h_{fc}$	$r_e + (1 - \alpha)r_b$
	$h_{rb} =$	$h_{ie}h_{oe}/(1 + h_{fe}) - h_{re}$	5×10^{-4}	$h_{rc} - 1 - h_{ic}h_{oc}/h_{fc}$	r_b/r_c
	$h_{fb} =$	$-h_{fe}/(1 + h_{fe})$	-0.98	$-(1 + h_{fc})/h_{fc}$	$-\alpha$
	$h_{ob} =$	$h_{oe}/(1 + h_{fe})$	0.6×10^{-6} mho	$-h_{oc}/h_{fc}$	$1/r_c$
Common collector	$h_{ic} =$	h_{ie}	$h_{ib}/(1 + h_{fb})$	1400 ohm	$r_b + r_e/(1 - \alpha)$
	$h_{rc} =$	$1 - h_{re}$	1	1.0	$1 - r_e/(1 - \alpha)r_c$
	$h_{fc} =$	$-(1 + h_{fe})$	$-1/(1 + h_{fb})$	-45	$-1/(1 - \alpha)$
	$h_{oc} =$	h_{oe}	$h_{ob}/(1 + h_{fb})$	27×10^{-6} mho	$1/(1 - \alpha)r_c$
T-equivalent	$\alpha =$	$h_{fe}/(1 + h_{fe})$	$-h_{fb}$	$(1 + h_{fc})/h_{fc}$	0.98
	$r_c =$	$(h_{fe} + 1)/h_{oe}$	$(1 - h_{rb})/h_{ob}$	$-h_{fc}/h_{oc}$	1.7×10^6 ohm
	$r_e =$	h_{re}/h_{oe}	$h_{ib} - (1 + h_{fb})h_{rb}/h_{ob}$	$(1 - h_{rc})/h_{oc}$	13 ohm
	$r_b =$	$h_{ie} - h_{re}(1 + h_{fe})/h_{oe}$	h_{rb}/h_{ob}	$h_{ic} + h_{fc}(1 - h_{rc})/h_{oc}$	830 ohm

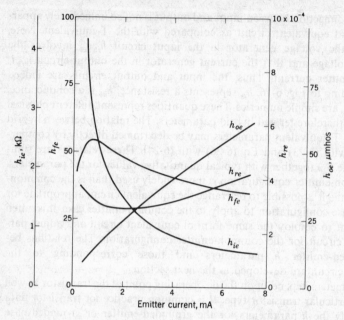

Figure 6-17 Variation of hybrid parameters with emitter current.

Illustrative values of the common-emitter h parameters and their variation with emitter current are given in Fig. 6-17. Note that the forward current gain h_{fe} and the reverse voltage amplification factor h_{re} are sensibly constant, while both the input resistance h_{ie} and output conductance h_{oe} vary considerably.

The two most sensitive measures of transistor quality are the emitter-collector current gain h_{fe} and the grounded-base output conductance h_{ob}. Large values of current gain are desirable to achieve maximum amplification. Note that h_{fe} and β as previously defined are two commonly used symbols for the same quantity. Small values of h_{ob} are desirable since then the output resistance is large and the feedback effect small. A small h_{ob} implies a large value of collector resistance (compare Table 6-3). Thus the two major parameters h_{fe} and h_{ob} reveal the quality of the emitter junction, base region, and collector junction.

JUNCTION-TRANSISTOR CIRCUITS

Common Emitter

The complete circuit of a practical common-emitter amplifier using an npn transistor is shown in Fig. 6-18. The input and output coupling capacitors C_1 and C_2 pass ac signal voltages and assure that the dc operating point of the transistor is

Figure 6-18 Practical common-emitter amplifier using *npn* type 2N930 transistor.

independent of the source and load conditions. The emitter bypass capacitor C_E shorts out the emitter bias resistor R_E for ac signals.

Assuming that the reactance of all three capacitors is negligible, the hybrid equivalent circuit appropriate for Fig. 6-18 is shown in Fig. 6-19. Note that it has the same form as for the grounded-base configuration, Fig. 6-16. The relation between grounded-emitter and grounded-base h parameters is developed by first writing the circuit equations pertaining to Fig. 6-19,

$$i_c = h_{fe} i_b + h_{oe} v_{ce} \tag{6-28}$$

$$v_{be} = h_{ie} i_b + h_{re} v_{ce} \tag{6-29}$$

where the h parameter subscripts are taken from Table 6-2. The corresponding relations for the grounded-base case, Eqs. (6-25) and (6-26), can be put into this form with the aid of the following general relations:

$$v_{be} + v_{ec} + v_{cb} = 0 \tag{6-30}$$

$$i_b + i_e + i_c = 0 \tag{6-31}$$

Figure 6-19 Hybrid equivalent circuit of amplifier in Fig. 6-18.

which apply to all configurations. Using (6-30) and (6-31) to eliminate i_e and v_{cb}, (6-25) and (6-26) may be put in the form

$$i_c = \frac{-h_{fb}}{1 + h_{fb}} i_b + \frac{h_{ob}}{1 + h_{fb}} v_{ce} \tag{6-32}$$

$$v_{be} = \frac{\cdot h_{ib}}{1 + h_{fb}} i_b + \left(\frac{h_{ib} h_{ob}}{1 + h_{fb}} - h_{rb} \right) v_{ce} \tag{6-33}$$

The approximation that h_{rb} and h_{ob} are small has been introduced in arriving at (6-32) and (6-33). Comparing these equations with (6-28) and (6-29) establishes the validity of the equivalent circuit in Fig. 6-19 and also gives the relations between the common-emitter h parameters and the common-base h parameters. These relationships are also summarized in Table 6-3.

The performance of the grounded-emitter amplifier is examined by analyzing the equivalent circuit as follows. The output voltage can be written directly as the current through the parallel combination of $1/h_{oe}$ and R_L

$$v_o = -h_{fe} i_b \frac{R_L}{1 + h_{oe} R_L} \tag{6-34}$$

Kirchhoff's rule applied to the input circuit yields

$$v_i = h_{ie} i_b + h_{re} v_o \tag{6-35}$$

Equation (6-35) is solved for i_b and this is substituted into Eq. (6-34). The result is arranged to give the voltage gain

$$a = \frac{v_o}{v_i} = -\frac{1}{h_{ie}(1 + h_{oe} R_L)/R_L h_{fe} - h_{re}} \tag{6-36}$$

By referring to Table 6-3 to compare the h parameters with the T-equivalent parameters, it may be seen that h_{re} is small and that $h_{oe} R_L$ may be neglected with respect to unity. Therefore Eq. (6-36) has the approximate form

$$a = -h_{fe} \frac{R_L}{h_{ie}} \tag{6-37}$$

According to Eq. (6-37), the voltage gain is approximately equal to the forward current gain of the transistor times the ratio of the load resistance to the input resistance. The voltage gain of the common-emitter amplifier is appreciable since both factors are large. The minus sign signifies that the input and output signals are 180° out of phase.

The transistor is a current-controlled device, and the current gain, which is the ratio of the output current to the input current, is also important. The output current is determined from Eq. (6-34)

$$i_o = \frac{v_o}{R_L} = -\frac{h_{fe} i_b}{1 + h_{oe} R_L} \tag{6-38}$$

For simplicity, the effect of R_{eq} in the bias network is assumed negligible, so the current gain is

$$g = \frac{i_o}{i_b} = -\frac{h_{fe}}{1 + h_{oe}R_L} \tag{6-39}$$

Here again, $h_{oe}R_L \ll 1$, so the approximate current gain is just h_{fe}.

The input resistance is the ratio of the input voltage to the input current, or

$$R_i = \frac{v_i}{i_b} = \frac{h_{ie}i_b + h_{re}v_o}{i_b}$$

Using Eq. (6-34),

$$R_i = h_{ie} - h_{re}\frac{h_{fe}R_L}{1 + h_{oe}R_L} \tag{6-40}$$

In arriving at Eq. (6-40), Eq. (6-35) has been used and the effect of R_{eq} has again been neglected. If necessary, R_{eq} may be included by calculating the parallel combination of R_{eq} and R_i. In Eq. (6-40), the second term is often negligible, so the input resistance is approximately h_{ie}. Note, however, that the exact input resistance depends upon the value of the load resistance R_L. This illustrates the coupling between input and output terminals inherent in transistors.

The output resistance of the amplifier includes the internal resistance of the input-signal source because of the coupling between input and output terminals. The effective internal resistance of the amplifier as viewed from the output terminals is determined using the Thévenin equivalent circuit. The equivalent internal resistance in the Thévenin circuit is the ratio of the open-circuit ($R_L = \infty$) voltage to the short-circuit ($R_L = 0$) current. Thus

$$R_o = \frac{(v_o)_{oc}}{(i_o)_{sc}} = -\frac{(h_{fe}/h_{oe})(i_b)_{oc}}{-h_{fe}(i_b)_{sc}} \tag{6-41}$$

The numerator uses Eq. (6-34) with $R_L = \infty$, and the denominator comes from the output loop of Fig. 6-19. Kirchhoff's rule applied to the input loop yields

$$v_g + i_b(R_g + h_{ie}) + h_{re}v_o = 0 \tag{6-42}$$

where v_g and R_g are the voltage and internal resistance of the signal source, respectively. Equation (6-42) may be solved for i_b and used to evaluate $(i_b)_{oc}$ by introducing v_o from Eq. (6-34). Similarly, $(i_b)_{sc}$ results from setting $v_o = 0$ in Eq. (6-42). Substituting these values into Eq. (6-41) gives the following expression for the output resistance of the amplifier,

$$R_o = \frac{1}{h_{oe} - h_{re}h_{fe}/(h_{ie} + R_g)} \tag{6-43}$$

To the same degree of approximation introduced previously, the output resistance is just $1/h_{oe}$, which is a fairly high value.

Table 6-4. Approximate parameters of transistor amplifiers*

Circuit configuration	Voltage gain	Current gain	Input resistance, Ω	Output resistance, Ω
Common emitter	280	47	1700	4×10^4
Common base	280	0.98	35	1.9×10^6
Common collector	1	48	5×10^5	22

* Using Fig. 6-17 for $I_E = 2$ mA; $R_L = 10^4 \ \Omega$.

In summary, the common-emitter transistor amplifier yields both voltage and current gain. It has a modestly high input resistance and large output resistance. Because of the favorable values of these four quantities, it is the most commonly used transistor amplifier circuit. Table 6-4 illustrates the magnitude of these important parameters, calculated from the h parameters of Fig. 6-17 at an emitter current of 2 mA. The various approximations introduced above have been used in preparing this table.

Grounded Base

A typical common-base amplifier circuit using an *npn* transistor is shown in Fig. 6-20. Careful comparison of this circuit with the common-emitter configuration, Fig. 6-18, reveals that the bias arrangements are identical. Therefore, bias considerations and the techniques for determining the operating point previously described for the common-emitter amplifier apply to the grounded-base circuit as well.

Capacitor C_B bypasses the base resistor R_2, and R_1 is shorted out for ac signals by the low impedance of the battery. Consequently, in the appropriate hybrid equivalent circuit, Fig. 6-16, these resistors are absent. The properties of this circuit can be found directly by comparing Fig. 6-16 with the hybrid equivalent of the common-emitter case, Fig. 6-19. The two are identical in form so that the previous results for the voltage and current gain, Eqs. (6-36) and (6-39), and for the input and output resistance, Eqs. (6-40) and (6-43), apply directly upon

Figure 6-20 Common-base transistor amplifier. Note bias circuit is identical to common-emitter case.

substituting the applicable h parameters. Note also that R_E replaces R_{eq} and i_e replaces i_b in converting the equations to the common-base configuration.

The approximate voltage gain in this case is $-h_{fb}R_L/h_{ib}$, which can be made large, but only if R_L is very great. This is so because the current gain is only $-h_{fb}$, which is approximately equal to unity. Note that h_{fb} is inherently a negative quantity (Table 6-3), so that the input and output voltage signals are in phase. The approximate input resistance h_{ib} is very low because it is essentially the resistance of the forward-biased emitter junction. Conversely, the output resistance $1/h_{ob}$ is the resistance of the reverse-biased collector junction and is therefore very large. This wide disparity between input and output resistance makes the grounded-base circuit less popular than the common-emitter amplifier, except for special applications. A summary of typical common-base circuit properties is presented in Table 6-4.

Emitter Follower

The common-collector amplifier, Fig. 6-21, is more often termed an emitter follower because of similarity to the source follower circuit. Here again, bias considerations are identical with those previously discussed, and the hybrid equivalent circuit also takes the same form as the one for the common-emitter and the grounded-base circuits. Relations between the common-collector h parameters and those of the other configurations are found by network analysis identical to that used above. The relationships are summarized in Table 6-3.

Circuit properties are obtained directly from the previous relations upon making the appropriate substitutions. Values of the common-collector h parameters are such that the voltage gain is unity. Furthermore, the output signal is in phase with the input. The approximate current gain $-h_{fc}$ is large. If the effect of R_{eq} can be neglected, the input resistance is simply $-h_{fc}R_L$. Thus, the input impedance is also large. The approximate output impedance reduces to $-(r_b + R_g)/h_{fc}$. Therefore, the output impedance of the amplifier itself is very small since the internal resistance R_g is associated with the signal source.

Figure 6-21 Common-collector transistor amplifier. This circuit is also called emitter follower.

Difference Amplifier

A very popular and versatile circuit often used as the input stage in laboratory instruments is the *difference amplifier*, Fig. 6-22, so named because it develops an output signal which is proportional to the difference between two input voltages. The two signals are introduced to the base terminals of a pair of similar transistors coupled together through a common emitter resistor. The output signal is taken between the collector terminals.

The useful properties of this amplifier are illustrated by analyzing the equivalent circuit, Fig. 6-23. Kirchhoff's voltage rule applied to both collector loops yields

$$i_{c1} R_L + (i_{c1} - h_{fe} i_{b1}) \frac{1}{h_{oe}} + R_E(i_{c1} + i_{c2}) = 0$$

$$i_{c2} R_L + (i_{c2} - h_{fe} i_{b2}) \frac{1}{h_{oe}} + R_E(i_{c1} + i_{c2}) = 0 \qquad (6\text{-}44)$$

Note that in arriving at Eq. (6-44) it has been assumed for simplicity that the emitter current is equal to the collector current in each transistor, that is, that the base currents are negligible in comparison to the collector currents. In particular, this means that the current through the common emitter resistor is just $i_{c1} + i_{c2}$.

Figure 6-22 Junction-transistor difference amplifier.

Figure 6-23 Equivalent circuit of difference amplifier.

The voltage equations around the base loops are

$$v_1 - i_{bs}h_{ie} - h_{re}v_{ec1} - R_E(i_{c1} + i_{c2}) = 0$$

$$v_2 - i_{b2}h_{ie} - h_{re}v_{ec2} - R_E(i_{c1} + i_{c2}) = 0 \qquad (6\text{-}45)$$

Note that the output signal is the sum of the voltages across both load resistors,

$$v_o = -R_L(i_{c1} - i_{c2}) \qquad (6\text{-}46)$$

The output signal may also be expressed as the difference between the emitter-collector voltages since the emitters are connected together,

$$v_o = v_{ec1} - v_{ec2} \qquad (6\text{-}47)$$

These circuit equations are solved by first subtracting the relations in Eq. (6-44),

$$\left(R_L + \frac{1}{h_{oe}}\right)(i_{c1} - i_{c2}) - \frac{h_{fe}}{h_{oe}}(i_{b1} - i_{b2}) = 0 \qquad (6\text{-}48)$$

and, similarly, subtracting those in Eq. (6-45),

$$(v_1 - v_2) - h_{ie}(i_{b1} - i_{b2}) - h_{re}(v_{ec1} - v_{ec2}) = 0 \qquad (6\text{-}49)$$

Equation (6-47) is inserted into Eq. (6-49) which is then used with Eq. (6-48) to eliminate $(i_{b1} - i_{b2})$. The result is rearranged to give

$$v_o = \frac{v_2 - v_1}{h_{ie}[(1 + h_{oe}R_L)/R_L h_{fe}] - h_{re}} \qquad (6\text{-}50)$$

This expression is identical to Eq. (6-36) for the amplification of a single grounded-emitter amplifier. Introducing the usual approximations for the small signal parameters,

$$v_o = h_{fe} \frac{R_L}{h_{ie}} (v_2 - v_1) \tag{6-51}$$

According to Eq. (6-51), the difference amplifier rejects any voltage signals that are common to both terminals, such that $v_1 = v_2$. At the same time a signal applied between the input terminals is amplified normally since in this case $v_1 = v_i/2$ and $v_2 = -v_i/2$,

$$v_o = -h_{fe} \frac{R_L}{h_{ie}} v_i \tag{6-52}$$

The same result holds if the signal is applied between either input terminal and ground so that, for example, $v_1 = v_i$ and $v_2 = 0$. Since the circuit rejects *common mode* signals resulting from, say, stray electric fields caused by the 60-Hz power mains, it is useful as the input stage of sensitive amplifiers. In addition, the input may be connected to signal sources that have neither terminal grounded, which often proves to be convenient.

The ability of the difference amplifier to reject common-mode signals rests on exact symmetry in the two halves of the circuit. Since this is rarely achieved in practical amplifiers, it is useful to examine the effect of small asymmetries. Suppose, for example, that the forward current gain parameters differ slightly, $h_{fe1} \neq h_{fe2}$. The subtraction in Eq. (6-44) yields

$$\left(R_L + \frac{1}{h_{oe}} \right) (i_{c1} - i_{c2}) - \frac{1}{h_{oe}} (h_{fe1} i_{b1} - h_{fe2} i_{b2}) = 0 \tag{6-53}$$

This equation is solved together with Eqs. (6-45) and (6-46) to determine the output signal analogous to that given by Eq. (6-51),

$$v_o' = h_{fe} R_L \frac{v_2 - (h_{fe2}/h_{fe1})v_1}{h_{ie} + (h_{fe1} - h_{fe2})R_E} \tag{6-54}$$

In arriving at Eq. (6-54) it is assumed that both h_{fe1} and h_{fe2} are not much different from their nominal value, h_{fe}. Suppose, now, $v_1 = v_2 = v_i$, then

$$v_o' = h_{fe} R_L \frac{1 - h_{fe1}/h_{fe2}}{h_{ie} + (h_{fe1} - h_{fe2})R_E} v_i \tag{6-55}$$

That is, an output signal results from circuit imbalance.

The ratio of the difference-signal output, Eq. (6-52), to the common-mode signal, Eq. (6-55),

$$\frac{v_o}{v_o'} = \frac{h_{fe}}{h_{ie}} \frac{h_{ie} + (h_{fe1} - h_{fe2})R_E}{h_{fe1} - h_{fe2}}$$

$$\cong \frac{h_{fe}}{h_{ie}} R_E \tag{6-56}$$

is called the *common-mode rejection ratio*. The magnitude of this ratio, and hence the ability of the difference amplifier to reject common-mode signals, is enhanced by large values of the common emitter resistor, R_E.

SPECIAL AMPLIFIERS

Complementary Symmetry

One of the more intriguing circuit applications of transistors is based on the combination of *npn* and *pnp* transistors, with their symmetrically inverted bias and signal-voltage polarities. Consider, for example, the *complementary-symmetry* circuit, Fig. 6-24, employing *npn* and *pnp* grounded-emitter amplifiers having common input and output connections. The base bias on both transistors is zero so in the absence of signal the transistors are cut off. Therefore, current is present in each transistor only when the input signal voltage biases its emitter junction in the forward direction. This happens on alternate half-cycles of the input voltage waveform because of the opposite polarities of the two transistors. Thus, the *npn* transistor delivers current to the load resistor when the *pnp* unit is cut off, and vice versa. The output signal is a replica of the input waveform, even though each transistor operates only half the time.

This particularly simple circuit is an efficient power amplifier since the quiescent current is zero and each transistor operates over the entire range of its characteristics. Furthermore, I^2R losses are small because the dc current in the load resistor is zero at all times. Unfortunately, it is not easy to fabricate *npn* and *pnp* transistors with identical characteristics, and the center-tapped collector supply voltage having neither terminal at ground is an awkward complication. A practical amplifier using complementary symmetry is described in Chap. 7.

Darlington Connection

The combination of two similar transistors in the so-called *Darlington connection*, Fig. 6-25, has many favorable circuit properties. Note that Q_2 is directly connected to Q_1 and the base-to-collector potential of Q_2 provides the emitter-

Figure 6-24 Complementary-symmetry amplifier employing *pnp* and *npn* transistors.

Figure 6-25 Darlington-connected amplifier.

collector voltage for Q_1. Furthermore, the output emitter current of Q_1 is the base input current of Q_2.

The circuit may be viewed as an emitter follower, Q_1, followed by a grounded-emitter amplifier, Q_2. The combination produces a very large current gain, $-h_{fc} \times h_{fe} = \beta^2$. The circuit also has the voltage gain of the common emitter with the large input impedance of the emitter-follower amplifier, both of which are desirable features.

For most purposes the combination may be considered to be a single device. Indeed, units having the two transistors in one case are commerciallly available. Proper bias potentials are applied by a conventional transistor bias circuit such as Fig. 6-13 in which the combination of Q_1 and Q_2 replaces the single grounded-emitter transistor.

Integrated-Circuit MOSFET

It is often advantageous to employ one transistor as the load resistor for another transistor in an amplifier circuit. This is particularly so in MOSFET integrated circuits because it then becomes possible to construct the entire amplifier from a suitable array of MOSFETs alone. This permits considerable flexibility in design with no particular increase in fabrication costs.

A useful example of this technique is the enhancement MOSFET amplifier shown in Fig. 6-26. In this circuit Q_2, which has the gate connected directly to the drain, acts as the drain load resistance for Q_1. Both V_{dd} and V_i are made large enough so that the gate voltage in each case is above threshold. The drain current is then, from Eq. (5-11),

$$I_d = I_{dss2}\left(1 - \frac{V_{dd} - V_o}{V_p}\right)^2 = I_{dss1}\left(1 - \frac{V_i}{V_p}\right)^2 \tag{6-57}$$

Taking the square root of both sides of the equation and solving for V_o,

$$V_o = (V_{dd} + V_p) - \left(\frac{I_{dss1}}{I_{dss2}}\right)^{1/2}(V_i + V_p) \tag{6-58}$$

Figure 6-26 Enhancement MOSFET amplifier.

According to Eq. (6-58), the output voltage is a linear function of the input signal and the gain is given by

$$a = \left(\frac{I_{dss1}}{I_{dss2}}\right)^{1/2} \qquad (6\text{-}59)$$

The gain may exceed unity by fabricating Q_1 with a wider channel than in Q_2 so that I_{dss1} is greater than I_{dss2}.

Note that Eq. (6-58) is linear over essentially the full operating range of the MOSFET. That is, the linearity between input and output voltages is not restricted to small signals. This useful property results from the compensating nonlinearities in the MOSFET amplifier and the MOSFET drain load resistance. Integrated circuit MOSFET amplifiers of this configuration are particularly appropriate in digital electronic applications, as discussed in Chap. 11.

SUGGESTIONS FOR FURTHER READING

E. J. Angelo, Jr.: "Electronics: BJT's, FET's, and Microcircuits," McGraw-Hill Book Company, New York, 1969.

"General Electric Transistor Manual," latest edition, General Electric Company, Semiconductor Products Department, Syracuse, N.Y.

Joseph A. Walston and John R. Miller (eds.): "Transistor Circuit Design," McGraw-Hill Book Company, New York, 1963.

EXERCISES

6-1 Determine the operating point of the FET grounded-source amplifier in Fig. 6-27. Use drain characteristics given in Fig. 5-21.

Ans.: −6 V, 4.5 mA, 1.8 V

Figure 6-27

6-2 Determine the small-signal parameters at the operating point of the amplifier of Exercise 6-1. Calculate the gain of the amplifier.

Ans.: 2.5×10^{-3} mho, 1.5×10^4 ohm; 2.5

6-3 Plot the transfer characteristics of the junction transistors whose collector characteristics are given in Figs. 5-12, 6-14, and 6-28. In each case calculate a value for h_{FE}.

Ans.: 500, 50, 300

Figure 6-28

6-4 Determine the operating point of the *npn* grounded-emitter amplifier, Fig. 6-13, using a 2N338 transistor, if $R_E = 470$ Ω, $R_1 = 16,000$ Ω, $R_2 = 6200$ Ω, $R_L = 700$ Ω, and $V_{cc} = 12$ V. Use the characteristic curves in Fig. 6-14.

Ans.: 75 μA, 4 mA

6-5 Determine the operating point of the 2N930 common-emitter amplifier, Fig. 6-18. Characteristic curves of the 2N930 are in Fig. 6-28.

Ans.: 4 μA, 4 mA

6-6 Calculate the voltage and current gain and the input and output impedances of the amplifier of Exercise 6-5. The *h* parameters are $h_{ie} = 3600$ Ω, $h_{re} = 3 \times 10^{-3}$, $h_{fe} = 150$, $h_{oe} = 1.4 \times 10^{-4}$ mho. Assume the reactances of all capacitors in the circuit are negligible and that the source resistance is 1000 Ω. Compare the calculated values with approximations introduced in the text.

Ans.: 362, 416; 62.4, 150; 1700 Ω, 3600 Ω; 2.4 × 10⁴ Ω, 7.1 × 10³ Ω

6-7 Determine the operating point of the 2N930 grounded-base amplifier, Fig. 6-20. Assuming the hybrid parameters given in Exercise 6-6, calculate the voltage and current gain and the input impedance for this circuit. Compare these with the approximations discussed in the text. Take $R_g = 100$ Ω.

Ans.: 200, 413; 0.99, 0.994; 27.5 Ω, 24 Ω

6-8 Determine the operating point of the type 2N930 emitter-follower amplifier, Fig. 6-21. Given the *h* parameters of Exercise 6-6, calculate the voltage and current gain and the input and output impedance for this circuit. Compare these results with the approximations introduced in the text. The source resistance is 10,000 Ω.

Ans.: 1, 1; 63, 151; 6.3 × 10⁵ Ω, 1.5 × 10⁶ Ω; 90 Ω

6-9 Determine the operating point of a grounded-emitter amplifier using a type 2N3521 transistor if $R_1 = 90$ kΩ, $R_2 = 45$ kΩ, $R_L = 10$ kΩ, $R_E = 6$ kΩ, and $V_{cc} = 36$ V. Use collector characteristics for the 2N3521 given in Fig. 5-12. Repeat if the ambient temperature is increased to 125°C using collector characteristics given in Fig. 5-13. Calculate the voltage gain of the amplifier in both cases. Is this circuit stabilized against changes in temperature?

Ans.: 5.9 V, 1.9 mA; 5.9 V, 1.9 mA; 750, 750; yes

6-10 Rearrange the circuit of Exercise 6-9 into an emitter-follower amplifier and calculate the input and output impedances. Suppose a type 2N930 transistor is inadvertently used in this circuit. Determine the operating point using the collector characteristics in Fig. 6-28 and calculate the input and output impedances. Is circuit performance sensitive to the change in transistors?

Ans.: 7.2 × 10⁶ Ω, 13 Ω; 5.5 V, 1.9 mA; 5.1 × 10⁶ Ω, 18 Ω; no

SEVEN

AMPLIFIER CIRCUITS

The principal applications of transistors are based on their ability to amplify electric signals. Some circuits amplify minute voltage signals by factors of many million, while others increase the electric power of a signal in order to operate a mechanical device such as an electric motor. Still other circuits amplify currents. In each of these applications the frequency range of the input signal is important. Different circuits have been developed for dc amplification and for use at high radio frequencies.

Most often the signal level is increased in several successive amplifier stages to attain the desired output-signal magnitude. In this case, the interaction between amplifier stages must be considered and fairly complicated networks are involved. Fortunately, the techniques of circuit analysis developed in previous chapters, in particular the ac equivalent circuits for transistors, are sufficient for a satisfactory understanding of complete amplifier circuits.

VOLTAGE AMPLIFIERS

Cascading

The transistor circuits discussed in previous chapters are ideally suited to amplify voltage signals with minimum waveform distortion. Gain factors greater than those possible with a single-stage amplifier are obtained by *cascading* several amplifier stages. The output of one amplifier stage is amplified by another stage or stages until the desired signal voltage level is achieved.

Consider, for example, the two-stage cascaded transistor amplifier, Fig. 7-1. Two individual circuits similar to those discussed in Chap. 6 are connected with the coupling capacitor C_{c2}. This capacitor passes the amplified ac signal from Q_1, to the base of Q_2. At the same time it blocks the collector voltage of Q_1 from the base of the second transistor. Similarly, capacitors C_{c1} and C_{c3} isolate the input and output circuits insofar as dc potentials are concerned.

Cascaded transistor voltage amplifiers most often employ the grounded-emitter configuration because of the combined voltage and current gain of this circuit. Neither the common-base nor the emitter-follower configuration achieves as great overall voltage amplification when cascaded. This is a result of the great impedance mismatch between the output impedance of one stage and the input impedance of the succeeding stage. Note that the bias resistors of the second stage in Fig. 7-1 are different from those of the first stage, even though both transistors are identical. The operating points are set at different places in order to obtain most favorable values of the h parameters in each stage.

The entire ac equivalent circuit of this amplifier may be drawn using the principles discussed in Chap. 6 and the performance of the system is determined from a complete ac-circuit analysis. Actually, this procedure is unwieldy because of the number of loops in the network and is rarely attempted. Rather, the circuit is analyzed in several separate steps, each of which has a minimum of mathemati-

Figure 7-1 Two-stage cascaded transistor amplifier.

Figure 7-2 Equivalent circuit of the two-stage transistor amplifier of Fig. 7-1.

cal complexity. This has the further advantage that the important effects can be isolated and more clearly examined.

For example, the reactances of both emitter bypass capacitors are assumed small enough to be negligible. Accordingly, these components are absent in the ac equivalent circuit of the amplifier, Fig. 7-2. Similarly, the reactances of the coupling capacitors are taken to be negligible. The output voltage may then be written immediately as

$$v_o = a_2 v' = a_1 a_2 v_i \tag{7-1}$$

where a_1 and a_2 are the gain factors of the first stage and the second stage, respectively. Evidently, the overall gain of a cascaded amplifier is just equal to the product of the gains of the individual stages.

Note that in calculating the gain of each stage the output load of the first stage involves the input impedance of the second. This means that the input impedance of this stage must be determined first. Since the input impedance of a transistor amplifier depends upon the output load impedance, it is necessary to work backward through the circuit starting from the output terminals. With the output load specified, the gain and input impedance of the second stage are calculated using results developed in the previous chapter. This input impedance is part of the load for the preceding stage, and its gain and input impedance are calculated accordingly. This procedure is not required in the case of cascaded FET stages because the very large input impedance inherent in the FET does not load the preceding stage. Thus the overall gain of an FET amplifier is just the product of the individual gains of isolated stages. In either case, the result is termed the *midband gain* of the amplifier because the reactances of the capacitors are taken to be negligible. This approximation applies to signal frequencies which are neither so low that the reactances cannot be ignored nor so high that other effects reduce the gain.

Low-frequency Gain

At sufficiently low frequencies the capacitive reactances may no longer be neglected. The effect of the coupling capacitors is usually of greater significance than that of the emitter or source bypass capacitors, although both tend to reduce the

Figure 7-3

gain at low frequencies. It is not practical to make the coupling capacitors overly large because large values of capacitance imply increased leakage current, which tends to upset the base or gate biases. The low input impedance of transistors permits coupling capacitors of a few tens of microfarads, whereas practical coupling capacitors are limited to values below about 0.5 μF in the case of FET circuits. Emitter and source bypass capacitors may be much larger, 100 μF or more, because leakage currents are less critical in this portion of the circuit.

Because the overall gain of cascaded stages is the product of individual stage gains, it is only necessary to examine the effect of the coupling capacitor reactance for an isolated amplifier stage. According to the ac equivalent circuits, Fig. 7-2, this effect can be treated by considering the simple RC circuit comprising the coupling capacitor and the input impedance of the amplifier. This portion is shown separately in Fig. 7-3 for clarity. The input impedance R_i is simply the gate resistor in the case of the FET amplifier, but in transistor stages it includes the input impedance of the transistor as well. In either case, Fig. 7-3 is a RC high pass filter already discussed in Chap. 2. The gain of this amplifier stage including the effect of the high pass filter is given by

$$a(f) = \frac{v_o}{v_i} = \frac{av'}{v_i} = \frac{a}{\sqrt{1 + (f_0/f)^2}} \tag{7-2}$$

where the frequency response characteristic of a high pass filter, Eq. (2-45), is used and f_0 is the half-power frequency. Note that the gain is reduced from the midband value, a, when the signal frequency is smaller than the half-power frequency.

At the same time, a phase shift is introduced between the input and output signal, which is also significant in determining waveform distortion of the amplifier. Recall that in the Fourier analysis of a complex signal waveform, both the amplitudes and relative phases of all frequency components must be preserved if the output wave is to be an amplified replica of the input signal. It is also important to recognize that Eq. (7-2) applies to an individual stage. The low-frequency response of the entire amplifier is always poorer than that of any individual stage because the gain of cascaded stages is the product of individual stage gains.

High-frequency Gain

The high-frequency gain of any amplifier is reduced by stray capacitive effects that are not purposely made part of the circuit. Referring to the simple FET amplifier stage, Fig. 7-4, these are the gate-source capacitance, C_1, the gate-drain capaci-

Figure 7-4 Stray capacitances in an FET amplifier.

tance, C_2, and the drain-source capacitance, C_3 of the FET. Also included in C_1 and C_3 are stray capacitances between the wires and components attached to the gate and drain terminals. All three of these capacitors shunt the signal at frequencies high enough for the capacitive reactances to be small.

The effect of the gate-drain capacitance is particularly important. Consider the pertinent equivalent circuit, Fig. 7-5, in which C_2 is connected between gate and drain terminals and the other stray capacitances are ignored. The input impedance of the amplifier stage is calculated assuming that the reactance of C_2 is the dominant factor,

$$\mathbf{Z}_i = \frac{v_i}{i_i} = \frac{v_i}{(v_i + av_i)/(1/j\omega C_2)}$$

$$= \frac{1}{j\omega(1 + a)C_2} \tag{7-3}$$

This result indicates that the effect of C_2 is as though a larger capacitance, $(1 + a)C_2$, is connected from gate to ground. The increase in effective shunt capacitance caused by the gain of the stage is called the *Miller effect* and is the dominating factor in determining the high frequency response of both FET and junction transistor amplifiers.

Figure 7-5

The appropriate high frequency equivalent circuit includes a shunt capacitance

$$C_s = C_1 + C_3 + (1 + a)C_2 \qquad (7\text{-}4)$$

from the gate or base terminal to ground. The effect of C_s is ascertained after isolating the input circuit, Fig. 7-6. In this circuit R_{eq} represents the entire shunt resistance to ground comprised of the output resistance of the preceding stage and the input resistance of the transistor, as well as the equivalent resistance of the bias resistors. Figure 7-6 is a RC low pass filter, so that the gain is, from Eq. (2-27),

$$a(f) = \frac{v_o}{v_i} = \frac{av'}{v_i} = \frac{a}{\sqrt{1 + (f/f_0)^2}} \qquad (7\text{-}5)$$

where, here again, f_0 is the half-power frequency and a is the midband gain. According to Eq. (7-5) the gain of the amplifier is reduced from its midband value when the signal frequency exceeds the half-power frequency.

Actually, the high-frequency amplification of many junction transistors is limited by the transit time of carriers diffusing across the base region. This effect results in a high-frequency gain given by an expression identical to Eq. (7-5) except that f_0 is determined by physical constants of the transistor, such as the width of the base. This characteristic frequency is usually specified by the transistor manufacturer. In specially designed high-frequency transistors the *alpha falloff frequency* is high enough that the gain is limited by the circuit parameters, as discussed above.

Using Eqs. (7-3) and (7-5), the *frequency response* of any voltage amplifier is similar to that illustrated in Fig. 7-7a. The *low-frequency cutoff* for each stage is determined from Eq. (7-3) while the *high-frequency cutoff* for each stage is found using Eq. (7-5). The *bandpass* of the complete amplifier is the frequency interval between the high- and low-frequency points where the gain falls to $1/\sqrt{2}$ of the midband gain. It is conventional to employ logarithmic scales on both axes of bandpass characteristics such as Fig. 7-7a because the range of gains and frequencies is so great. The vertical scale is often put in terms of a unit called the *bel*, named after Alexander Graham Bell, the inventor of the telephone. Actually, a unit one-tenth as large, the *decibel*, abbreviated dB, proves more convenient in practice. The decibel is defined as

$$\text{dB} = 20 \log \frac{a(f)}{a} \qquad (7\text{-}6)$$

Figure 7-6

Figure 7-7 (a) Frequency-response characteristic of single amplifier stage. Characteristic of two-stage amplifier is shown dashed. (b) Phase-shift characteristic of single amplifier stage.

Correspondingly, the midband gain is often quoted in terms of decibels using the definition

$$dB = 20 \log \frac{v_o}{v_i} \tag{7-7}$$

The advantage of this unit is that the total gain in dB of several amplifier stages is simply the sum of the individual gains in terms of decibels. Note that, according to Eq. (7-7), the amplifier gain is down 3 dB at the upper and lower half-power points.

The phase-shift characteristics of a single-stage amplifier are illustrated in Fig. 7-7b. The output signal leads the input at frequencies below the low-frequency cutoff and lags at frequencies above the high-frequency cutoff. The phase-shift characteristics of an entire amplifier are determined by adding the contributions from each stage.

Figure 7-8 Small inductance as load impedance improves high-frequency response because gain is increased by higher load impedance.

Often an amplifier must have a wide bandpass in order to minimize waveform distortion. A number of minor circuit alterations have been developed to accomplish this end. For example, source or emitter bypass capacitors may purposely be made small so that the capacitive reactance is appreciable except at frequencies near the high-frequency cutoff. This reduces the midband gain but increases the gain at high frequencies where the capacitive reactance becomes small. The net result is an extended high-frequency response, although at the expense of smaller overall amplification. If necessary, the loss in gain can be made up by adding another stage.

A second useful way of extending the high-frequency response is to include a small inductance as part of the load, Fig. 7-8. The load impedance increases at high frequencies and the gain of the amplifier is larger, according to Eq. (6-14). This technique is called *peaking* since the resulting frequency-response characteristic tends to be peaked at the high-frequency end.

Figure 7-9 Combination $R_3 C_3$ improves low-frequency response of amplifier because reactance of C_3 becomes large.

Improved low-frequency response can be achieved by adding a series resistor-capacitor combination R_3 and C_3 to the output circuit, Fig. 7-9. The gain at frequencies where the reactance of C_3 is small includes the effect of R_3 in the ac load resistance. At low frequencies the gain increases as the reactance of the capacitor increases and removes R_3 as part of the output load. The result is an extended low-frequency response at the expense of midband gain.

Decoupling

When three or more stages of amplification are cascaded, it is usually necessary to *decouple* the power supply of the input stage from the remainder of the amplifier. The reason for this is that the supply voltage changes with current because of the effective internal impedance of the power supply. Any small change in the power supply voltage alters the bias on the first stage, and this change is amplified in the same fashion as an input signal. If the change in bias increases the current in the first stage, current in the second stage is reduced because of the 180° phase shift in the input amplifier. The current in the third stage is increased, however, because of the second 180° phase shift in the second stage. The change in the third stage is much larger than the original disturbance because of the gain of the amplifier. The additional load causes a decrease in the power supply voltage. This, in turn, further alters the bias on the first stage and the process is cumulative. The changes continue until one transistor is driven to cutoff or into saturation, which reduces the overall gain to zero. The power supply voltage then returns to normal and the process repeats itself. The result of this *feedback* from output to input is that the amplifier rapidly oscillates from cutoff to saturation at a rate which is a function of the circuit components.

A low-pass RC filter inserted in the power supply lead to the first stage, Fig. 7-10, circumvents this difficulty. The time constant of this decoupling filter is selected so that power supply variations are sufficiently attenuated and feedback is eliminated. Actually, the characteristic filter frequency is put below the low-frequency cutoff of the amplifier where the gain is insufficient to support feedback oscillations.

Figure 7-10 Simple RC filter eliminates instability in amplifier by reducing feedback effect resulting from common power supply.

POWER AMPLIFIERS

Transformer Coupling

When transistor amplifiers deliver appreciable amounts of power, it is no longer feasible to use resistors in the collector circuit. The I^2R losses become significant at the high currents associated with large powers. Instead, a transformer couples the circuit to the load, Fig. 7-11. The dc collector current in the winding resistance

Figure 7-11 Power amplifiers use transformer to couple transistor to load to reduce dc power lost in load resistance.

Figure 7-12 Location of operating point and dc and ac load lines for power amplifier in Fig. 7-11.

introduces only a small power loss, yet the reflected resistance of R_L into the primary circuit provides the proper ac load impedance for the amplifier. Furthermore, the output impedance of the amplifier is matched to the load by the transformer and the actual load resistance can be any convenient value.

The dc load line is essentially vertical on the collector characteristic curves, Fig. 7-12, because of the small winding resistance of the transformer primary. The quiescent operating point is determined exactly as outlined in Chap. 6. The ac load line corresponding to the reflected load resistance R'_L as seen from the primary side of the output transformer passes through the operating point. As usual, the slope of the ac load line is $-1/R'_L$, as shown in Fig. 7-12.

The operating point of a power amplifier is chosen to maximize the efficiency of the amplifier and to minimize the possibility of thermal runaway. Power dissipation in the transistor due to collector current is limited by the allowable temperature rise of the collector junction. The permissible operating range of collector current and voltage is to the left of a *maximum-power hyperbola*, Fig. 7-13. Power

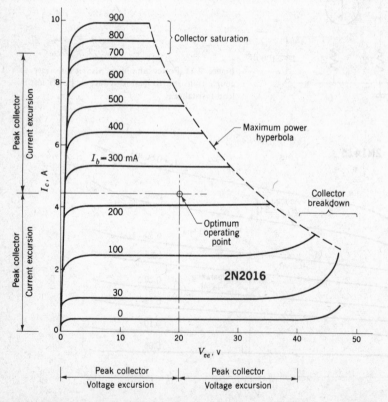

Figure 7-13 Optimum location of operating point for power amplifier is determined by collector saturation and collector breakdown.

transistors are often firmly mounted on a good heat conductor. This moves the maximum-power hyperbola farther away from the origin and extends the permissible operating current and voltage range. In addition, cooling fins are often provided to maximize heat conduction away from the transistor.

The operating point is located so that the largest possible ac signals can be developed in order to maximize the power output without distortion. The maximum instantaneous collector potential is limited by reverse breakdown at the collector junction. Similarly, the maximum instantaneous transistor current corresponds to collector saturation, where the collector current no longer increases with emitter-junction current. The output waveform is badly distorted if either of these limits is exceeded because the peaks of the signal wave are clipped. Therefore, the optimum position for the operating point is in the center of the rectangle bounded by collector breakdown, collector saturation, zero collector current, and zero collector voltage, Fig. 7-13. Here the collector current and voltage excursions on either side of the operating point are maximized without distortion.

The efficiency of a power amplifier is equal to the ratio of the ac signal power to the dc or average power from the power supply. The average power is simply the product of the quiescent current times the quiescent collector voltage, $I_c V_c$. If the operating point is located at the optimum position, the peak output signal current is equal to I_c and the peak output signal voltage is equal to V_c. Con-

Figure 7-14 Dynamic transfer characteristic is used to determine amplified waveform of 2N2016 power amplifier. Note distortion in output current caused by nonlinear transfer characteristic.

sequently, the maximum efficiency of this power amplifier is 50 percent. Practical transistor amplifiers approach the ideal quite closely, even though for minimum distortion the signal excursions must be somewhat smaller than the ideal case. Efficiencies of the order of 48 percent are achieved in practice.

An equivalent-circuit representation of power amplifiers is not feasible because of the large signal voltage excursions. Consequently, all analyses are carried out graphically. It is most useful to determine the dynamic transfer characteristic of the amplifier, which is a plot of the collector output current as a function of base input current, Fig. 7-14. The transfer characteristic is determined from intersections of the collector characteristic curves, Fig. 7-13, with the dynamic load line. For minimum distortion the transfer characteristic should be a straight line since any curvature introduces irregularities into the output waveform.

Push-Pull Amplifier

The two-transistor *push-pull* amplifier, Fig. 7-15, has increased power output, efficiency, and less distortion than a single-transistor circuit. The center-tapped input transformer drives each of the transistors with signals 180° out of phase, which accounts for the name of the circuit. The amplified collector currents combine in the center-tapped output transformer to produce a load current waveform that is a replica of the input signal. The input transformer also matches the driver stage to the input impedance of the amplifier.

Increased efficiency results when the push-pull amplifier is biased nearly to cutoff. This is known as *class B* operation to distinguish the circuit performance from that previously described in which current is present in the transistor during the entire input-signal cycle. Because each transistor is biased near cutoff, the quiescent current is very small and the signal-voltage and current excursions can be equal to the maximum permissible collector voltage and current, Fig. 7-16. Each tube delivers one-half of a sine-wave signal to the output transformer and the output waveform is preserved even though signal currents in each transistor represent only one-half of the input signal. This action has already been noted in

Figure 7-15 Push-pull power amplifier.

Figure 7-16 Operating point for class B push-pull operation permits voltage to swing over maximum range of collector characteristics.

the complementary-symmetry amplifier discussed in Chap. 6. In push-pull operation the peak output voltage can equal the maximum collector potential, which is the same as the dc collector supply voltage (see Fig. 7-16). Correspondingly, the peak signal current is equal to the maximum collector current. The average power of the stage is equal to the power of a half-sine wave, since only one transistor conducts at a time. The maximum efficiency of a class B push-pull amplifier turns out to be 78 percent, a considerable improvement over the single-transistor amplifier.

A small quiescent base bias current minimizes *crossover* distortion resulting from nonlinearity in the transfer characteristic of each transistor at small currents. This is illustrated in Fig. 7-17, where the transfer characteristics of the two transistors are plotted in opposite quadrants corresponding to their reversed signal polarities. The composite transfer characteristic of the entire amplifier is the average of the individual curves and is much more linear than either one. In particular, the nonlinearities cancel each other near the origin where both transistors are active. The cancellation effect means that the push-pull circuit has much less distortion than a single-ended stage.

Figure 7-17 Composite transfer characteristic of push-pull power amplifier is more linear than that associated with each transistor. Compare output-signal current amplitude and waveform with single-transistor case, Fig. 7-14.

Special Circuits

The frequency response of power amplifiers is seriously limited by transformer characteristics. In particular, good low-frequency response requires large (and expensive) transformers. This implies poor high-frequency operation because of unavoidable stray capacitances. The small output impedance of the emitter-follower configuration may often be used advantageously to eliminate the output transformer and to achieve satisfactory performance over a wide frequency interval. A typical circuit, Fig. 7-18, also illustrates the flexibility inherent in transistor designs.

Transistor Q_1 is a grounded-emitter input stage which is directly coupled to a complementary-symmetry pair of transistors Q_2 and Q_3. This pair is biased for class B operation and each transistor produces an output signal on alternate halves of the input wave, as previously discussed in connection with Fig. 6-24.

Actually, Q_2 and Q_3 have a small quiescent bias to minimize crossover distortion. This bias is set by the voltage drop across the forward-biased 1N536 silicon

Figure 7-18 Single-ended push-pull power amplifier.

diode. The emitter resistance of a transistor decreases significantly with increasing temperature. The forward resistance of the diode similarly decreases with temperature and reduces the base bias of Q_2 and Q_3 correspondingly. Thus the quiescent current in Q_2 and Q_3 is temperature compensated.

The complementary-symmetry pair is in the Darlington connection with respect to the output transistors Q_4 and Q_5. This produces large current gain, as described in the previous chapter. Both Q_4 and Q_5 operate class B except that a small bias provided by the voltage drop across the 390-Ω resistors minimizes crossover distortion. Note that the push-pull signals developed by Q_2 and Q_3 result in activating Q_4 and Q_5 on alternate half-cycles of the input waveform. This may be seen most clearly by assuming the input signal is a square wave. Suppose the input square wave is on the positive cycle so that Q_1 is nearly cut off. This means Q_2 and Q_4 conduct, while Q_3 and Q_5 are cut off, the output signal is negative, and C_3 is charged through R_L. On the alternate half-cycle Q_3 and Q_5 discharge C_3 through the load while Q_2 and Q_4 are cut off. Since each output transistor acts independently of the other, the circuit is, in effect, a single-ended class B push-pull arrangement.

An integrated-circuit power amplifier is illustrated in Fig. 7-19. The circuit produces a gain of 58 dB and a power output of $\frac{1}{2}$ W. As previously described in Chap. 5, the entire amplifier is contained within one small monolithic block of silicon.

The circuit of the amplifier, Fig. 7-19a, consists of an input difference amplifier, Q_1 and Q_2. The output signals from Q_1 and Q_2 are properly phased to feed Q_3 and

(a)

(b)

Figure 7-19 (a) Wiring diagram of type CA3020 integrated amplifier circuit and (b) practical power amplifier.

Q_4 in push-pull. Q_3 and Q_4 are directly connected to the output transistors Q_5 and Q_6. Transistor Q_7 is a separate unit which may be connected to the remainder of the amplifier or ignored, as desired.

The combination of R_{10} and R_{11} together with the three diodes makes up a temperature-compensated voltage regulator for the input stage. Since the voltage drop across a forward-biased silicon diode is approximately 0.6 V, the total base bias supply for Q_1 and Q_2 is 1.2 V, while the collector supply voltage is 1.8 V. Furthermore, the magnitudes of these voltage sources change with temperature to compensate for corresponding changes in transistor properties with temperature. Note that R_5 and R_7 set the bias on Q_1 and Q_2 from the base potentials of Q_5 and Q_6. Thus if the bias on the base of Q_5 increases, say, because of a temperature change, the bias on Q_1 also increases. This produces a decreased voltage on the base of Q_3 and, in turn, a decreased bias on Q_5. In effect, the original change has been partially compensated. Such *feedback* helps stabilize the operating points of the transistors against drifts caused by temperature changes and slow changes in component characteristics.

A practical power amplifier using this integrated circuit is shown in Fig. 7-19b. In this connection Q_7 is arranged as an emitter follower for a high impedance input and the 10-kΩ potentiometer acts as a variable gain control. The base of Q_2 is grounded for ac signals through the 5-μF capacitor, and a standard push-pull output transformer couples Q_5 and Q_6 to the load. Note how many fewer components are required in using the integrated-circuit amplifier compared to the number necessary if the entire circuit were assembled from separate components.

TUNED AMPLIFIERS

Tuned Coupling

Resonant circuits couple the output of one stage to the input of the next when it is only necessary to amplify signals of a single frequency or of a narrow band of frequencies. The impedance of parallel resonant circuits is very great at resonance, as discussed in Chap. 3. Therefore, appreciable gain is achieved at the resonant frequency when a *tuned circuit* is the load impedance of a transistor amplifier. A tuned amplifier also rejects signals far from the resonant frequency, which is often a considerable advantage. In addition, stray circuit capacitances are incorporated into the resonant circuit and do not shunt the signal at high frequencies.

The elementary two-stage transistor tuned amplifier, Fig. 7-20, uses parallel resonant circuits for the input circuit and output load of each transistor. The coupling capacitor C_c carries the signal from one stage to the next. The operating point for each transistor is determined in the standard fashion. It is common practice to make the tuning capacitors C_1, C_2, C_3, and C_4 adjustable so that each circuit can be brought to the same resonant frequency including the effect of all stray capacitances in each stage. Tuned circuits are resistive at resonance, which means that the amplifier can be analyzed by the methods previously developed. Stray capacitances can be neglected, however, and values of the h parameters appropriate at the frequency of interest must be used. Circuit analysis at other than the resonant frequency is rather complicated because of reactance effects, but can be treated straightforwardly using the ac equivalent circuit.

If all four resonant circuits are tuned to the same frequency, the response

Figure 7-20 Two-stage tuned amplifier.

characteristic is sharply peaked at the resonant frequency. Such a characteristic is useful when signals having one specific frequency are amplified. Alternatively, each circuit can be tuned to a slightly different frequency, Fig. 7-21, in which case the response characteristic becomes flat-topped. This permits amplification over a band of frequencies such as for the modulated sine wave discussed in Chap. 4. The midband gain of such a *stagger-tuned* amplifier is less than that of the single-frequency circuit since the maximum amplification of each stage occurs at different frequencies.

If L_2 and L_3 are wound on the same core the response characteristic may be double-peaked and approximate the flat-top response in Fig. 7-21, even though

Figure 7-21 Stagger-tuned amplifier has response characteristic with relatively flat top.

both primary and secondary windings are tuned to the center frequency. This is caused by mutual inductance between the windings, which are said to be *overcoupled*. The mutual inductance is altered by changing the distance between the two coils, and in this way the frequency-response curve may be either sharply peaked, or relatively flat-topped. The overcoupled case is particularly useful because the midband gain is greater than in the stagger-tuned amplifier and the sides of the response curve rise much more steeply. The circuit thus rejects signals at frequencies immediately outside of the passband. Note that the coupling capacitor is no longer necessary since the signal is coupled from one stage to the next by the mutual inductance. In fact, the combination $L_2 C_2$ and $L_3 C_3$ is looked upon as a tuned transformer.

Neutralization

The collector-junction capacitance in many transistors is large enough to cause an undesirable feedback effect between the collector and base. At high frequencies the capacitive reactance becomes small and the amplifier oscillates because the amplified collector signal is returned to the input where it is reamplified, etc. The circuit is useless as an amplifier when this occurs. The effect of the collector capacitance can be *neutralized* by feeding the base a signal of the same amplitude as that produced by the feedback capacitance but 180° out of phase so that the two feedback signals cancel each other. One technique for accomplishing this is illustrated in Fig. 7-22. Here, a portion of the output signal is fed back to the input by the neutralizing capacitor C_n. The magnitude of the return signal is determined by the relative values of the small series inductance L_n and the capacitor C_p, as well as the value of C_n. The return signal is 180° out of phase with that

Figure 7-22 Effect of collector-base capacitance is neutralized by signal returned from output circuit through C_n.

due to the collector capacitance because of the inductive phase shift introduced by the series inductance. Other circuit configurations are also used. In transistor design, every effort is made to minimize collector capacitance so that neutralization is unnecessary.

A similar effect occurs in field-effect transistors because of the drain-gate capacitance. The magnitude of the drain-gate capacitance of MOSFETs is markedly reduced by interposing a second gate between the control gate and the drain, Fig. 7-23, so that neutralization becomes unnecessary. The second gate is grounded for ac signals to provide the shielding effect.

The feedback effect in transistor amplifiers can also be circumvented by employing the grounded-base configuration, Fig. 7-24. In this circuit the collector-emitter capacitance is so small that neutralization is unnecessary. Furthermore, the grounded-base h parameters are relatively independent of frequency so that the transistor is a useful amplifier at frequencies very close to the α-cutoff frequency. The low input impedance of the grounded-base stage can be matched using a tuned transformer. The winding L_2 has only a few turns, and it is usually not advantageous to resonate the secondary of the transformer.

In the circuit of Fig. 7-24, the collector is tapped down on the inductance of the resonant circuit $L_3 C_3$. This reduces the loading of the transistor output impedance upon the resonant circuit, thereby increasing the Q and making the resonance curve sharper. Furthermore, the collector capacitance is much less significant in determining the resonant frequency. This is important because of the change in collector capacitance with temperature. If the collector is connected across the entire resonant circuit, the collector capacitance is effectively in parallel with the tuning capacitor C_3, and the resonance frequency varies with temperature.

Figure 7-23 Dual-gate MOSFET amplifier.

Figure 7-24 Grounded-base tuned amplifier.

PULSE AMPLIFIERS

Signals which consist of a series of rapid transitions from one voltage level to another as, for example, square waves, are equally important as sinusoidal waveforms. Although such pulse waveforms may be resolved into their sinusoidal harmonic components using the method of Fourier analysis, it usually proves more convenient to consider the signal pulses themselves. An ideal pulse amplifier introduces no change in the pulse shape other than increasing its amplitude, and practical circuits can come close to this performance.

Rise Time

The important characteristics of pulse amplifiers, gain, frequency response, stability, and distortion, are similar to those already discussed in connection with conventional amplifiers. To illustrate, in particular, the significance of frequency response, consider the ideal square pulse shown in Fig. 7-25 and the nonideal pulse shape resulting from passing this waveform through a practical amplifier. Note that the amplified pulse does not rise instantaneously to the maximum value. This is a result of the high-frequency cutoff of the amplifier caused by shunt capacitance. Consider the case of a single-stage transistor or vacuum-tube amplifier, which can be represented for present purposes by the input circuit of Fig. 7-26, in which C_s represents the total shunt capacitance (compare Fig. 7-6), R_i is the input impedance, and R_s is the output impedance of the preceding stage. The waveform of the output pulse v_g due to a square voltage pulse v_i starting at $t = 0$ is found by the transient circuit analysis technique described in Chap. 2.

Figure 7-25 Comparison of ideal square voltage pulse with amplified, nonideal pulse.

According to Eq. (2-59), the output signal increases exponentially at a rate given by the circuit time constant

$$\tau = \frac{C_s}{1/R_s + 1/R_i} \tag{7-8}$$

It is desirable to make the time constant as small as possible to minimize distortion of the leading edge of the pulse.

At the end of the input pulse, the output signal decays exponentially to zero at a rate again determined by the circuit time constant. Note, however, that the preceding stage may be completely cut off at the end of the pulse so that R_s becomes large enough to be ignored in Eq. (7-8). This means that the leading and trailing edges of the pulse waveform are likely to have different time constants; the rise time is usually more rapid. This disparity is minimized in junction transistor pulse amplifiers because the low input impedance of transistors dominates in Eq. (7-8).

It is convenient to define the *rise time* t_r of a pulse as the time required for the voltage to go from 10 to 90 percent of its final value, as indicated in Fig. 7-25. A useful approximate expression for t_r may be developed from Eq. (2-59) by setting

$$0.9 = 1 - e^{-t_r/\tau} \tag{7-9}$$

Figure 7-26

Solving for t_r and introducing the high frequency cutoff, $2\pi f_o = 1/\tau$, of the amplifier from Eq. (2-46), the result is

$$t_r = \frac{\ln 10}{2\pi f_o} \cong \frac{1}{3 f_o} \qquad (7\text{-}10)$$

According to Eq. (7-10) an amplifier with a good high-frequency response is necessary if the pulse rise time is short. For example, if the pulse rise time is 1 μs, the amplifier bandwidth must extend to 370 kHz. In this connection it should be remembered that the bandwidth of a multiple-stage amplifier is always poorer than that of any individual stage.

Tilt

A second form of pulse distortion is departure of the top of the pulse from a constant value, called *tilt*, in Fig. 7-25. This is a result of low-frequency cutoff of the amplifier. In effect, the pulse length represents a dc signal of short duration. Therefore, the appropriate single-stage equivalent circuit, Fig. 7-27, includes the interstage coupling capacitor C_2 as in Fig. 7-3. This is a high pass filter and the voltage transient is an exponential waveform characterized by the time constant

$$\tau = (R_i + R_s)C_c \qquad (7\text{-}11)$$

Clearly, the circuit time constant τ must be long for minimum decay of output voltage during the pulse. This means that a large value of C_c is required.

The tilt is small in most practical situations, so the exponential can be approximated by the first two terms of a series expansion,

$$v_g = v_i\left(1 - \frac{t}{\tau}\right) \qquad (7\text{-}12)$$

The percentage tilt $P = \Delta v_g / v_i$ at the end of the pulse duration t_w is then

$$P = \frac{\Delta v_g}{v_i} = \frac{t_w}{\tau} \qquad (7\text{-}13)$$

Introducing the low-frequency cutoff $2\pi f_0 = 1/\tau$ from Eq. (7-4),

$$t_w = \frac{P}{2\pi f_0} \qquad (7\text{-}14)$$

Figure 7-27

Notice that a long pulse requires an amplifier with an excellent low-frequency response. For example, suppose that tilt distortion must be less than 1 percent. Then the low-frequency cutoff of an amplifier required to handle 1-ms pulses must extend down to 1.6 Hz.

The results of these two sections provide a convenient technique for determining response characteristics of amplifiers. A square wave is applied to the input, and the output waveform is displayed on an oscilloscope. The frequency of the square wave is reduced until tilt is measurable in the output waveform and the low-frequency cutoff determined using Eq. (7-14). Similarly, the square-wave frequency is increased until the rise time of the amplified pulse is observed on the oscilloscope. Then Eq. (7-10) gives the high-frequency cutoff of the amplifier. Of course, it is assumed that the square-wave signal source and the response characteristics of the oscilloscope are sufficiently good so that they introduce negligible pulse distortion.

Such *square-wave testing* is particularly advantageous in adjusting amplifiers for optimum response because the results of any changes are immediately apparent in the output pulse waveform. This makes repeated laborious determinations of the sine-wave frequency-response characteristic unnecessary. Waveforms similar to those of Figs. 2-28 and 2-30 are often seen, depending upon the relative values of the square-wave frequency and amplifier cutoff frequency.

DC AMPLIFIERS

Direct-Coupling

All the amplifier circuits discussed to this point have zero gain for dc signals because of the infinite reactance of the coupling networks at zero frequency. Amplification of dc or very slowly varying signals is achieved by eliminating the coupling networks entirely. In addition to the dc response of such a *direct-coupled* amplifier, the high-frequency performance is also enhanced. This comes about because stray capacitances associated with the coupling networks are eliminated as well.

The Darlington connection already discussed is an elementary two-stage dc amplifier. More elaborate circuits are necessary to achieve larger values of voltage gain. Consider, for example, the three-stage direct-coupled amplifier in Fig. 7-28 which consists of two grounded-emitter transistors preceded by a FET input stage. Note that the two grounded-emitter stages use complementary symmetry to simplify interconnections and keep the dc voltage levels in each stage to reasonable values. A major difficulty with such cascaded dc amplifiers is *drift* caused by small variations in transistor characteristics with time or temperature and changes in the supply potentials. The input transistor is particularly susceptible because any small changes in the input are amplified by succeeding stages in the amplifier.

It is common practice to stabilize the quiescent point of direct-coupled amplifiers by returning a portion of the output voltage to the input stage in such a

Figure 7-28 Three-stage dc amplifier.

fashion that drifts tend to be compensated. This is accomplished in Fig. 7-29 by the resistor R_1 which sets the base bias on the input transistor. Note that if the emitter voltage at the second stage, V_E, increases for some reason, the base current of the first stage increases correspondingly. This results in a decrease in base current of the second transistor which tends to return V_E to its original value. The stabilizing action of R_1 can be examined best with the equivalent circuit of

Figure 7-29 Direct-coupled ac amplifier.

the input stage base circuit, which is similar to Fig. 6-13b. The voltage drops around the equivalent circuit are

$$\frac{R_2}{R_1 + R_2} V_E - I_b \frac{R_1 R_2}{R_1 + R_2} - V_{be} = 0$$

where V_{be} is the emitter-base voltage. Solving for V_E

$$V_E = \left(1 + \frac{R_1}{R_2}\right) V_{be} + R_1 I_b \tag{7-15}$$

Now V_{be} is a relatively constant quantity. If the circuit is arranged so that the first term in Eq. (7-15) is much larger than the second, V_E is constant and the amplifier is stabilized. After calculating V_E from Eq. (7-15), the currents in each transistor may be determined from the transistor characteristics and the quiescent point of the amplifier is completely determined.

Note, however, that in achieving stability against drift the amplifier does not respond to dc signals. If the second term in Eq. (7-15) is negligible, an input signal I_b does not cause any change in V_E; hence the output signal is zero. This is not so for ac signals where the 100-μF emitter bypass capacitor shorts out V_E, and no signal is returned to the input, so that the circuit is an effective ac amplifier.

Figure 7-30 Difference amplifier is balanced dc amplifier.

Drifts caused by variations in device characteristics and supply voltages are minimized by using a balanced difference amplifier, Fig. 7-30. Changes in one side of the circuit tend to be compensated by similar changes on the other side. Furthermore, the output terminals are at the same potential when the input signal is zero. Note the similarity of Fig. 7-30 to Fig. 5-29a and Fig. 6-22.

The purpose of transistor Q_3 is to provide a constant-current source for the difference amplifier to further reduce drift tendencies. In addition, the ac collector impedance of Q_3 is very large, essentially equal to $1/h_{oe}$, which greatly reduces the effect of circuit assymetry, according to Eq. (6-56). This is more convenient than a large emitter resistor since the voltage drop across Q_3 is much smaller than that associated with a resistance equal to $1/h_{oe}$. Note that the base bias resistors for Q_1 and Q_2 are returned to the opposite collectors. This return tends to stabilize the operating point without reducing gain since the collector potentials change in opposite directions in response to an input signal.

The output signal of the difference amplifier is, using Eq. (6-51),

$$v_0 = g_m R_L (v_2 - v_1) \tag{7-16}$$

Figure 7-31 Balanced FET millivoltmeter.

It is instructive to evaluate the transconductance of the transistor using the rectifier equation, Eq. (5-4),

$$g_m = \frac{dI_c}{dV_e} = \alpha \frac{dI_e}{dV_e} = \frac{\alpha e}{kT} I_o e^{eV_e/kT} = \frac{\alpha e}{kT} I_e \tag{7-17}$$

The circuit is balanced so that $2I_e = I_3$, and the output signal is then

$$v_o = \frac{\alpha e}{2kT} I_3 R_L (v_2 - v_1) \tag{7-18}$$

According to Eq. (7-18) the gain of the amplifier is set by the current in Q_3. This current, in turn, is determined by the base bias on Q_3 and may be adjusted to suit gain requirements.

The stability and balance of the difference amplifier is such that several stages may be cascaded to achieve large values of amplification, as considered in the following chapter. A useful two-stage amplifier which is a practical dc millivolt-meter is illustrated in Fig. 7-31. A FET stage is followed by a grounded-emitter transistor stage. The FET amplifier provides large input impedance and is sta-bilized against voltage changes through the 2N726 emitter transistor and the 651-C4 zener diode and also against temperature changes as a result of the 1N461 diode in the base bias circuit. The overall gain of the amplifier is such that full-scale deflection of the meter is achieved for an input signal of 50×10^{-3} V, so that the equivalent sensitivity is 40 MΩ/V. Alternatively, the circuit may be con-sidered to be a microammeter having a full-scale deflection of $(50 \times 10^{-3})/(2 \times 10^6) = 25$ nA.

Chopper Amplifiers

To circumvent the drift and instabilities inherent in direct-coupled amplifiers, it is useful to convert the dc input signal to an ac voltage which can be amplified by a standard ac-coupled circuit. Subsequently, the amplifier output is rectified to recover the amplified dc input signal. The superior stability features of ac-coupled amplifiers provide much greater gains than are practical with dc amplifiers.

A convenient technique for converting a dc voltage to an ac signal is with a mechanical *chopper*, which is a rapidly vibrating switch driven by an electromag-net, Fig. 7-32. As the switch alternately opens and closes, a square wave with an amplitude equal to the dc signal is produced. The frequency of the square wave corresponds to the chopping frequency, which is most often 60 Hz, since it is convenient to drive the chopper from the power line. The amplified ac output signal is converted back to a dc signal using a peak-reading voltmeter circuit similar to Fig. 4-31.

The sensitivity of chopper amplifiers to very low-level dc signals is limited by switching noise effects at the contacts. The mechanical chopper has a very low resistance when the contact is closed and a very high resistance when the contact is open and is, therefore, quite an effective chopper. In effect the chopper resistance

Figure 7-32 Chopper amplifier using electromagnetically driven switch to convert dc input voltage to ac signal.

in combination with the series resistance R_s in Fig. 7-32 represents an alternating voltage divider. A large change in the dividing ratio caused by the opening and closing of the switch means that even feeble input signals can override contact noise effects.

In many applications the chopper amplifier must accurately reflect changes in the input signal. As a rule of thumb, the upper frequency limit of a chopper amplifier is about one-fourth of the chopping frequency. Mechanical choppers are limited to 60 Hz, although 400-Hz units are occasionally used. To achieve higher chopping frequencies nonmechanical choppers employing diodes, transistors, photoconductors, etc., are used in various circuits.

It is possible, for example, to replace the chopper in Fig. 7-32 with a MOSFET which is driven alternately into saturation or cutoff by a square-wave signal. A more satisfactory approach is the integrated-circuit MOSFET chopper in Fig. 7-33 wherein Q_1 and Q_2 are the two elements of the input voltage divider.

Figure 7-33 Integrated-circuit MOSFET chopper.

Figure 7-34 Synchronous chopper amplifier preserves polarity of input signal.

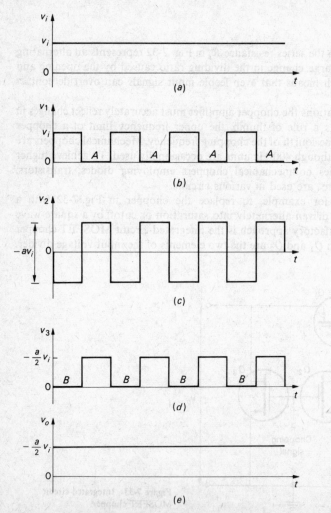

Figure 7-35 Waveforms in synchronous chopper amplifier.

The purpose of Q_3 is to introduce a 180° phase shift of the chopping signal applied to Q_1 so that when Q_2 is conducting Q_1 is cut off and vice versa. Since, therefore, both arms of the input voltage divider are changed simultaneously, the chopping action is enhanced.

A major advantage of chopper amplifiers is their very low internal noise level which means that even minute input signals can be amplified. As discussed in Chap. 10, interference caused by random-noise effects is proportional to the bandwidth of the amplifier, so that it proves advantageous to employ the smallest possible bandwidth. In the case of the chopper amplifier the effective bandwidth is set by the characteristic frequency of the RC filter of the output peak-reading voltmeter. In effect, the overall bandwidth is equal to the frequency interval from dc (zero Hz) to $1/2\pi RC$. If, for example, the RC time constant is 10 s, the effective bandwidth is 0.016 Hz, a very small value indeed. Of course, when the RC time constant is 10 s, a time interval of approximately 30 s is required for the output signal to reach its final value. Thus the amplifier responds only very slowly to changes in the input signal. This reciprocity between bandwidth and response time is a general property of all measuring systems.

A difficulty with the simple chopper amplifier is that the output signal is independent of the polarity of the input signal. This is so because the ac amplifier yields an ac output signal for either polarity of input voltage. The defect is corrected in chopper amplifiers which employ a second contact to rectify the output, Fig. 7-34. As the mechanical chopper alternately closes contacts A and B, the input signal is converted to a square wave, Fig. 7-35a and b, and the amplified ac signal, Fig. 7-35c, is converted back to a dc signal, Fig. 7-35d and e.

Comparing the waveforms in Fig. 7-35 shows that if v_i is positive, v_o is also positive. Similarly, if the input signal is negative, the output signal is also negative. An electronic version of this *synchronous rectifier* is analyzed in the next section.

Lock-in Amplifier

The principles of the synchronous chopper amplifier are employed in an electronic circuit which has found wide use in instrumentation systems. The version illustrated in Fig. 7-36 uses a diode modulator to convert the input signal to a form suitable for the amplifier and a diode demodulator to recover the amplified version of the input signal. For generality, we consider the input signal to be a slowly varying signal; if dc signals are important, the input transformer can be replaced by a center-tapped resistor.

The input circuit is analyzed using techniques discussed in Chap. 4. It is assumed that the diode characteristic can be represented by a quadratic expression. Then the current in the diode D_1 is, from Eq. (4-16),

$$i_1 = a_1 v(t) + a_1 V_2 \sin \omega_2 t + a_2 v^2(t)$$
$$+ a_2 V_2^2 \sin^2 \omega_2 t + 2a_2 v(t)V_2 \sin \omega_2 t \quad (7-19)$$

where a_1 and a_2 are constants related to the diode characteristic, $v_2 = V_2 \sin \omega_2 t$ corresponds to the chopping signal, and $v(t)$ is the input signal. An identical

Figure 7-36 Electronic Synchronous chopper amplifier.

expression applies to the current in D_2, except that the polarity of $v(t)$ is reversed with respect to v_2. The voltage signal applied to the amplifier is

$$v_a = R(i_1 - i_2) \tag{7-20}$$

Substituting for i_1 and i_2, many terms cancel because of the reversed polarity of $v(t)$ in D_2. The result is

$$v_a = 2Ra_1 v(t) + 4Ra_2 v(t)V_2 \sin \omega_2 t \tag{7-21}$$

The first term in Eq. (7-21) is not transmitted by the amplifier since it is assumed that ω_2 is much larger than any of the frequency components associated with $v(t)$. Note that the second term is simply a modulated sine wave of frequency ω_2. The amplitude variations correspond to the input signal.

Ignoring the capacitors C for a moment, the current in D_3 is, again using Eq. (4-16),

$$i_3 = a_1 v_b + a_1 V_2 \sin \omega_2 t + a_2 v_b^2 + a_2 V_2^2 \sin^2 \omega_2 t + 2a_2 v_b V_2 \sin \omega_2 t \tag{7-22}$$

The current in D_4 is similar except that the polarity of v_b with respect to v_2 is reversed. Consequently, the output voltage becomes

$$v_o = R(i_3 - i_4) = 2a_1 Rv_b + 16a_2^2 kR^2 v(t)V_2^2 \sin^2 \omega_2 t \tag{7-23}$$

Inserting the standard trigonometric identity $2 \sin^2 \omega t = 1 - \cos 2\omega t$,

$$v_o = 2a_1 Rv_b + 8a_2^2 kR^2 V_2^2 v(t)(1 - \cos 2\omega_2 t) \tag{7-24}$$

The filter capacitors eliminate the high-frequency terms in ω_2 and $2\omega_2$. Therefore, the filtered output signal is simply

$$v_o = 8a_2^2 kR^2 V_2^2 v(t) \tag{7-25}$$

According to Eq. (7-25) the output voltage is an amplified replica of the input signal.

The amplifier passband is conveniently peaked at ω_2 to minimize noise effects. Actually, however, the output filter circuit determines the effective bandwidth in the same way as for the chopper amplifier, and very-low-noise performance is possible.

In instrumentation applications it is often possible to produce the initial modulation in some way associated with the physical quantity being measured. For example, the infrared beam of an infrared spectrometer is chopped by means of a rotating shutter before it strikes the infrared detector. The amplified signal from the detector is demodulated by a circuit similar to Fig. 7-36 using a voltage v_2 derived from the shaft of the rotating shutter. The low-noise performance of the circuit permits extremely weak infrared signals to be detected. In this form the circuit is usually called a *lock-in amplifier*, since the detector is locked in step with the input signal. The system is also called a *phase-sensitive detector* because the circuit can recognize the phase of the input signal.

SUGGESTIONS FOR FURTHER READING

E. J. Angelo, Jr.: "Electronic Circuits," McGraw-Hill Book Company, New York, 1964.
Edwin R. Jones: "Solid-State Electronics," Intext Educational Publishers, Scranton, Penn., 1974.
Jack J. Studer: "Electronic Circuits and Instrumentation Systems," John Wiley & Sons, Inc., New York, 1963.

EXERCISES

7-1 Plot the frequency-response characteristics of the two-stage transistor amplifier in Fig. 7-1. Assume the output load resistance is 5000 Ω. The common-base h parameters for Q_1 are: $h_{ib} = 50 \ \Omega$, $h_{rb} = 300 \times 10^{-6}$, $h_{fb} = -0.99$, $h_{ob} = 0.2 \times 10^{-6}$ mho; for Q_2: $h_{ib} = 11 \ \Omega$, $h_{rb} = 600 \times 10^{-6}$, $h_{fb} = -0.99$, $h_{ob} = 0.72 \times 10^{-6}$ mho. Assume $C_1 + C_3 = 50$ pF and $C_2 = 8$ pF.

7-2 Plot the midband and high-frequency response of the FET amplifier in Fig. 7-8. Compare with the response of the amplifier without the inductance. Assume $C_1 + C_3 = 10$ pF and $g_m = 8 \times 10^{-3}$ mho.

7-3 Plot the low-frequency response of the transistor amplifier of Fig. 7-9. Compare with the response of the amplifier with R_3 and C_3 removed. Use values of h parameters given in Exercise 6-6.

7-4 Consider the power amplifier in Fig. 7-11. Calculate the maximum power output for load resistances of 5, 10, 20, and 40 Ω, using the characteristic curves of Fig. 7-12. Calculate the power efficiency for each load.

Ans.: 0.08, 0.05, 0.025, 0.013 W; 0.41, 0.25, 0.13, 0.06.

7-5 Plot the dynamic transfer characteristic for each load resistance of the power amplifier studied in Exercise 7-4. Which load yields minimum distortion?

Ans.: 5 Ω

7-6 Determine the required reflected load impedance R'_L for a class B push-pull power amplifier using 2N1415 type transistors if the output power is 0.3 W and $V_{cc} = 6$ V.

Ans.: 60 Ω

7-7 Plot the composite transfer characteristic of the amplifier of Exercise 7-6. Repeat if each transistor is biased to a quiescent base current of 0.2 mA.

7-8 Sketch the equivalent circuit of the transistor tuned amplifier, Fig. 7-20. Calculate the peak gain assuming that the reactances of the tuned circuit are infinite at resonance. Use h parameters given in Exercise 6-6.

Ans.: 3.7×10^5

7-9 Draw the equivalent circuit of the MOSFET tuned amplifier, Fig. 7-23. Calculate the peak gain, assuming that the reactance of the tuned circuit is infinite at resonance. Take $g_m = 9500$ μmhos, $r_d = 5 \times 10^5$ Ω.

Ans.: 4.75×10^3

7-10 Select appropriate square-wave test frequencies to confirm that an amplifier has a passband extending from 20 to 20,000 Hz. Sketch the expected output waveforms at both frequencies.

Ans.: 314 Hz; 3 kHz.

EIGHT

OPERATIONAL AMPLIFIERS

The performance of transistor amplifiers is enhanced in many respects by returning a fraction of the output signal to the input terminals. This process is called feedback. The feedback signal may either augment the input signal or tend to cancel it, and the latter, called negative feedback, is the primary concern of this chapter. Improved frequency-response characteristics and reduced waveform distortion are attained with negative feedback. In addition, amplifier performance is much less dependent upon changes in transistor parameters caused by aging or temperature effects.

A particular form of negative feedback, known as operational feedback, is used in amplifiers which perform mathematical operations such as addition or integration on an input signal. Commercially available high-performance dc-coupled units have come to be called operational amplifiers. Operational amplifiers are widely used in measurement and control applications, as well as in electronic analog computers.

NEGATIVE FEEDBACK

Voltage Feedback

The circuit alterations of a standard amplifier which return a portion of the output signal to the input may be analyzed by the techniques developed in previous chapters. It is more illustrative, however, to isolate the feedback portion of the circuit and treat it separately. Consider the feedback amplifier, Fig. 8-1, comprising a standard amplifier with a gain a and a feedback network indicated by the box marked β. According to this circuit, a voltage βv_o is added to the input signal v_i so that the total input signal to the amplifier is

$$v_1 = v_i + \beta v_o \tag{8-1}$$

Introducing the fact that $v_o = av_1$,

$$v_o = av_1 + a\beta v_o \tag{8-2}$$

so that

$$v_o = \frac{a}{1 - \beta a} v_i \tag{8-3}$$

According to Eq. (8-3) the overall gain of the amplifier with feedback,

$$a' = \frac{a}{1 - \beta a} \tag{8-4}$$

may be greater or smaller than that of the amplifier alone, depending upon the algebraic sign of βa.

The condition of greatest interest in this chapter is *negative feedback*, when βa is a negative quantity. In this case, Eq. (8-4) shows that the overall gain is reduced because, in effect, the feedback voltage cancels a portion of the input signal. If the amplifier gain is very large, so that $-\beta a \gg 1$, the overall gain reduces to

$$a' = \frac{1}{\beta} \tag{8-5}$$

This shows that the gain depends only upon the properties of the feedback circuit.

Figure 8-1 Block diagram of feedback amplifier.

Most often, the feedback network is a simple combination of resistors and/or capacitors. Therefore, the gain is independent of variations in transistor parameters in the amplifier. In addition to this desirable improvement in stability, the gain may be calculated from circuit values of the feedback network alone. Thus it is not necessary to know, for example, the h parameters of all transistors in the circuit.

Negative feedback is also effective in reducing waveform distortion in amplifiers. Waveform distortion results from a nonlinear transfer characteristic, which may be interpreted as a smaller gain where the slope of the transfer characteristic is less, and as a larger gain where the slope of the transfer characteristic is greater. According to Eq. (8-5), however, the gain of an amplifier with feedback is essentially independent of variations caused by nonlinearities in tube or transistor characteristics. Therefore, the transfer characteristic is more linear and distortion is reduced.

The benefits of negative feedback are obtained at the expense of reduced gain. This is not a serious loss, however, because large amplifications are easily obtained in transistor amplifiers. As discussed in Chap. 10, the maximum usable gain is limited by random-noise effects anyway and it is not difficult to achieve the maximum amplification that can be effectively used, even with feedback included.

In the foregoing, a portion of the output voltage is returned to the input terminals, a condition referred to as *voltage feedback*. The two-stage transistor amplifier, Fig. 8-2, uses voltage feedback introduced by the resistor R_F connecting the output terminal with the input circuit of the first stage. The feedback voltage is introduced into the emitter circuit of the first stage because the output voltage of a

Figure 8-2 Two-stage feedback amplifier. Feedback is determined by ratio R_1/R_F.

two-stage amplifier is in phase with the input signal. The feedback factor is the result of the resistor divider made up of R_F and R_1, so that

$$\beta = \frac{R_1}{R_F + R_1} \cong \frac{R_1}{R_F} \qquad (8\text{-}6)$$

Capacitor C_1 isolates the dc components of the two stages. Also, note that dc interstage coupling is used in this amplifier, which is quite independent of negative-feedback considerations. Resistor R_2 sets the bias on the first stage and provides dc feedback which helps stabilize the direct-coupled amplifier against slow drifts, as discussed in Chap. 7. It is not part of the ac feedback circuit.

The effect of the feedback ratio on the frequency-response characteristic is illustrated in Fig. 8-3. Without feedback (resistor R_F removed) the midband gain of the amplifier is 1000 and the high-frequency cutoff is 100 kHz. As feedback is increased by using smaller values of R_F the midband gain is reduced and the frequency response is extended to lower and higher frequencies. With a 1200-Ω feedback resistor the gain is 10 and the upper frequency cutoff is extended to 15 MHz. Under this condition $\beta = 120/1200 = 0.1$, so $a\beta = 100$. According to Eq. (8-5) the midband gain is $1/\beta$, in agreement with the experimental response curves. Thus, a very stable wideband amplifier is possible through the use of feedback.

Note that in Fig. 8-3 the gain increases at the extreme of the response curve when a large feedback ratio is used. This is a result of phase shift in the amplifier at these frequencies. In effect, the negative feedback is reduced since the feedback voltage is no longer exactly 180° out of phase with the input signal. The phase characteristic of feedback amplifiers at frequencies outside of the passband is very

Figure 8-3 Effect of feedback on response characteristic of amplifier in Fig. 8-2.

important. Such irregularities in the response curve should be small, particularly when large feedback ratios are employed. It is possible for the phase shifts to become great enough to cause positive feedback at the frequency extremes, a condition which must be avoided if the amplifier is to remain stable. This matter is considered further in a later section.

If the feedback circuit is frequency-selective, it is possible to develop a specific frequency-response characteristic for the amplifier. Consider, for example, an amplifier having a bridged-T feedback network in one stage, Fig. 8-4. In this circuit source followers are used to provide a high input impedance and a low output impedance so that the frequency-selective stage is isolated from disturbing influences of input and output loads. A difference amplifier is used in the first two stages to simplify coupling between stages in the presence of the feedback network.

Since the bridged-T filter is connected from drain to gate, the feedback voltage is 180° out of phase with the input signal, as required for negative feedback. According to the response curve of the bridged-T filter, the feedback voltage is a minimum at the characteristic frequency of the filter. Accordingly, the amplifier gain is a maximum at this frequency and the result is a tuned amplifier. The feedback ratio β is a function of frequency because of the characteristics of the bridged-T filter, and the feedback effect considerably enhances the selectivity of the filter, as illustrated in Fig. 8-5.

Feedback tuned amplifiers are commonly used at audio frequencies where high-Q inductances are difficult to construct because large values of inductance

Figure 8-4 Tuned amplifier using a frequency-selective feedback network consisting of bridged-T filter.

Figure 8-5 Frequency-response characteristic of the bridged-T amplifier in Fig. 8-4.

are required. Furthermore, it is a simple matter to tune the amplifier by making the resistors variable. This approach is also useful in integrated circuits where inductances are difficult to produce. Other frequency-selective feedback networks, such as the twin-T or Wien bridge, are also used in tuned feedback amplifiers.

Negative feedback alters the input and output impedance of an amplifier. To see how this comes about, replace the amplifier by its Thévenin equivalent, Fig. 8-6. The output voltage and the input signal are then

$$v_o = av_1 - i_o r_o \tag{8-7}$$

and

$$v_1 = v_i + \beta v_o \tag{8-8}$$

Equation (8-8) is introduced into Eq. (8-7),

$$v_o = av_i + a\beta v_o - i_o r_o$$

Solving for v_o,

$$v_o = \frac{a}{1 - a\beta} v_i - \frac{r_o}{1 - a\beta} i_o \tag{8-9}$$

Figure 8-6 Circuit to determine effective input and output impedance of amplifier with voltage feedback.

Equation (8-9) is the Thévenin expression for the amplifier with negative feedback. Note that the coefficient of v_i is the gain which is in agreement with Eq. (8-3). The coefficient of i_o is the effective output impedance. According to Eq. (8-9) the output impedance is reduced, and may become very small if $-a\beta$ is large. The effective input impedance is found by replacing v_o by $R_L i_o$ and v_1 by $r_i i_i$ in Eqs. (8-7) and (8-8). The result is solved for the ratio v_i/i_i, or the input impedance,

$$R_i = r_i\left(1 - \frac{a\beta}{1 + r_o/R_L}\right) \tag{8-10}$$

According to Eq. (8-10) the input impedance is increased by negative feedback. Both reduced output impedance and increased input impedance are favorable features in amplifiers, as discussed in previous chapters.

Current Feedback

It is possible to develop a feedback signal proportional to the output current rather than to the output voltage and this is called *current feedback*. Both voltage and current feedback can be employed in the same amplifier, Fig. 8-7. Voltage relations in the output circuit of Fig. 8-7 are

$$v_o = av_1 - i_o(r_o + R_F) \tag{8-11}$$

Similarly, Kirchhoff's voltage equation for the input circuit yields

$$v_1 = v_i + \beta_v v_o + \beta_i i_o R_F \tag{8-12}$$

Substituting Eq. (8-12) into Eq. (8-11) and solving for the output voltage,

$$v_o = \frac{a}{1 - a\beta_v} v_i - i_o \frac{r_o + (1 - a\beta_i)R_F}{1 - a\beta_v} \tag{8-13}$$

Figure 8-7 Combined voltage and current feedback.

Note that this expression incorporates Eq. (8-9) for voltage feedback alone, $\beta_i = 0$.

Combined current and voltage feedback permits a unique adjustment of the output impedance of the amplifier. The output impedance is given by the coefficient of i_o in Eq. (8-13) and can be made to vanish if

$$r_o + (1 - a\beta_i)R_F = 0 \tag{8-14}$$

or

$$a\beta_i = 1 + \frac{r_o}{R_F} \tag{8-15}$$

If Eq. (8-15) is satisfied, the amplifier has zero internal impedance and therefore can deliver maximum power to any load impedance. Note that this requires positive current feedback, according to Eq. (8-15). Positive current feedback is permissible in this case because of the stabilizing effect of negative voltage feedback which is also present.

Stability

The amplifier gain a and the feedback factor β are inherently complex numbers. That is, phase shifts associated with coupling capacitors and stray capacitance effects are present, particularly at frequencies outside of the passband of the amplifier. These phase shifts cause a departure from the 180° phase shift necessary for the feedback voltage to interfere destructively with the input signal. It can happen that the overall phase shift becomes zero (that is, 360°) so that $a\beta$ is positive and the feedback voltage augments the input signal. This is called *positive feedback* and leads to serious instability effects in feedback amplifiers.

Note, particularly, that if $a\beta = +1$, the output voltage, Eq. (8-3), becomes very large, even in the absence of an input signal. This means that positive feedback may cause an amplifier to oscillate, as previously discussed in connection with the feedback effect of the collector capacitance in transistors. Oscillation is deleterious in amplifiers since the output voltage is not a replica of the input signal. Positive feedback is, however, a useful condition in oscillator circuits, as discussed in the next chapter.

Great care is taken in the design of feedback amplifiers to make them stable in the face of phase shifts leading to positive feedback at the frequency extremes of the amplifier bandpass. Fortunately there exists a rather straightforward criterion which can be applied to establish the stability of feedback amplifiers. According to Eq. (8-4), instability occurs if $a\beta$ is positive and equal to unity. It follows that if the absolute value of $a\beta$ drops below unity before the phase shift reaches $-360°$, the amplifier will be stable. It is convenient to plot the magnitude and phase of $a\beta$ as a function of frequency, Fig. 8-8, in order to establish the relative positions at which $a\beta = 1$ and the phase shift equals zero. By this standard, the amplifier with characteristics represented by Fig. 8-8 is stable.

In order to apply this criterion for stability, the amplifier must be stable in the absence of feedback. Also, there exist certain situations in which a feedback

Figure 8-8 Feedback amplitude and phase characteristics of stable amplifier.

amplifier can be conditionally stable, that is, stable under some conditions of, say, loading, and not under others. For most practical circuit analysis, however, the stated condition is sufficient.

It is useful to consider the $a\beta$ characteristics of several amplifier types in the light of this stability requirement. Since in most cases the feedback ratio is independent of frequency, it is sufficient to examine the amplitude and phase characteristics of the amplifier gain separately. In the case of a single RC-coupled amplifier stage, Fig. 8-9a, the phase shift never exceeds 90°. This means the phase of $a\beta$ is always less than $90 + 180 = 270°$. Since this is smaller than 360°, a single-stage negative feedback amplifier cannot be unstable.

Two identical RC-coupled stages in cascade have the ideal characteristic sketched in Fig. 8-9b. The phase shift reaches 180° at very low and very high frequencies, but the gain is very small at these extremes. Therefore, it is unlikely that $a\beta$ can be equal to unity, where the total phase shift is $180 + 180 = 360°$. Practical amplifiers always have stray capacitive effects that can introduce additional phase shift at high frequencies, however. It is possible that the additional phase shift caused by stray capacitances may result in unstable conditions at high frequencies if the gain is large.

The characteristics of three identical RC-coupled stages, Fig. 8-9c, are such that the extreme phase shift approaches 270°. Therefore, 180° phase shift is encountered at both low and high frequencies, and a three-stage feedback amplifier is certain to be unstable if the midband gain is large enough. It can be shown that the maximum value of $a\beta$ permitted at midband in this case is equal to 8, although even this value puts the amplifier on the verge of oscillation.

Many possibilities exist to remove this unfavorable difficulty other than minimizing the gain. For example, the bandwidth of two stages in a three-stage amplifier can be made very much greater than that of the remaining stage. This means that the phase characteristics are determined primarily by the single stage which is unconditionally stable. Alternatively, dc coupling can be used between two stages to eliminate the phase shift associated with one interstage coupling network. It is common practice to alter the feedback and gain characteristics,

Figure 8-9 Gain and phase characteristics of (a) one-stage amplifier, (b) two-stage amplifier, and (c) three-stage amplifier.

particularly at high frequencies, by including small capacitances to change the phase and gain characteristics in a way that improves stability. Usually such changes must be made empirically on the actual amplifier because of the unavoidable and unknown stray wiring capacitances.

OPERATIONAL FEEDBACK

The Virtual Ground

A particularly versatile negative-feedback connection is called *operational feedback* because the circuit is capable of performing a number of mathematical operations on input signals. High-gain dc-coupled amplifiers are universally used

Figure 8-10 Block diagram of operational amplifier.

in this application. Such *operational amplifiers* exhibit high input impedance, low output impedance, and conventionally introduce a 180° phase shift between the input signal and the output voltage.

To illustrate the features of an operational amplifier, consider the feedback circuit, Fig. 8-10, in which negative voltage feedback is produced by the resistor R_f connected between the output and input. Note that the feedback is negative because of the phase inversion in the amplifier. The feedback ratio in operational feedback can vary from unity for a high-impedance source to $R/(R + R_f)$ for a low-impedance source since the feedback voltage is effectively connected in parallel with the input signal source. It is convenient to analyze the operational feedback circuit by applying Kirchhoff's current rule to the branch point S. Since the amplifier input impedance is large, the current in this branch is negligible, which means the current in R equals the current in R_f, or

$$\frac{v_i - v_1}{R} = \frac{v_1 - v_o}{R_f} \tag{8-16}$$

Introducing $v_1 = -v_o/a$ and rearranging,

$$v_o\left(1 + \frac{1}{a} + \frac{R_f}{aR}\right) = -\frac{R_f}{R}v_i \tag{8-17}$$

Since the gain is very large

$$v_o = -\frac{R_f}{R}v_i \tag{8-18}$$

which means that the output voltage is just the input signal multiplied by the constant factor $-R_f/R$. If precision resistors are used for R_f and R, the accuracy of this multiplication operation is quite good.

The branch point S has a special significance in operational amplifiers. This may be illustrated by determining the effective impedance between S and ground, which is given by the ratio of v_1 to the input current,

$$Z_s = \frac{v_1}{i_i} = \frac{v_1 R_f}{v_1 - v_o} = \frac{R_f}{1 - v_o/v_1} = \frac{R_f}{1 + a} \tag{8-19}$$

where the right side of Eq. (8-16) has been inserted for the input current. According to Eq. (8-19) the impedance of S to ground is very low if the gain is large. Typical values are $R_f = 10^5 \ \Omega$ and $a = 10^4$, so that the impedance is 10 Ω. The

Figure 8-11 Conventional circuit symbol for operational amplifier.

low impedance results from the negative feedback voltage, which cancels the input signal at S and tends to keep the branch point at ground potential. For this reason the point S is called a *virtual ground*. Although S is kept at ground potential by feedback action, no current to ground exists at this point. The virtual ground at S shows immediately that the impedance viewed from the input terminals is equal to R.

Mathematical Operations

The operational feedback circuit, Fig. 8-10, multiplies the input signal by the constant $-R_f/R$. It is conventional to simplify the circuit diagram, as in Fig. 8-11, by not showing the ground terminals specifically. The operational amplifier is indicated by a triangle pointing toward the output terminal. It is understood that the simplified circuit diagram, Fig. 8-11, actually implies the corresponding complete circuit, Fig. 8-10. If $R_f = R$ in the multiplier circuit, the signal is simply multiplied by -1. This is often useful in obtaining a signal multiplied by a positive constant wherein this operational amplifier precedes one which determines the multiplier. Multiplicative factors ranging from -0.1 to -10 are possible, and the precision is determined principally by the accuracy of the two resistors.

The operational amplifier can also be used to sum several signals by connecting individual resistors to the branch point S, Fig. 8-12. In this circuit, the sum of the currents in R_1, R_2, and R_3 equals the current in R_f, since no current to ground exists at S. Furthermore, S is at ground potential, so that [compare Eq. (8-16)]

$$\frac{v_1}{R_1} + \frac{v_2}{R_2} + \frac{v_3}{R_3} = -\frac{v_o}{R_f} \tag{8-20}$$

$$v_o = -R_f\left(\frac{v_1}{R_1} + \frac{v_2}{R_2} + \frac{v_3}{R_3}\right) \tag{8-21}$$

Figure 8-12 Summing circuit using operational amplifier.

Figure 8-13 Integrating circuit.

Figure 8-14 Differentiating circuit.

Many signals can be added in this way. There is no interaction between individual signal sources since S is a virtual ground. Because the addition appears to take place at this point, S is commonly called the *summing point*. Note that each signal may be multiplied by the same factor (or -1), if $R_1 = R_2 = R_3$, etc., or that individual factors may be selected, as convenient.

If the feedback resistor is replaced by a capacitor, Fig. 8-13, the circuit performs the operation of integration. Using the fact that S is at ground potential,

$$v_o = \frac{q}{C} = \frac{1}{C} \int_0^t i \, dt = -\frac{1}{RC} \int_0^t v_i \, dt \qquad (8\text{-}22)$$

According to Eq. (8-22) the output voltage is the integral of the input signal. Note that there is no restriction placed on the frequency components of the input signal, as in the case of the simple RC integrator discussed in Chap. 2. It is only necessary that the amplifier bandwidth be large enough to handle all signal frequencies. The integral of the sum of several signals is obtained by introducing several input resistors, as in Fig. 8-12.

Interchanging R and C, Fig. 8-14, results in a differentiating circuit which gives the time derivative of the input signal. Again equating currents at the summing point,

$$-\frac{v_o}{R} = \frac{dq}{dt} = \frac{d}{dt} C v_i = C \frac{dv_i}{dt} \qquad (8\text{-}23)$$

$$v_o = -RC \frac{dv_i}{dt} \qquad (8\text{-}24)$$

The differentiator circuit is equally independent of frequency restrictions, assuming only that the gain of the amplifier is sufficiently large at all signal frequencies.

Simple Amplifier

An operational amplifier makes an effective high-gain amplifier for use with small signals. If, however, the input impedance R, Fig. 8-10, is 10^5 Ω and a gain of 10^3 is desired, the feedback resistor must be 10^8 Ω. Stable resistors of this magnitude are not readily available, so an alternate connection, Fig. 8-15, is used. The voltage-divider action of R_1 and R_2 returns only a fraction of the output signal to the input while maintaining the operational-feedback connection. The circuit is

Figure 8-15 High-gain operational amplifier.

analyzed by writing voltage drops around the feedback loops, remembering that S is at ground potential,

$$v_o - (R_2 + R_1)i_1 + R_1 i_2 = 0$$

$$(R_f + R_1)i_2 - R_1 i_1 = 0 \qquad (8\text{-}25)$$

Since no current flows to ground at S, $v_i/R = -i_2$. This value is inserted into Eqs. (8-25) and the result solved for v_o/v_i,

$$\frac{v_o}{v_i} = - \frac{R_f}{R}\left[1 + \frac{R_2}{R_1}\left(1 + \frac{R_1}{R_f}\right)\right] \qquad (8\text{-}26)$$

Most often, $R_f \gg R_1$ and $R_2 \gg R_1$ so that Eq. (8-27) reduces to the previous expression for the gain of an operational amplifier, Eq. (8-18), multiplied by the ratio R_2/R_1. A gain of 10^3 can thus be achieved at an input impedance of $10^5\ \Omega$ with a feedback resistor, $R_f = 10^6\ \Omega$, if, say, $R_2 = 10^4$ and $R_1 = 10^2\ \Omega$.

OPERATIONAL AMPLIFIER CIRCUITS

Practical Amplifiers

The circuit of a useful but relatively simple operational amplifier, Fig. 8-16, includes a FET difference-amplifier input stage employing a constant-current source transistor to improve balance, as discussed in the previous chapter. The second stage is an *npn* grounded-emitter amplifier using a *pnp* transistor as the collector load resistor. This is directly coupled to an emitter-follower output stage. Negative voltage feedback is provided by the 680-kΩ resistor connected from the output to the emitter of the second stage. In addition, positive current feedback is obtained by connecting the output emitter resistor to a portion of the emitter resistor of the collector load transistor.

This circuit has a large gain, 10^5, and a large input impedance, approximately 5 MΩ, by virtue of the FET input stage. Additionally, the output impedance is low because of the emitter-follower output and the effect of negative feedback. Positive current feedback is employed to achieve some gain in the third stage and to increase the maximum permissible output-voltage swing without distortion. Note

Figure 8-16 Operational amplifier with FET input stage.

that the differential input terminals include polarity markings. The negative terminal, usually referred to as the *inverting* input, causes an output signal of opposite polarity to the input signal. The positive terminal, called the *noninverting* input, produces an output signal in phase with the input. The inverting input is the normal input terminal since the 180° phase shift is necessary in operational feedback.

Integrated-circuit operational amplifiers are very effective because of their small size and useful properties. The circuit of a typical unit, Fig. 8-17a, is compared with the actual silicon wafer in Fig. 8-17b. The entire unit is mounted in a small package, such as Fig. 8-17c or Fig. 5-29. The five-stage circuit consists of a Darlington-connected difference amplifier with a temperature-compensated emitter transistor. This is followed by a second difference-amplifier stage coupled to an emitter-follower amplifier which uses another transistor for the emitter resistor. A grounded-emitter amplifier feeds the output emitter follower. The emitter follower also employs a transistor for an emitter resistor. Note that temperature compensation is provided in the input, middle, and output stages by diodes in the base bias circuits of the transistors acting as emitter resistors.

As indicated in Fig. 8-17, access to the collectors of the input-stage transistors and to the internal negative feedback network at the third stage is provided in addition to input, output, and power-supply terminals. Terminal pairs 7, 8 and 9, 10 permit the addition of external phase-compensation networks to prevent positive feedback at the extremes of the amplifier passband. The compensation required depends somewhat upon the particular operational-amplifier application and, accordingly, is not fabricated as part of the integrated circuit. Examples of phase compensation networks are illustrated in a later section.

Even with very careful design and well-stabilized supply voltages, it is difficult to prevent drifts in dc amplifiers. Drift voltages at the amplifier output mean that operational feedback no longer keeps the branch point S at virtual ground and the error in the virtual ground is referred to as *offset*. Typically, offset voltages are of the order of 1 mV, which may be serious in critical applications. Improved performance is achieved by using a chopper amplifier to stabilize the virtual ground voltage, Fig. 8-18. Here, the chopper amplifier, which is drift-free because it is ac-coupled, measures the offset voltage at S and provides an amplified error signal to the positive input of the operational amplifier. This signal counteracts drifts in the operational amplifier and, in practice, offset can be reduced to about 10 μV.

Note that a capacitor is interposed between S and the operational amplifier so that offset is controlled only by the chopper amplifier. This prevents the opera-

Figure 8-17 (a) Integrated-circuit operational amplifier.

(b)

(c)

Figure 8-17 (b) Actual silicon wafer, and (c) flat-pack case. (*Motorola Semiconductor Products, Inc.*)

tional amplifier itself from responding to dc signals, but the chopper amplifier does so and, in feeding its output signal to the operational amplifier, becomes part of the operational feedback circuit. In effect, the operational amplifier is not dc-coupled and handles the high-frequency signals, while the chopper amplifier handles the dc and very-low-frequency signals. The dc gain is very large since it is the product of the gains of both amplifiers. This means that the overall response curve is nonuniform in that the low-frequency gain is larger than the gain at high frequencies. According to Eq. (8-18) this is unimportant since by feedback action the gain does not appear in the expression for the output signal, so long as the gain remains large at all frequencies of interest.

Figure 8-18 Chopper-stabilized operational amplifier.

The combination of an operational amplifier stabilized by a chopper amplifier is often itself referred to as an operational amplifier. Thus the dashed line in Fig. 8-18 encloses a chopper-stabilized operational amplifier. Chopper amplifiers specifically designed for stabilizing operational amplifiers are also available as separate units. A typical circuit, shown in Fig. 8-19, is a rather conventional ac-coupled amplifier and uses a synchronous chopper to recover the dc signal. Note that unbypassed emitter and cathode resistors provide current feedback to improve phase characteristics so that the synchronously chopped output is in phase with the chopped input signal. In fact, dc coupling is used for convenience for the same reason. Input and output capacitors are present, so that the amplifier is actually ac-coupled. Integrated-circuit chopper-stabilized operational amplifiers employing MOSFET choppers similar to Fig. 7-33 are also commercially available.

Figure 8-19 Typical chopper amplifier.

The properties required of an operational amplifier depend upon the particular application envisioned. Certain parameters, however, are characteristic of many applications and these serve as a basis for comparison between different units. Consider the characteristics of the several types of operational amplifiers summarized in Table 8-1. This listing compares two typical transistor amplifiers assembled from discrete components (one with FET input similar to Fig. 8-16), with two integrated-circuit units similar to Fig. 8-17, and a chopper-stabilized unit. Many variations of these basic types are also available for different applications, but the range of parameters represented in the table is quite representative.

The foremost characteristic of an operational amplifier is, of course, the gain. This parameter is usually specified as the *open-loop* voltage gain, which simply means the gain of the amplifier itself without any operational feedback connection. A large value is desirable in order for Eq. (8-18) to apply accurately. According to the second column in Table 8-1, values in excess of 1000 and ranging to more than 10 million are typical.

Bandwidth is also significant since it specifies the range of signal frequencies over which appreciable gain is available. Note that a uniform frequency response is not necessary since the actual value of gain does not appear in the operational feedback expression. For this reason it is not appropriate to cite the bandwidth at the half-power frequency, as used in Chap. 7. A useful basis for comparison is the open-loop bandwidth at unity gain, which is the interval from direct current to the frequency for which the gain falls to unity. Although this parameter is useful for comparison purposes, an operational amplifier is not useful over this entire bandwidth, since the gain near the extreme frequency is not sufficient for satisfactory

Table 8-1. Properties of operational amplifiers

Type	Open-loop voltage gain	Open-loop bandwidth, unity gain, MHz	Input impedance, Ω	Input offset voltage, mV	Offset drift, $\mu V/°C$
Transistor, discrete component	3×10^4	1	2×10^5	adjust	25
FET, discrete component	10^6	10	10^{12}	adjust	2
MC1431, integrated circuit	3500	20	6×10^5	5	10
μA702A, integrated circuit	3600	30	4×10^4	0.5	2.5
Chopper-stabilized	3×10^7	15	5×10^5	0.01	0.2

performance. The third column shows that very high signal frequencies can be accommodated, however.

The input impedance of an operational amplifier must be great enough for the input current to be negligible, according to the analysis leading to Eq. (8-18). This is easily achieved with an FET stage, and the Darlington connection is often employed for the same reason. Although the input impedance of a simple grounded-emitter difference amplifier is smaller than these alternatives (fifth column), it is sufficiently large for most applications.

If the input difference amplifier is not exactly balanced, an output voltage exists even in the absence of an external input signal. This spurious signal can be troublesome, as, for example, in the integrator circuit since, according to Eq. (8-22), a constant voltage results in an ever-increasing output signal which eventually drives the amplifier into saturation. Although, as previously noted, a constant-current transistor in the emitter or source circuit minimizes the effect of circuit imbalance, it usually proves necessary to provide an adjustable balance control in discrete component amplifiers (compare Fig. 8-16). The balance control may require periodic adjustments as components age. Integrated-circuit amplifiers are sufficiently well balanced so that this is unnecessary, but it is desirable to know the extent of any residual assymmetry. The voltage that must be applied to the input to obtain zero output voltage, the offset, is a convenient measure. Values of offset in the millivolt range or less are common, according to Table 8-1.

The *drift* of the input offset voltage, particularly with temperature, is significant since a slow change in offset voltage is indistinguishable from a slowly varying signal. According to the last column in Table 8-1, temperature changes of tens of degrees are permissible before the drift signal is comparable with the offset signal.

The superior properties of the chopper-stabilized operational amplifier are clearly evident from the parameters listed in Table 8-1. Certain applications may require the very large input impedance of the FET circuit, however, while for other uses the small size and lower cost of the integrated-circuit design may be more appropriate. Thus the parameters of the several types make the amplifiers complementary rather than competitive.

In addition to those listed in Table 8-1, a number of other parameters are commonly used to specify operational amplifiers. Some of these additional properties are output impedance, input bias current and current offset, common-mode rejection ratio, and internal noise. For a specific application any one of these, as well as others, may prove more important than the common parameters indicated in the table.

Integrator-Differentiator Circuit

A practical integrator and differentiator circuit is illustrated in Fig. 8-20. The integrated-circuit operational amplifier used here is similar to that in Fig. 8-17. The switch in the input circuit interchanges the positions of the integrating capaci-

Figure 8-20 Practical integrator-differentiator circuit.

Figure 8-21 (a) Square-wave input (b) integrated and (c) differentiated by operational-amplifier circuit of Fig. 8-20.

tor and resistor, as may be noted by tracing through the circuit and comparing with Figs. 8-13 and 8-14. Thus the same circuit can be used either to integrate or differentiate the input signal. Typical experimental output waveforms for the case of a square-wave input signal are shown in Fig. 8-21.

The purpose of the 390-kΩ resistor shunting the integrating capacitor in Fig. 8-20 is to eliminate the effect of long-time integration of the input offset voltage. In effect the resistor keeps the capacitor discharged for slow changes in signal and therefore the integrator does not integrate properly at frequencies below about $f = 1/2\pi RC = 20$ Hz. The two series combinations of a small capacitor and resistor are phase compensation networks. As previously discussed, these networks ensure that the overall phase shifts do not produce positive feedback at the high-frequency limit of the amplifier bandpass.

Logarithmic Amplifier

Nonlinear feedback components in operational-amplifier circuits are used to develop special input-output characteristics. Consider, for example, the grounded-base transistor feedback element, Fig. 8-22, that results in an amplifier in which the output voltage is the logarithm of the input signal. Such a logarithmic amplifier operates over a very wide range of input-signal magnitudes without saturating.

The logarithmic input-output characteristic comes about in the following way. The collector current in a transistor is just the emitter current times the current gain, according to Fig. 5-10, and the emitter current is given by the rectifier equation. Ignoring the unity in Eq. (5-4) since the exponential term dominates, the collector current is

$$I_c = \alpha I_0 e^{eV/kT} \tag{8-28}$$

In the operational-feedback connection appropriate for Fig. 8-22, the collector current may be expressed in terms of the input signal

$$\frac{v_i}{R} = \alpha I_0 e^{ev_o/kT} \tag{8-29}$$

where the output signal v_o is the base-to-emitter voltage. Rearranging and taking the logarithm of both sides,

$$v_o = \frac{kT}{e} \ln \frac{v_i}{\alpha I_0 R} \tag{8-30}$$

Figure 8-22 Elementary logarithmic amplifier.

Figure 8-23 Input-output characteristic of logarithmic amplifier.

Experimental input-output characteristics, Fig. 8-23, illustrate the logarithmic properties predicted by Eq. (8-30) and the extremely wide range of input signals over which the amplifier operates. In practical circuits it is usually necessary to supplement the elementary circuit in Fig. 8-22 by including a collector bias supply and to compensate for offset voltages using the noninverting input in order to obtain the logarithmic characteristic over a wide range of input voltages.

In addition to its unique signal-handling capability, the logarithmic amplifier can be used to multiply two arbitrary input signals. The outputs of two logarithmic amplifiers are summed and then the output of the combination is passed through an inverse logarithmic amplifier. The result is a signal proportional to the product of the two input signals.

Active Filter

Consider the operational amplifier in Fig. 8-24 in which both the inverting and noninverting inputs have the operational-feedback connection. Because of the large input impedance of the amplifier, the current into the amplifier input terminals is negligible so that

$$i_1 = -\frac{v_o}{r_1} \quad \text{and} \quad i_2 = -\frac{v_o}{r_2} \tag{8-31}$$

which means

$$i_1 = \frac{r_2}{r_2} i_2 \tag{8-32}$$

Figure 8-24 Operational amplifier connected as a negative impedance converter.

Writing Kirchhoff's voltage equation around the outer loop,

$$v_1 - i_1 r_1 + i_2 r_2 - v_2 = 0 \qquad (8\text{-}33)$$

Inserting Eq. (8-32), the relationship between the input and output voltages is simply

$$v_1 = v_2 \qquad (8\text{-}34)$$

Consider now the apparent input impedance of the circuit when an impedance $\mathbf{Z}_2$ is connected to the output terminals. To focus on the impedance transformation, use the block diagram, Fig. 8-25, in which the portion in the dashed rectangle in Fig. 8-24 is labeled *NIC*, for reasons justified below. According to Fig. 8-25, the impedance $\mathbf{Z}_2$ is given by

$$\mathbf{Z}_2 = \frac{-i_2}{v_2} \qquad (8\text{-}35)$$

The input impedance is

$$\mathbf{Z}_1 = \frac{v_1}{i_1} = \frac{v_2}{r_2 i_2 / r_1}$$

$$\mathbf{Z}_1 = -\frac{r_1}{r_2} \mathbf{Z}_2 \qquad (8\text{-}36)$$

where Eqs. (8-32) and (8-34) have been used. According to Eq. (8-36) the circuit has the interesting property that the apparent input impedance is the negative of the impedance connected to the output terminals. That is, the circuit is a *negative impedance converter*, or *NIC*. If, for example, the output impedance is capacitive,

Figure 8-25 Block diagram of a negative impedance converter.

Figure 8-26 Active filter.

the input impedance is inductive. Furthermore, the magnitude of the inductance depends upon the scale factor r_1/r_2.

A simple application of this useful property is the *active filter*, Fig. 8-26, in which capacitor C_2 at the output terminals is connected to an inductance to form a resonant circuit with C_1. The input-output characteristic of this circuit is just

$$\frac{v_o}{v_i} = \frac{v_2}{i_1 \mathbf{Z}_1 + v_1} = \frac{1}{1 + (i_1/v_1)\mathbf{Z}_1} \tag{8-37}$$

where Eq. (8-34) has been used. Inserting Eq. (8-36),

$$\left(\frac{v_o}{v_i}\right)^{-1} = 1 - \frac{r_2}{r_1}\frac{\mathbf{Z}_1}{\mathbf{Z}_2} \tag{8-38}$$

The ratio $\mathbf{Z}_1/\mathbf{Z}_2$ is

$$\frac{\mathbf{Z}_1}{\mathbf{Z}_2} = \left(R_1 - \frac{j}{\omega C_1}\right)\frac{R_2 - j/\omega C_2}{-jR_2/\omega C_2} = \left(\frac{R_1}{R_2} + \frac{C_2}{C_1}\right) + j\omega\left(R_1 C_2 - \frac{1}{\omega^2 C_1 R_2}\right) \tag{8-39}$$

The circuit is in resonance when the imaginary part vanishes,

$$\omega_0^2 = \frac{1}{R_1 C_1 R_2 C_2} \tag{8-40}$$

Inserting Eq. (8-39) and Eq. (8-40) into Eq. (8-38), the result is

$$\left(\frac{v_o}{v_1}\right)^{-1} = \left[1 - \frac{r_2}{r_1}\left(\frac{R_1}{R_2} + \frac{C_1}{C_2}\right)\right] - j\frac{r_2}{r_1}\sqrt{\frac{R_1 C_2}{R_2 C_1}}\left(\frac{\omega}{\omega_0} - \frac{\omega_0}{\omega}\right) \tag{8-41}$$

This expression should be compared with Eq. (3-49) for an *RLC* circuit. In the simple case where $R_1 = R_2$ and $C_1 = C_2$, the input-output characteristic is

$$\left(\frac{v_o}{v_i}\right)^{-1} = \left(1 - 2\frac{r_2}{r_1}\right) - j\frac{r_2}{r_1}\left(\frac{\omega}{\omega_0} - \frac{\omega_0}{\omega}\right) \tag{8-42}$$

Note that the equivalent Q of the circuit is just r_2/r_1, which means that the Q can be adjusted independently of the resonant frequency. Alternatively, it is possible to adjust the resonant frequency by changing either R_1 (and R_2) or C_1 (and C_2) without changing Q. The output signal may exceed the input voltage at resonance if the real term in Eq. (8-42) is less than unity, so that an active filter may also have gain.

Figure 8-27 Operational-amplifier comparator.

Comparator

Operational amplifiers are useful in comparing the magnitudes of two signals. If one signal is applied to the inverting input and the other to the noninverting input, the output of the amplifier is zero only when the two signals are equal.

A practical *comparator* circuit, Fig. 8-27, employing an integrated-circuit operational amplifier yields a zero output voltage whenever the instantaneous input signal is equal to V. The large common-mode rejection ratio and excellent input-stage balance characteristic of operational amplifiers are important in this application.

ANALOG COMPUTERS

Simulation

Operational-feedback circuits can be assembled into *analog computers* in which voltage signals are analogous to variables in physical systems. Many physical systems can be described by mathematical equations based on the laws of nature. Analog computors are used to solve these equations and thereby to display the behavior of the system.

The voltages and other circuit parameters of an analog computer correspond to the variables and properties of the real system. The parameters of the computer are easily altered and adjusted, however, so that the behavior of the system may be examined over a wide range of conditions. The computer is, in fact, looked upon as *simulating* the physical system, and its response to external stimuli corresponds to the actual system to the extent that the mathematical description is accurate.

Damped Harmonic Oscillator

A great many physical systems are described by differential equations, for example, mechanical motions in accordance with Newton's laws of motion. Consider the *damped harmonic oscillator*. This is the mechanical vibration of a body of mass m on the end of a spring with force constant k in the presence of viscous damping described by the damping constant b and driven by an arbitrary force $F(t)$. The differential equation for the position x of the body is

$$m\frac{d^2x}{dt^2} + b\frac{dx}{dt} + kx = F(t) \tag{8-43}$$

Rearranging,

$$\frac{d^2x}{dt^2} = -\frac{b}{m}\frac{dx}{dt} - \frac{k}{m}x + \frac{1}{m}F(t) \tag{8-44}$$

The design of an analog computer to solve this equation begins by assuming a voltage signal corresponding to d^2x/dt^2 is available. This is integrated to yield $-dx/dt$, where for convenience the RC time constant in Eq. (8-22) is made equal to unity. Next, $-dx/dt$ is integrated again to obtain x. A fraction b/m of $-dx/dt$ is obtained from a voltage-divider potentiometer across the output of the first integrator; this is inverted and added to the fraction k/m of x from the output of the second integrator and to a voltage signal corresponding to $(1/m)F(t)$ (Fig. 8-28). The sum is equal to d^2x/dt^2, according to Eq. (8-44), and is returned to the input, where the second-derivative signal was assumed to be originally. This computer, therefore, continuously solves the original differential equation Eq. (8-43) and voltages corresponding to x and, if desired, dx/dt can be measured at appropriate points in the circuit.

It is necessary to set voltages corresponding to dx/dt, and x for their initial values at the time when the solution begins, as in solving any differential equation.

Figure 8-28 Analog computer for vibrating-mass problem.

This is most effectively accomplished by opening switches s_1 and s_2 at $t = 0$. The voltages across the integrating capacitors represent the velocity and displacement at all times, as can be seen by the fact that the summing points are at ground potential. These switches must be opened simultaneously with the beginning of $F(t)$ and in practice this is most often done with electronic switches such as diodes.

Note that Eq. (8-43) has the same form as Eq. (3-20) for the current in an *RLC* circuit. This means that the analog computer in Fig. 8-28 also can be used to investigate, say, resonance effects in this simple circuit. Although the computer is much more complicated than the circuit it simulates, study of the circuit is facilitated by the ease with which circuit parameters can be altered. Furthermore, the computer can be rewired to simulate other circuits as the need arises.

SUGGESTIONS FOR FURTHER READING

Jerald Graeme: "Applications of Operational Amplifiers," McGraw-Hill Book Company, New York, 1973.

Brian Jones: "Circuit Electronics for Scientists," Addison-Wesley Publishing Company, Inc., Reading, Mass., 1974.

R. D. Middlebrook: "Differential Amplifiers," John Wiley & Sons, Inc., New York, 1963.

EXERCISES

8-1 Draw the equivalent circuit of the bridged-T feedback amplifier, Fig. 8-4. Plot the characteristic curve of the filter alone, that is, the frequency variation of β, and determine the response curve of the feedback amplifier. Assume $g_m = 5 \times 10^{-3}$ mho for all FETs.

8-2 Analyze the source follower, Fig. 6-8, as a feedback amplifier and develop expressions for the gain, and for the input and output impedances.

Ans.: $a/(1 + a)$; $(R_g + R_2)(1 + a)$; $R_2/(1 + a)$

8-3 Calculate the input and output impedances of the transistor feedback amplifier in Fig. 8-2. The h parameters are $h_{ie} = 3600\ \Omega$, $h_{fe} = 150$, $h_{re} = 3 \times 10^{-3}$, $h_{oe} = 1.4 \times 10^{-4}$ mho.

Ans.: $1.61 \times 10^6\ \Omega$, $12\ \Omega$

8-4 Analyze the circuit of Fig. 6-18 as an amplifier with current feedback if the emitter bypass capacitor is omitted. Calculate β and the gain.

Ans.: 1, 3.3

8-5 Derive an expression for the input impedance of a feedback amplifier with both voltage and current feedback. Use this expression to show that both current and voltage negative feedback increase the input impedance. Using Eq. (8-15) find the expression for the input impedance when the output impedance is zero.

8-6 Calculate the voltage and current feedback ratios β_v and β_i in the output stages of the operational-feedback amplifier of Fig. 8-16.

Ans.: 10^{-2}, 1

8-7 Use Fig. 8-23 to develop a numerical expression corresponding to Eq. (8-30). What is the approximate value of I_0 for the transistor used in the circuit of Fig. 8-22? Take $e/kT = 38\ \text{V}^{-1}$.

Ans.: 3.8×10^{-13} A

8-8 Using the principles employed in Fig. 8-22, sketch a circuit for which the output voltage is either the square or the cube of the input signal.

8-9 Design an analog computer to solve the falling-body problem, for which the equation is

$$\frac{d^2x}{dt^2} = g \qquad (8\text{-}44)$$

where g is the acceleration due to gravity.

8-10 Write down the two simultaneous equations in x and y that the analog-computer circuit in Fig. 8-29 is set up to solve.

    Ans.: $a_1 x + b_1 y = c_1$; $a_2 x + b_2 y = c_2$

Figure 8-29 Analog-computer circuit to solve simultaneous linear equations.

NINE

OSCILLATORS

Electronic circuits can generate ac signals of a variety of waveforms over a wide range of frequencies. In fact, transistor and vacuum-tube oscillators are the only convenient way of generating high-frequency voltages. They are widely used in radio and TV transmitters and receivers, for dielectric and induction heating, and in electronic instruments for timing and testing purposes. An oscillator, in effect, converts power delivered by the dc supply voltages into ac power having the desired characteristics. In addition to the frequency and waveform of the oscillations, the conversion efficiency and frequency stability are important in the design of oscillator circuits.

POSITIVE FEEDBACK

Oscillation is achieved through positive feedback which produces an output signal without any input signal. According to Eq. (8-4), the gain of a feedback amplifier is given by

$$a' = \frac{a}{1 - a\beta} \tag{9-1}$$

If circuit conditions are arranged so that

$$a\beta = 1 \tag{9-2}$$

Eq. (9-1) indicates that the gain becomes infinite. The physical interpretation is that an output signal exists even when the input signal is zero. For sinusoidal oscillations the feedback network is designed so that Eq. (9-2), called the *Barkhausen criterion*, is satisfied at only one frequency, and the circuit oscillates at that frequency. The Barkhausen criterion requires that the overall phase shift of the feedback signal is 360°, and this is the significant factor in determining the frequency of oscillation. In addition, the amplifier gain must be large enough to assure that the $a\beta$ product is equal to unity, in order for the oscillations to persist.

The amplitude of oscillation is determined indirectly by Eq. (9-2). The gain of any amplifier is reduced at large-signal amplitudes upon the onset of cutoff or saturation conditions. Accordingly, the quiescent amplitude is such that the absolute value of gain is $1/\beta$. Since the feedback network is most often a passive circuit, the amplitude depends primarily upon amplifier characteristics.

It is not necessary to supply an input signal in order to initiate oscillations. Random noise voltages or transients accompanying application of the supply voltages are sufficient to start the feedback process. Since the amplitude of the feedback signal depends upon amplifier gain, the rapidity with which the oscillations reach the steady-state magnitude increases when the gain is large. It is usually desirable for the small-signal gain to be significantly larger than required by the Barkhausen criterion. This produces strong oscillations unaffected by minor circuit changes. On the other hand, if the gain is very great, nonsinusoidal oscillations may result from nonlinearities accompanying large-signal amplitudes.

RC OSCILLATORS

Phase-shift Oscillator

A simple but useful oscillator circuit employing a conventional amplifier stage and an *RC* feedback network is the *phase-shift oscillator*, Fig. 9-1. The grounded-emitter stage has an inherent phase shift of 180°, and the three cascaded *RC* circuits shift the phase an additional 180° in order to satisfy the Barkhausen criterion. At some particular frequency the phase shift in each *RC* section is 60°, so

Figure 9-1 Phase-shift oscillator.

the total phase shift in the feedback network is 180° and the circuit oscillates at this frequency, provided the amplification is great enough. Note that the maximum phase shift in one RC section is limited to 90°. This means that a two-section feedback network is not possible because an infinite gain would be required to overcome the attenuation in the feedback network at a total phase shift of 180°. Conversely, there is no particular advantage to having more than three RC sections in the feedback network, although it is possible to design such an oscillator.

The phase-shift oscillator is analyzed by first ignoring the loading effect of the amplifier upon the network. Considering that a voltage v_o is applied to the feedback network, the signal v_i applied to the transistor can be calculated by straightforward network analysis. The result is

$$\beta = \frac{v_i}{v_o} = \frac{1}{1 - 5/(\omega RC)^2 + j[1/(\omega RC)^3 - 6/\omega RC]} \tag{9-3}$$

In order for the phase shift of the feedback network to be 180°, the imaginary part of Eq. (9-3) must vanish, or

$$\frac{1}{(\omega_0 RC)^3} = \frac{6}{\omega_0 RC} \tag{9-4}$$

The oscillation frequency is found by solving for $f_0 = \omega_0/2\pi$,

$$f_0 = \frac{1}{2\pi\sqrt{6}} \frac{1}{RC} \tag{9-5}$$

Inserting Eq. (9-5) into Eq. (9-3), $\beta = 1/(1 - 5 \times 6) = -1/29$, which means that the gain must be

$$a = \frac{1}{\beta} = -29 \tag{9-6}$$

in order to satisfy the Barkhausen criterion. According to this result the amplification must be at least 29 or the circuit cannot oscillate.

Actually the amplifier gain must be somewhat larger than 29 in order to assure stable oscillations in the face of circuit losses and component aging effects. The amplitude of oscillations increases until limited to a value of 29 by nonlinearities in the transistor. Most often, the onset of cutoff at the peak of the wave is the limiting factor, and a peak signal amplitude nearly equal to the quiescent collector potential is expected.

The simplicity of the phase-shift oscillator makes it attractive for noncritical applications, particularly at medium and low frequencies down to about 1 Hz. The frequency stability is not as good as can be obtained with other RC oscillators, however. In addition, in order to change frequency it is necessary to vary all three capacitors (or the three resistors), and this is inconvenient.

Wien bridge Oscillator

The frequency-selective properties of the Wien bridge discussed in Chap. 3 are very appropriate for the feedback network of an oscillator. Actually, the *Wien bridge oscillator*, Fig. 9-2, is widely used for variable-frequency laboratory instruments called *signal generators*. In Fig. 9-2 a conventional integrated-circuit amplifier provides positive feedback at the resonant frequency of the bridge, where, according to Eq. (3-68), the phase shift in the feedback network is zero. The characteristic frequency is, from Eq. (3-64),

$$f_0 = \frac{1}{2\pi RC} \tag{9-7}$$

Figure 9-2 Integrated-circuit Wien bridge oscillator.

The output voltage of the Wien bridge is zero at exact balance when $R_2 = 2R_1$ (refer to Fig. 3-15), so it is necessary to unbalance the bridge slightly by adjusting the ratio R_2/R_1. This provides sufficient feedback voltage to maintain stable oscillations, since the feedback ratio is determined by the relative values of R_1 and R_2. The combination of the two 1N914 diodes and the 47-kΩ resistors shunting R_2 provides amplitude stability by automatically adjusting the feedback ratio. Suppose the output signal is small so that the signal voltage across the diodes is less than the characteristic diode forward-voltage drop, approximately 0.6 V. Then neither diode conducts and the shunt arm is open-circuited. If the output signal is large enough for the diodes to conduct, the 47-kΩ resistor is, in effect, connected in parallel with R_2 and feedback is reduced. Therefore, the output amplitude is stabilized at that level at which the diodes just conduct at the peaks of the waveform.

The excellent frequency stability of the Wien bridge oscillator compared with the phase-shift oscillator is a result of the rapid change of phase of the feedback voltage with frequency. The feedback ratio is, from Eq. (3-68),

$$\beta = \frac{v_o}{v_i} = \frac{1}{3 + j(\omega/\omega_0 - \omega_0/\omega)} - \frac{1}{1 + R_2/R_1} \tag{9-8}$$

The phase shift is found by rationalizing Eq. (9-8) and computing the tangent of the phase angle. For frequencies near to the resonant frequency, the result is

$$\tan \theta \cong \frac{1 + R_2/R_1}{9 - 3(1 + R_2/R_1)} \left(\frac{\omega}{\omega_0} - \frac{\omega_0}{\omega} \right) \tag{9-9}$$

If, for example, $R_2 = 1.9R_1$, the coefficient of the frequency term in Eq. (9-9) is equal to 9.7. By comparison, the corresponding expression for the phase-shift oscillator is, after rationalizing Eq. (9-3),

$$\tan \theta \cong -\frac{\sqrt{6}}{5} \left(\frac{\omega}{\omega_0} - \frac{\omega_0}{\omega} \right) \tag{9-10}$$

where Eq. (9-5) has been used to introduce ω_0. The numerical factor in Eq. (9-10) is only 0.49, which is smaller by a factor of 20 than in the case of the Wien bridge.

This comparison is illustrated more clearly in Fig. 9-3 where the phase angle of the feedback signal is plotted as a function of ω/ω_0 for both oscillators. The phase angle changes much more rapidly with frequency in the Wien bridge feedback circuit. This means that the oscillation frequency is quite stable since only feedback signals with a near-zero phase angle are effective.

The excellent frequency stability together with the relative ease in changing frequency (only the two capacitors need be variable) makes the Wien bridge oscillator popular. In most practical circuits the variable capacitors provide a frequency range of about 10 to 1. In addition, fixed decade values of the resistors are selected by a switch so that a wide frequency range can be covered in a single instrument. A typical commercial Wien bridge oscillator can have a frequency range extending from 5 Hz to 1 MHz in decade steps.

Figure 9-3 Phase-shift characteristics near resonant frequency for two *RC* oscillators.

RESONANT CIRCUIT OSCILLATORS

LC Oscillators

Resonant *LC* circuits are often used in the feedback network of oscillators to select the frequency of oscillation. Consider the so-called *Hartley oscillator*, Fig. 9-4, in which a parallel resonant circuit is connected between gate and drain. The inductance is tapped so that, in effect, the portion L_1 is part of the drain load while the remainder L_2 is in the gate circuit. The resonant frequency involves the series inductance of L_1 and L_2, so that, as in Chap. 3,

$$\omega_0 = \frac{1}{\sqrt{(L_1 + L_2)C}} \tag{9-11}$$

The feedback ratio is determined by first calculating the feedback voltage across L_2,

$$v_i = ij\omega L_2 = \frac{v_o \, j\omega L_2}{j\omega L_2 + 1/j\omega C} \tag{9-12}$$

Introducing the resonance condition $\omega_0(L_1 + L_2) = 1/\omega_0 C$,

$$v_i = v_o \frac{\omega_0 L_2}{-\omega_0 L_1} \tag{9-13}$$

so that the feedback ratio is just

$$\beta = -\frac{L_2}{L_1} \tag{9-14}$$

Note that the ratio is negative, which means that the additional 180° phase shift of the amplifier produces positive feedback as required. Equation (9-14) and the

Figure 9-4 Hartley oscillator.

Barkhausen criterion specify the gain of the amplifier needed to sustain oscillations.

The bias conditions in the Hartley oscillator are worthy of note. Operating bias is supplied by the $R_g C_g$ combination in the gate circuit. When oscillations start, the gate bias is zero. As oscillations build up, the gate-source diode rectifies the feedback signal, thereby charging C_g to nearly the peak value of the input signal. The $R_g C_g$ time constant is much longer than the period of oscillation so the voltage across C_g is constant and represents the necessary dc gate bias. In effect, the gate is clamped at ground potential (compare Figs. 4-27 and 4-28). The gate bias therefore automatically adjusts itself to the amplitude of the feedback signal and this action stabilizes the amplitude of oscillation.

This form of bias means that the transistor is cut off during most of the cycle, a condition labeled *class C* operation. The gate voltage produces drain current at the peak of the feedback voltage cycle. The large impedance of the resonant circuit allows only the fundamental component of the output signal to have any appreciable amplitude, however, so that the output waveform is sinusoidal. Another way of explaining the action of the resonant circuit is to say that the circuit is excited by periodic pulses from the class C amplifier and that a continuous wave is produced by ringing in the resonant circuit. These two explanations are equivalent.

The *Colpitts oscillator*, Fig. 9-5, is similar to the Hartley circuit except that the feedback ratio is determined by the relative values of C_1 and C_2. Base bias in this circuit is such that the transistor operates as a class A amplifier, but class C operation equivalent to the Hartley oscillator is equally possible. The output load is coupled to the resonant circuit with a secondary winding L_o, which is particularly useful in feeding low-impedance loads.

Figure 9-5 70-MHz Colpitts oscillator.

Another way to develop feedback voltage is by including a secondary winding, or *tickler* winding, coupled to the inductance. Consider, for example, the grounded-base oscillator, Fig. 9-6a. Mutual inductance between L and the tickler winding induces a feedback signal of the proper amplitude and phase to sustain oscillations. From the definition of mutual inductance in Chap. 3, the feedback voltage is just the ratio of the mutual inductance M between coils to the total inductance L times the voltage across L. Therefore

$$v_i = \frac{M}{L} v_o \tag{9-15}$$

or

$$\beta = \frac{v_i}{v_o} = \frac{M}{L} \tag{9-16}$$

The equivalent circuit of the oscillator, Fig. 9-6b, is analyzed in the following way. The impedance of the parallel resonant circuit is very large at resonance, so the gain of the grounded-base amplifier is, from Eq. (6-36),

$$a = \frac{-1}{h_{ib}h_{ob}/h_{fb} - h_{rb}} = \frac{1}{r_e/\alpha r_c + r_b/r_c} \cong \frac{r_c}{r_b} = \frac{1}{h_{rb}} \tag{9-17}$$

Applying the Barkhausen criterion, the condition for oscillation is obtained by comparing Eqs. (9-17) and (9-16),

$$h_{rb} = \frac{M}{L} \tag{9-18}$$

In a good-quality transistor h_{rb} is of the order of 10^{-4}, which means that the mutual inductance may be quite small. Accordingly, a tickler winding consisting of only a few turns is sufficient.

A major advantage of the tickler feedback circuit is that the amplitude of the feedback voltage can be easily adjusted by choosing the number of turns on the

Figure 9-6 (a) Grounded-base tickler oscillator and (b) equivalent circuit.

feedback winding. The proper phase is obtained by interchanging the leads of the winding, if necessary, to obtain positive feedback. Thus, although Fig. 9-6 is a grounded-base circuit, the grounded-emitter and grounded-collector configurations are equally useful. Furthermore, the resonant circuit can be put in the emitter or base circuits as well. The specific configuration illustration by Fig. 9-6 is particularly appropriate for good frequency stability, as discussed below.

Class C operation of transistor oscillators is often used in circuits designed to develop appreciable power at high frequencies. Figure 9-7 is the circuit diagram of a typical power oscillator. Overall power efficiencies of the order of 70 percent are possible. In this circuit, base bias is developed by the resistor-capacitor combination in the base-emitter circuit analogous to the action in the Hartley oscillator.

Crystal Oscillators

The frequency stability of LC oscillators is determined primarily by the Q factor of the resonant circuit. The resonance curve is sharply peaked and the rate of change of phase with frequency is rapid when the Q is large, and both factors contribute to frequency stability of the oscillator. In this connection, any equivalent resistance connected in parallel with the resonant circuit lowers the effective Q. Therefore,

Figure 9-7 100-kHz 10-W power oscillator.

the loading effect on the resonant circuit should be minimized to improve frequency stability. According to the discussion in Chap. 6, the output impedance of the grounded-base amplifier is larger than for any other configuration. This is why the grounded-base oscillator with the resonant combination in the collector circuit is the most satisfactory transistor oscillator configuration. Correspondingly, FET oscillators most often have the resonant circuit in the gate circuit. In either case, practical values of Q from 100 to 500 can be obtained, and quite stable oscillations result.

Many applications require a higher order of frequency stability than can be obtained with LC resonant circuits, and *crystal oscillators* are widely used to fill this need. Certain crystalline materials, most notably quartz, exhibit piezoelectric properties; that is, they deform mechanically when subjected to an electric field. Piezoelectricity also implies that the inverse is also true: when the crystal is forcibly deformed, an electric potential is developed between opposing faces of the crystal. As a result of this piezoelectric property, a thin plate of quartz provided with conducting electrodes vibrates mechanically when the electrodes are connected to an alternating voltage source. The vibrations, in turn, produce electrical signals which interact with the voltage source. The vibrations and electrical signals are a maximum at the natural mechanical resonant frequency of the crystal.

The equation for the motion of a vibrating body has already been written in Eq. (8-43),

$$m\frac{dx^2}{dt^2} + b\frac{dx}{dt} + kx = F(t) \tag{9-19}$$

where in the present case m is the mass of the crystal, b is the internal mechanical loss coefficient, and k is the elastic constant of the crystal. As has already been noted, this expression is identical in form with that for the current in a series resonant circuit, Eq. (3-20),

$$L\frac{d^2i}{dt^2} + R\frac{di}{dt} + \frac{1}{C}i = F(t) \tag{9-20}$$

Comparing Eqs. (9-19) and (9-20) it can be seen that the vibrating mass is analogous to inductance, mechanical losses are equivalent to resistance, and the elasticity corresponds to the reciprocal of capacitance. Because of the identical form of the two equations, mechanical resonance is expected, and it is useful to define a mechanical Q factor by analogy with Eq. (3-47),

$$Q = \frac{\omega m}{b} \qquad (9-21)$$

It turns out that the internal losses in quartz crystals are very small and that Q values reaching as high as 100,000 can be achieved. Furthermore, the elastic constants are such that resonant frequencies ranging from 10 kHz to several tens of megahertz are possible, depending upon the mechanical size and shape of the crystal.

The piezoelectric properties of quartz result in electrode potentials corresponding to the mechanical vibrations. This suggests that the electrical characteristics can be represented by an equivalent circuit. Comparing Eq. (9-19) and (9-20), the appropriate circuit is a series combination of a resistance, inductance, and capacitance. To this must be added the electric capacitance resulting from the parallel-plate capacitance of the electrodes with the crystal as a dielectric. Therefore, the complete equivalent circuit of a quartz crystal is the series-parallel combination shown in Fig. 9-8a. In this equivalent circuit L, C, and R are related to the properties of the quartz crystal and C' is the electrostatic capacitance of the electrodes. Appropriate values for a 90-kHz crystal are $L = 137$ H, $C = 0.0235$ pF, $R = 15,000$ Ω, and $C' = 3.5$ pF. The conventional circuit symbol for a crystal is a parallel-plate capacitor with the crystal between the plates, Fig. 9-8b.

The series-parallel equivalent circuit of a quartz crystal shows that there will be a series resonant frequency (zero impedance) and a parallel resonant frequency (infinite impedance). The frequency of series resonance is $\omega_S = 1/\sqrt{LC}$. The parallel resonance occurs when the reactance of C' equals the net inductive reactance of the combination of L and C, $\omega_P = \sqrt{1/L(1/C + 1/C')}$. Accordingly, the parallel resonant frequency is always greater than the series resonant frequency, although,

(a)   (b)

Figure 9-8 (a) Equivalent circuit of a quartz crystal and (b) its circuit symbol.

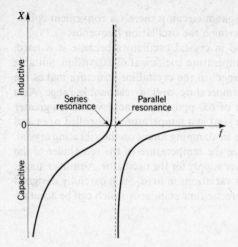

Figure 9-9 Impedance of quartz crystal.

since $C' \gg C$, the two are very close. The reactance is capacitive both below and above the resonant frequencies, Fig. 9-9.

Extremely stable oscillator circuits are possible with the large Q of a quartz resonator, and a variety of circuits have been designed. Either the series or parallel resonance frequency may be used, although parallel resonance is more common. Consider, for example, the *Pierce oscillator*, Fig. 9-10, in which the crystal is connected between base and collector. This circuit is identical to the Colpitts oscillator, Fig. 9-5, with the crystal replacing the resonant circuit. The feedback ratio is determined by the relative values of C_1 and C_2. The inductance RFC (for *radio-frequency choke*) in the collector lead is a useful way of applying collector potential without shorting the collector to ground at the signal frequency. It can be replaced by a 10,000-Ω resistor with some loss in circuit performance. A parallel resonant LC combination can also be used with attendant gain in circuit

Figure 9-10 Quartz-crystal oscillator.

performance. In the latter case the resonant circuit is merely a convenient collector load impedance and does not determine the oscillation frequency.

Quartz is almost universally used in crystal oscillators because it is hard, reasonably strong, and has a small temperature coefficient of expansion. Suitable orientation of the plane faces with respect to the crystalline structure makes the resonant frequency independent of temperature over a reasonable range. As a result, frequency stability of the order of 100 ppm can be achieved. Even greater accuracy is obtained by placing the crystal in a temperature-controlled oven and evacuating the crystal holder to reduce air damping forces on the vibrating crystal. It is also common practice to stabilize the temperature of the remainder of the circuit and to employ a regulated power supply for the oscillator. Amplifier stages are used to isolate the oscillator from variations in load. Such carefully designed crystal oscillators provide extremely precise time standards which can be accurate to 1 part in 100 million.

NEGATIVE RESISTANCE OSCILLATORS

Stability Analysis

The current-voltage characteristics of several devices, most notably the tunnel diode and the unijunction transistor, exhibit *negative resistance* properties. That is, over a portion of the characteristic the current decreases as the applied voltage increases. The physical mechanisms leading to negative resistance properties which are discussed in Chap. 5 prove to be very useful in a variety of oscillator circuits.

Analysis of negative resistance oscillation is best carried out by examining the complex impedance of the ac small-signal linear equivalent circuit of the oscillator. Consider, for example, the series resonant circuit in Fig. 9-11 where r represents the negative resistance of a tunnel diode at the operating point. The impedance of the circuit is

$$\mathbf{Z} = R + j\omega L + \frac{1}{1/r + j\omega C} = R + j\omega L + \frac{r}{1 + j\omega C}$$

$$= R + \frac{r}{1 + (\omega r C)^2} + j\omega \left[L - \frac{r^2 C}{1 + (\omega r C)^2} \right] \tag{9-22}$$

Figure 9-11 Equivalent circuit of negative resistance oscillator.

As usual, the circuit current is given by Ohm's law

$$i = \frac{v}{Z} \tag{9-23}$$

Equation (9-23) predicts an infinite current under conditions for which the impedance vanishes. This situation is analogous to Eq. (9-2) and implies that the circuit oscillates. Analysis of circuit oscillation is effected, therefore, simply by searching for zeros of the impedance function.

The impedance vanishes when both the real and imaginary parts of Eq. (9-22) are equal to zero,

$$R + \frac{r}{1 + (\omega rC)^2} = 0 \quad \text{and} \quad L - \frac{r^2 C}{1 + (\omega rC)^2} \tag{9-24}$$

These relations are each solved for the angular frequency

$$\omega^2 = -\left(\frac{1}{rC}\right)^2 \left(1 + \frac{r}{R}\right) \quad \text{and} \quad \omega^2 = \frac{1}{LC} - \left(\frac{1}{rC}\right)^2 = \omega_0^2 - \left(\frac{1}{rC}\right)^2 \tag{9-25}$$

where ω_0 is the resonant frequency. Since both expressions must yield the same value for the oscillation frequency,

$$-\left(\frac{1}{rC}\right)^2 \left(1 + \frac{r}{R}\right) = \frac{1}{LC} - \left(\frac{1}{rC}\right)^2 \tag{9-26}$$

Solving for the value of negative resistance necessary to sustain oscillations,

$$r = -\frac{L}{RC} \tag{9-27}$$

Inserting this value into Eq. (9-25), the frequency of oscillation is

$$\omega^2 = \omega_0^2 - \left(\frac{R}{L}\right)^2 \tag{9-28}$$

This expression is very similar to the natural frequency of a series resonant circuit. In effect, the diode negative resistance cancels out the positive resistances in the circuit and the circuit rings continuously at its natural frequency.

According to Eq. (9-27) the diode resistance at the operating point is negative, as expected. Note, too, that the absolute value of r must be greater than the circuit resistance R in order for the frequency calculated from Eq. (9-25) to have a real value. Useful circuit operation is also achieved when $|r| < R$, however. The fact that Eq. (9-25) indicates that the oscillation frequency is imaginary under this condition suggests that the small-signal linear equivalent-circuit analysis does not apply.

Circuit performance in the two cases is best compared with the aid of load lines plotted on the tunnel-diode current-voltage characteristic, Fig. 9-12. When $|r| > R$ the load line intersects the curve only at the operating point on the negative resistance portion and the circuit oscillates at its resonant frequency as

Figure 9-12 Load lines for tunnel-diode oscillator and switching circuit.

analyzed above. By comparison, when $|r| < R$ three intersections occur. In this case both point 1 and point 2 are stable (nonoscillatory) circuit conditions. Point 3 is a condition of unstable equilibrium such that if the circuit is brought to this point the current immediately begins to change and finally settles at either point 1 or point 2. In effect the circuit has two stable states. It may be switched from one state to the other by temporary application of voltages. Suppose, for example, that the circuit is at point 1, and that a temporary increase in applied voltage increases the current to the peak. When the temporarily greater voltage is removed, the circuit traverses the *IV* characteristic and ends up at point 2. Similarly, the circuit may be switched from point 2 to point 1 by temporarily removing the applied voltage. When the voltage is reapplied, the current increases from the origin and stops at point 1. Such *switching* properties prove to be very useful in digital computers, for example.

The variety of operating conditions is not exhausted by these two cases which are specified by the dc load line. The oscillatory condition, $|r| > R$, actually has several different modes of operation that depend upon the ac load line. If, for example, the inductance L is large, the frequency calculated from Eq. (9-24) is imaginary even though the magnitude of the diode negative resistance is proper for oscillation. This condition leads to nonsinusoidal *relaxation* oscillations which are treated later in this chapter. On the other hand, if the capacitance C is large, the frequency is zero, according to Eq. (9-24). This means that the circuit oscillations are completely damped out.

The relation between circuit conditions for sinusoidal oscillation, relaxation oscillation, and damped behavior may be appreciated from the current-voltage characteristic, Fig. 9-13, and the ac load lines. According to this analysis sinusoidal oscillations at the natural frequency of the circuit are obtained when the magnitude of the negative resistance at the operating point is properly matched to circuit parameters, Eq. (9-27). A large value of inductance and small capacitance

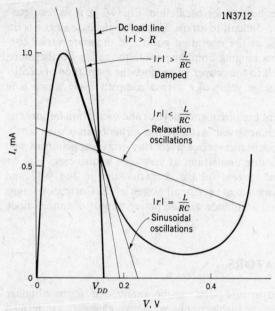

Figure 9-13 Combination of dc and ac load lines determine oscillating mode of tunnel-diode oscillator.

lead to relaxation oscillations and large C and small L produce complete damping. This latter condition is useful in experimentally tracing out the entire current-voltage characteristic which, clearly, cannot be accomplished with the circuit oscillating.

Tunnel-diode Oscillator

The circuit of a practical tunnel-diode oscillator is shown in Fig. 9-14. In this configuration the diode is effectively in parallel with a parallel resonant circuit but the principle of operation is identical with the analysis of the preceding section. The resistive voltage divider biases the diode to an operating point on the negative resistance portion of the current-voltage characteristic.

Figure 9-14 Tunnel-diode oscillator.

It may appear that the condition for oscillation, Eq. (9-27), is so restrictive that sinusoidal oscillations are difficult to attain. This is not the case since r is the slope of the IV characteristic at the operating point and therefore varies considerably with operating points ranging from the peak to the valley. Furthermore, the circuit tends to adjust itself to the proper value since the amplitude of oscillation changes to yield an average value of r over a complete cycle which is in agreement with Eq. (9-27).

Tunnel diodes are excellent high-frequency devices and oscillation frequencies as large as 10^{11} Hz have been achieved. Actually this virtue often proves troublesome in experimental circumstances since even tiny stray capacitances and inductances are sufficient to cause oscillations at very high frequencies.

According to the typical current-voltage characteristics in Fig. 9-12 and Fig. 9-13, tunnel diodes can operate at very small power levels. Correspondingly, however, the maximum signal amplitude and ac power output of tunnel-diode oscillators are limited.

RELAXATION OSCILLATORS

Oscillator circuits considered to this point can be analyzed in terms of linear elements. Circuits which employ highly nonlinear active elements are termed *relaxation oscillators* for reasons which become clear in the following discussion. Very often relaxation oscillators are based upon negative resistance properties of the active element. Although, as discussed in the previous section, it is possible to generate sinusoidal waveforms by means of negative resistance characteristics, relaxation oscillators characteristically produce nonsinusoidal signals.

Sawtooth Generators

Consider the unijunction relaxation-oscillator circuit in Fig. 9-15. The operation of this circuit can be understood after examining the current-voltage characteristic of a unijunction transistor, Fig. 9-16. This characteristic exhibits a negative resistance region between the peak voltage V_p and the valley point V_v, as discussed in Chap. 5.

The unijunction oscillator operates as follows: Capacitor C charges through resistor R and the voltage across C increases exponentially until the potential V_p is attained. At this point the emitter junction becomes biased in the forward direction and the capacitor rapidly discharges through the emitter junction. When the capacitor potential drops to a low value (essentially equal to V_v), the emitter junction is again reverse-biased and the capacitor begins to recharge. The waveform at output terminal 1, Fig. 9-17, is a series of RC charging curves with a peak-to-peak amplitude equal to $V_p - V_v$. Note that following every charging period the circuit "relaxes" back to the starting point. This is the origin of the terminology for this type of oscillator. The output waveform at terminal 2 is a series of sharp positive pulses, Fig. 9-17.

Figure 9-15 Unijunction relaxation oscillator.

The period of oscillation is found from the expression for the capacitor voltage in a simple RC circuit,

$$v = V(1 - e^{-t/RC})$$

The voltage reaches V_v at t_1,

$$V_v = V(1 - e^{-t_1/RC}) \qquad (9\text{-}29)$$

Solving for t_1

$$t_1 = -RC \ln (1 - V_v/V) \qquad (9\text{-}30)$$

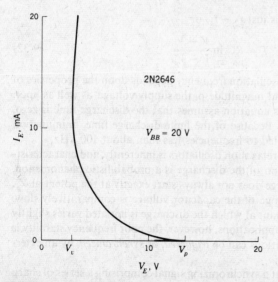

Figure 9-16 Current-voltage characteristic of unijunction transistor.

Figure 9-17 Output waveforms of unijunction oscillator at (a) output terminal 1 and (b) output terminal 2.

Similarly, the voltage reaches V_p at t_2, so that

$$t_2 = -RC \ln (1 - V_p/V) \qquad (9\text{-}31)$$

The period of the oscillation is just $t_2 - t_1$, or

$$T = RC \ln \frac{V - V_v}{V - V_p} \qquad (9\text{-}32)$$

According to Eq. (9-32) the oscillation frequency depends upon the properties of the uninjunction transistor and magnitude of the supply voltage, as well as upon the circuit time constant. This equation assumes that the discharge time is zero, which is not true in practice. Because of the finite discharge time, uninjunction relaxation oscillators are limited to frequencies less than about 100 kHz.

The frequency stability of relaxation oscillators is inherently, and characteristically, very poor. The initiation of the discharge is a probabilistic phenomenon, which means that the discharge does not always start exactly at the potential V_p. Furthermore, the rate of change of the capacitor voltage is comparatively slow. This means that the exact instant at which the discharge is initiated varies slightly from cycle to cycle. In many applications, however, the poor frequency stability is an advantage since the oscillations can be *triggered*, or *synchronized*, by an external signal.

Suppose, for example, that a synchronizing signal comprising a series of sharp voltage pulses is applied across the capacitor, Fig. 9-18. The sharp voltage rise of

Figure 9-18 Synchronizing frequency of a relaxation oscillator.

one of the pulses causes the capacitor voltage to exceed V_p at some point and the transistor is triggered into conduction. This repeats on successive cycles with the result that the frequency of oscillation is synchronized to the period of the pulse signal. Note that the period of the relaxation oscillator is a multiple or submultiple of the period of the pulses. In this way a relaxation oscillator acts as a frequency multiplier or divider. Frequency division by a factor as large as 10 or so is easily possible.

The free-running frequency of the relaxation oscillator must be reasonably close to a multiple or submultiple of the sync frequency. Since, however, the frequency stability of a relaxation oscillator is poor, the disparity may be as large as 20 percent or so without loss of synchronization. Triggering by a sinusoidal synchronizing voltage is also possible. Sharp pulses are more effective, however, since the time at which the breakdown voltage is attained is more definite.

The current-voltage characteristic of an SCR, Fig. 4-20, is similar in form to that of the unijunction transistor. However, the breakdown voltage is easily adjusted by varying the gate current. This means that the amplitude and frequency of oscillation can be changed easily in a SCR relaxation oscillator. Furthermore, synchronizing signals can be introduced into the gate circuit to take advantage of the inherent amplification of the SCR.

A useful relaxation oscillator employing an SCR is shown in Fig. 9-19. The gate potential is determined by the variable voltage divider comprising R_1 and

Figure 9-19 Relaxation sawtooth generator.

R_2. Initially, when C is uncharged, the cathode of the SCR is at a potential of 20 V with respect to ground, so the gate is biased in the reverse direction. As the capacitor voltage increases, the SCR remains switched off until the cathode potential becomes slightly less positive than the gate. At this point the SCR switches on, the capacitor is discharged, and the cycle repeats. The purpose of the 5-Ω resistor in series with the cathode is to limit the discharge current to a safe value for the SCR. This is necessary because the internal resistance of the SCR in the on state is so small that the peak current can be destructive.

A linear sawtooth waveform is produced by the constant-current transistor. Since the charging current is constant, the capacitor voltage increases linearly with time according to

$$v = \frac{q}{C} = \frac{I}{C} t \qquad (9\text{-}33)$$

where I is the current value and t is the time measured from when the capacitor voltage is zero.

A transistor is an excellent constant-current source since the collector current is essentially equal to the emitter current quite independent of the collector potential. The emitter current is determined by the base bias resistor R_b. It can be made adjustable in order to vary the charging current and consequently the period of oscillation. The peak amplitude of the sawtooth can be nearly as large as the dc supply voltage, which is advantageous in most applications.

Multivibrators

Two-transistor feedback circuits called *multivibrators* are important relaxation oscillators. Additionally, suitably modified versions have exceedingly versatile properties other than oscillation.

It is easiest to begin with a study of the so-called *astable* multivibrator, Fig. 9-20, the terminology for which will become clear later. This circuit can be recognized as a two-stage RC amplifier with the output coupled back to the input. The feedback ratio is unity and positive because of the 180° phase shift in each

Figure 9-20 Astable multivibrator.

stage; thus, the circuit oscillates. Because of the very strong feedback signals, the transistors are driven into either cutoff or saturation, and nonsinusoidal oscillations are generated.

Suppose that at some instant the feedback voltage drives V_1 into cutoff; this implies that V_2 is conducting because of the relative phase shift between the two stages. The voltage drop across the collector resistor of V_2 puts the collector at nearly ground potential, and C_1 charges through R_1 toward the collector supply potential. When the voltage across C_1 increases sufficiently to bias the emitter junction of V_1 in the forward direction, V_1 begins to conduct. The collector voltage of V_1 drops and V_2 is driven into cutoff through coupling capacitor C_2. Now, C_2 charges through R_2 until V_2 becomes forward-biased and the cycle repeats.

Notice that the circuit alternates between a state in which V_1 is conducting and simultaneously V_2 is cut off and a state in which V_1 is cut off and V_2 is conducting. The transition between these two states is rapid because of the strong feedback signals. The time in each state depends upon the coupling capacitor and bias resistor time constant. Since each transistor is driven alternately into cutoff and saturation, the voltage waveform at either collector is essentially a square wave with a peak amplitude equal to the collector supply voltage.

This picture is confirmed by actual collector voltage waveforms, Fig. 9-21. The triangular base voltage waveforms illustrate the alternate charging and discharging of the coupling capacitors. Of particular interest is the very low

Figure 9-21 Collector and base voltage waveforms in transistor multivibrator.

Figure 9-22 Equivalent circuits of the V_2 portion of Fig. 9-20 showing the origin of overshoot.

voltage drop across the transistors when saturated (≈ 0.1 V) and the fact that the transition from one state to another is initiated when the base voltage of the cutoff transistor just slightly exceeds zero.

A small voltage spike, or *overshoot*, is present in the base voltage waveform at the transition from one state to the other. The reason for the overshoot may be seen by focusing attention on capacitor C_2 when V_1 goes from conducting to cutoff. Consider that portion of the circuit shown in Fig. 9-22a where the on-to-off transition of V_1 is represented by a switch. Further simplification in the circuit is effected by ignoring R_2 since $R_L \ll R_2$; in addition, the forward-biased emitter junction of V_2 is represented approximately by a simple resistor r, Fig. 9-22b. According to the waveforms of Fig. 9-21, V_1 switches from saturation to cutoff at transition B when the base voltage of V_2 (and hence the voltage across C_2) is essentially zero. Thus, when the switch in the circuit of Fig. 9-22b opens, C_2 begins to charge through R_L and r and the charging current results in an overshoot voltage drop across the emitter junction resistance.

The charging current through R_L also prevents the collector voltage of V_1 from rising immediately to V_{cc}. This is the source of the rounding of the collector waveform pulse edge shown in Fig. 9-21. A typical value for r is 200 Ω, so that the peak overshoot in the multivibrator of Fig. 9-20 is $10(200/10,200) = 0.2$ V. This small value, plus the relatively short time constant, means the overshoot introduces only a minor departure from a square collector waveform.

The pulse width in a multivibrator depends upon the time constant of C_1 (or C_2) charging through R_1 (or R_2). A value for t_w can be written down immediately from the expression for the period of a relaxation oscillator, Eq. (9-32). Comparing the waveforms in Fig. 9-21 with those of Fig. 9-17,

$$t_w = R_1 C_1 \ln \frac{V_{cc} - (-V_{cc})}{V_{cc} - 0}$$

$$t_w = R_1 C_1 \ln 2 \tag{9-34}$$

When $R_1 = R_2$ and $C_1 = C_2$ the waveform is a square wave of frequency $f = 1/2t_w$. It is equally possible to generate asymmetrical waveforms by choosing nonequal values for, say, the coupling capacitors. The result is alternating pulses of widths given by Eq. (9-34) for each time constant.

Astable multivibrators are relaxation oscillators and can therefore be synchronized by application of external signals. Positive synchronizing pulses can be applied to either base to trigger the other transistor into conduction. Multivibrators are often used in this way as frequency dividers. The circuit is also a useful square-wave generator, and, in conjunction with a differentiating circuit, yields sharp, spike waveforms.

A dc-coupled multivibrator can exist in either of its two states indefinitely and can be caused to make a transition from one state to the other by an external trigger pulse. Because of its two stable states, a multivibrator with dc coupling is called a *binary*. A binary is also commonly referred to as a *flip-flop*.

Consider the typical transistor binary in Fig. 9-23. Note that conventional dc coupling using resistors R_1 and R_2 is used. In other respects the circuit is similar to the ac-coupled, astable multivibrator. Trigger pulses are introduced across the emitter resistor R_E. The purpose of capacitors C_1 is to improve triggering action, as discussed subsequently. Output is conventionally taken from either, or both, collectors.

It is important to verify that a stable state of the binary exists when one transistor is cut off and the other is saturated. To begin, assume that V_1 in Fig. 9-23 is cut off so that its collector current is negligible. Collector current in the saturated transistor V_2 is found by drawing a load line corresponding to the total load resistance $R_L + R_E$ on the collector characteristics. The intersection of

Figure 9-23 Binary circuit is dc-coupled multivibrator.

Figure 9-24 Equivalent circuits of transistor binary used to show that binary has only two stable states.

this load line with the current axis gives the collector current, 9.3 mA, and also shows that a base current of 125 μA is needed to reach saturation. The bias on V_1 may now be calculated using the portion of the circuit shown in Fig. 9-24a, where the emitter voltage $V_e = i_c R_E = 9.3 \times 10^{-3} \times 220 = 2.1$ V. The emitter-collector voltage of V_2 is taken to be 0.1 V when saturated. Actually, a more accurate value for the emitter-collector voltage may be determined from transistor manufacturer's data, but it is usually small enough to be ignored in any event. Using Fig. 9-24a, the emitter base voltage of V_1 is

$$V_{eb} = V_e \frac{R_1}{R_1 + R_2} = 2.1 \frac{42}{42 + 33} = 1.2 \text{ V}$$

This value of reverse bias at the emitter junction is much more than required to keep V_1 cut off, as assumed.

To show that V_2 is biased to saturation, focus attention on that portion of the circuit shown in Fig. 9-24b. The resistance R_2 may be ignored in comparison with the resistance of the emitter junction, and the current through V_1 is negligible because it is cut off. Therefore,

$$i_b = \frac{V_{cc} - V_e}{R_L + R_1} = \frac{22.5 - 2.1}{2.2 + 42} 10^{-3} = 460 \text{ } \mu\text{A}$$

This is much greater than the 125 μA needed to keep V_2 in saturation. Thus, it is established that V_1 is cut off and V_2 is saturated. Because of the symmetry of the circuit, it is obvious that an equally stable state is the one in which V_1 is saturated and V_2 cut off. When the binary switches from one configuration to the other the collector voltage changes from essentially the emitter potential in the saturated condition to a voltage equal to V_{cc} less the drop across R_L resulting from the base bias current. Therefore the output signal is

$$v_o = (22.5 - 2.2 \times 10^3 \times 460 \times 10^{-6}) - 2.1 = 19.4 \text{ V}$$

The circuit is triggered into switching from one state to the other by a positive voltage pulse applied across the emitter resistor. This drives the saturated transistor out of saturation and also tends to turn on the cutoff transistor. As current in the saturated transistor decreases, the collector signal is coupled to the cutoff transistor, further turning it on. Once initiated, the circuit flips itself over because of the strong regenerative feedback. Each successive trigger pulse causes the binary to assume its alternate state. The trigger-pulse amplitude must be somewhat greater than the quiescent voltage across R_E in order to initiate the transfer; in the circuit of Fig. 9-23, a positive pulse of 2.5 V is sufficient.

It is of interest to inquire if the binary has any other stable states than the two considered so far. According to the above analysis, it is not possible for both transistors to be cut off simultaneously, for then the base bias conditions on one transistor would be inappropriate. Similarly, both transistors cannot be saturated at the same time. A state in which equal, but not saturation, currents are in each transistor is possible in principle, but this is a state of unstable equilibrium. Any slight departure from exact balance upsets the equilibrium and the circuit reverts to one of the stable states. Suppose, for example, that a random-noise voltage decreases the current in V_1 slightly. The signal at the collector of V_1 tends to increase the current in V_2, and the collector signal of V_2 further reduces the current in V_1. This regenerative action continues until V_1 is cut off and V_2 is saturated. We conclude that the only two stable states of a binary are with one transistor cut off and the other saturated, and vice versa. Note that minor noise voltages cannot cause a transition out of one of the stable states because the transistors are cut off and saturated. Therefore, the amplifier gain is very small and regeneration is not effective.

The speed with which a binary makes the transition from one state to the other limits the trigger pulse rate for reliable switching. Stray shunt capacitances reduce the transition speed, and the same considerations that apply to the high-frequency response of amplifiers are important in binary circuits. This is not surprising since during the transition the circuit really operates as a feedback amplifier. In particular, the total shunt capacitance from base to ground is the limiting factor. This is so because this capacitance must be charged through the relatively high resistance of R_1 in order to change the base voltage corresponding to one state to that of the other state. The time required to charge the input capacitance is reduced by shunting both R_1 resistors with small capacitances C_1, called *commutating capacitors*. When commutating capacitors are used, the time between states is reduced to essentially that required for C_1 to charge through the parallel combination R_1 and R_2. Thus, the resolving time of the binary in Fig. 9-23 is

$$t = \frac{C_1}{1/R_1 + 1/R_2} = 220 \times 10^{-12} \frac{42 \times 33}{75} \times 10^3 = 4.1 \ \mu s$$

This means that the binary can respond reliably to sequential trigger pulses only if they are separated by about 5 μs. With special design and additional circuit complications binaries having resolving times of the order of 0.01 μs are possible.

The versatility of a binary is enhanced by the variety of inputs which may cause a change of state. A particularly flexible and popular arrangement, Fig. 9-25, employs three separate inputs, R, S, and T, and four diodes to steer input pulses to the desired transistor. Suppose, for example, input S is at ground potential, input R is at $+5$ V, and a positive pulse is applied to terminal T. The

Figure 9-25 RS/T flip-flop.

leading and trailing edges of the pulse are differentiated by the 47-pF capacitors. The positive leading edge is blocked by both D_1 and D_2 and shunted to ground by D_3; thus, nothing happens. The negative trailing edge is passed by D_2 and turns transistor V_2 off. The negative pulse is not transmitted through D_1 because of the positive potential at R. The result is that V_2 is cutoff and V_1 is conducting and the binary is said to be *set*.

If the R and S inputs are reversed so that R is at ground potential and S is at $+5$ V, a positive pulse applied to terminal T results in the state in which V_1 is cutoff and V_2 is conducting. The binary is then said to be *reset*, or *cleared*. Note that a positive potential at S causes the binary to be set by an input pulse while a positive potential at R causes it to be reset. In effect these control potentials steer input pulses at T to the corresponding transistor and turn it off. The transitions between states are timed or "clocked" by the signal at T. Therefore, the T input is sometimes referred to as the *clock* connection; the entire configuration in Fig. 9-25 is known as an RS/T flip-flop.

For many applications it is desired that the binary change state upon successive input pulses at terminal T. This is easily accomplished by connecting S to output terminal Q and R to output terminal $\bar{Q}$, Fig. 9-25. Suppose, for example that V_2 is conducting and V_1 is cut off so that S is essentially at ground potential while R is essentially at $+5$ V. According to the preceding discussion, the negative trailing edge of a positive pulse applied to T will be steered to V_2, turning it off so that the binary changes state. Now S is at $+5$ V and R is grounded so that the next pulse is steered to V_1 and the original state is restored. This action is much like that of a mechanical toggle switch and for this reason T is also called the *toggle* input.

Figure 9-26 Schmitt trigger uses emitter feedback.

A useful binary circuit called the *Schmitt trigger* (after its inventor) results if one collector-to-base coupling is omitted and replaced by feedback through a common-emitter resistor, Fig. 9-26. This circuit has two stable states and the magnitude of the input voltage determines which of the two is possible. Suppose the input voltage is zero and so that V_1 is cut off while V_2 is conducting but not saturated. Since V_1 is cut off, the quiescent current of V_2 is found by determining the operating point. In the circuit of Fig. 9-26, for example, the emitter bias is 6.6 V.

Suppose now the input voltage in Fig. 9-26 is increased from zero. Nothing happens until the input voltage just exceeds the base bias, 6.6 V, at which point V_1 begins to conduct. The bias on V_2 is reduced and the emitter bias voltage decreases; this, in turn, increases current in V_1. Regenerative action continues until V_1 is conducting and V_2 is cut off. This state is maintained so long as the input voltage is greater than 6.6 V. As the input voltage is decreased, the circuit does not regain its original state until an input voltage significantly smaller than 6.6 V is attained. The reason for this is that the base bias on V_2 is much lower in this state than previously because the collector of V_1 is at a lower potential. Therefore the current in V_1 must be considerably reduced before the smaller emitter voltage and increased collector potential of V_1 combine to increase the base current of V_2 and it begins to turn on. In the circuit of Fig. 9-26 it is necessary to reduce the input voltage to 5.2 V before regeneration returns the circuit to the original state.

An interesting application of the Schmitt trigger is as a pulse regenerator or *squaring circuit*, illustrated in Fig. 9-27. The input signal, a series of degraded pulse waveforms, is converted to a square pulse output as the circuit is triggered back and forth between its two states by the input waveform. Note that even a sinusoidal signal may be converted to a square wave in this way.

Figure 9-27 Waveforms in Schmitt trigger used as squaring circuit.

The Schmitt trigger is also useful as a *pulse-height discriminator* to measure the voltage pulse amplitudes of an input signal. Each time an input pulse exceeds the trigger threshold the circuit generates an output pulse. Thus, by varying, say, a dc voltage in series with the input signal, the range of pulse sizes in a signal can be determined. A precision of the order of 0.1 V is possible with this circuit. Note that the output pulses are all of uniform amplitude, independent of the input trigger pulse. This is useful in that subsequent circuits need not be capable of handling a wide range of pulse sizes.

A final form of multivibrator is the partial combination of a binary and an astable multivibrator. Both dc and ac coupling are employed, Fig. 9-28. As might be anticipated, this circuit has only one stable state: with V_2 cut off and V_1 saturated. When triggered out of this state by a negative pulse applied to the base of V_1, V_2 becomes conducting and V_1 is cut off. This state lasts for a time determined by the $R_1 C_1$ time constant, after which the circuit reverts spontaneously to the stable configuration. Because of this action the circuit is called a *one-shot* or *monostable* multivibrator. This completes the roster of multivibrator

Figure 9-28 Monostable multivibrator uses both ac and dc coupling between stages.

Figure 9-29 Operational amplifier (a) astable, (b) monostable, (c) bistable multivibrators.

types: the binary, which has two stable states and is therefore bistable; the monostable multivibrator with its one stable state; and the astable circuit which is a free-running oscillator having no stable states.

The versatile operational amplifier makes an effective multivibrator and exceedingly simple circuits are possible, particularly when integrated-circuit operational amplifiers are used. Analysis of an astable multivibrator, Fig. 9-29a, begins by assuming the output signal is positive. Feedback to the noninverting input is large enough to assure that the output is saturated. The capacitor C charges positively through R until the positive potential at the inverting input exceeds the feedback signal at the noninverting input. This net positive input signal causes the output to go negative and, again, the feedback signal drives the amplifier into saturation. The capacitor now begins to charge negatively until the signal at the inverting input is more negative than that at the noninverting input. Now the net negative input signal drives the output positive and the cycle repeats. The result is a square-wave output signal, the frequency of which is determined by the RC time constant.

Shunting the capacitor with a diode yields a monostable multivibrator, Fig. 9-29b. When the output is saturated positively, the capacitor cannot charge to a potential greater than the forward voltage drop of the diode, say, 0.6 V. The circuit rests stably in this condition until a negative input signal applied to the noninverting input causes the output to go negative. Feedback drives the amplifier into saturation, and the capacitor charges negatively through R. Since the diode is reverse-biased, the voltage at the inverting terminal eventually exceeds the feedback signal and the net negative input signal drives the output positive again. Thus the circuit remains at positive saturation until driven negatively by an input signal and the duration of the negative state is determined by the RC time constant.

Feedback to the noninverting input in the bistable multivibrator, Fig. 9-29c, holds the circuit either saturated positively or saturated negatively. Suppose the output is positive; then a positive pulse applied to the inverting input causes the output to go negative and saturate. Only a subsequent negative input pulse can return the circuit to the original condition. Thus, the circuit has two stable states which are determined by the polarity of the input signal.

WAVEFORM GENERATORS

Nonsinusoidal waveforms can be generated by relaxation oscillators and multivibrator circuits. The output signals from these generators can be further modified through the use of diode clippers and clamps to select a portion or set the dc level of a waveform. Also, integrating and differentiating circuits can be used to modify waveforms in specific ways. A number of additional waveform generators are described in this section.

Figure 9-30 Diode pump.

Diode Pump

Consider the action of the so-called *diode-pump* circuit, Fig. 9-30, in response to negative-going signal pulses. On each negative pulse, diode D_1 conducts and C_1 charges to the peak value of the input pulse. Since the input pulse biases D_2 in reverse, no charge reaches C_2. Thus the charge on C_1 is

$$Q_1 = C_1 v_i \qquad (9\text{-}35)$$

During the time that the input voltage is zero, the voltage on C_1 biases D_2 in the forward direction and C_2 becomes charged. Since the two capacitors are now effectively connected in parallel, the charge on C_2 is

$$Q_2 = C_2 v_o = C_2 \frac{Q_1}{C_1 + C_2} = \frac{Q_1}{1 + C_1/C_2} \qquad (9\text{-}36)$$

If $C_2 \gg C_1$, $Q_2 = Q_1$ and the charge has effectively been pumped from the source to C_1 and thence to C_2. Note that under this condition C_1 is essentially discharged and the process can be repeated on the next cycle.

The output voltage after the first cycle is

$$v_o = \frac{Q_2}{C_2} = \frac{v_i}{1 + C_2/C_1} \cong \frac{C_1}{C_2} v_i \qquad (9\text{-}37)$$

Figure 9-31 Staircase waveform produced by diode pump when input is negative square wave.

where Eqs. (9-35) and (9-36) have been used. On each subsequent cycle a voltage increment given by Eq. (9-37) appears across C_2 as long as the total voltage remains small compared with the input pulse amplitude. The result is a *staircase* waveform, Fig. 9-31. In most applications the voltage across C_2 is rapidly discharged by an auxiliary circuit after a finite number of steps and the staircase waveform then repeats.

Ramps

A voltage that increases linearly with time, called a *ramp*, is widely used for the horizontal sweep voltage in oscilloscopes. Recurring ramps such as those produced by relaxation oscillators are referred to as *sawtooth* waves. Other ramp generators are designed to produce a single ramp waveform when triggered by an external signal.

Figure 9-32 is the circuit diagram of a ramp generator called the *Miller sweep*, which yields a linear ramp with a peak amplitude nearly equal to the dc supply potential. The quiescent condition has the trigger transistor Q_1 saturated and the Miller sweep transistor Q_2 turned off so that the output potential is equal to V_{cc} and the capacitor C is fully charged.

A negative input pulse turns Q_1 off and allows Q_2 to operate as an amplifier. Thus C charges through R_1 and Q_2 and the output voltage decreases toward zero. The decrease is linear because of the feedback between collector and base provided by the capacitor. At the end of the input pulse Q_1 becomes saturated, Q_2 is turned off, and the capacitor charges rapidly to V_{cc} through R_2. The circuit is then ready for another input pulse.

Figure 9-32 Miller sweep ramp generator.

Figure 9-33 Block diagram and typical waveforms of triggered-sweep circuit used in oscilloscopes.

Actually the Miller sweep circuit may be thought of as an operational-amplifier integrator, Fig. 8-13, in which Q_2 is the amplifier. The input signal is the constant voltage V_{cc}, which is applied through R_1. Since the integral of a constant value increases linearly, the output signal is a ramp. The function of Q_1 is to start and stop the integration interval.

Circuits similar to the Miller sweep are often used in *triggered-sweep* oscilloscopes, which are particularly useful for examining waveforms of transient signals. A typical block diagram of the sweep circuit for such an oscilloscope, Fig. 9-33, starts with a Schmitt trigger activated by the input signal applied to the vertical deflection amplifier. The output of the Schmitt trigger is differentiated and clipped to produce a sharp negative pulse suitable for triggering a monostable multivibrator. The multivibrator generates a square pulse input to the Miller integrator. The output ramp is clamped at a voltage level such that the sweep starts at the desired place on the oscilloscope screen.

A number of adjustable controls are included to increase the flexibility of the circuit. For example, bias on the Schmitt trigger can be adjusted to permit the circuit to trigger on input signals of different amplitudes (*trigger amplitude*, in Fig. 9-33). Similarly, the bias of the multivibrator is variable (*sweep stability*) so that it may also be operated as a synchronized astable multivibrator to produce recurring sweeps. The sweep speed is altered by changing the charging capacitor (and/or resistor) in the Miller sweep circuit. At the same time, it is necessary to change the time constant of the multivibrator (by switching capacitors) so that the sweep duration is commensurate with the sweep speed.

Pulses

Pulse waveforms of endless variety can be generated using combinations of multivibrator circuits. Consider the two cascaded monostable multivibrators in Fig. 9-34. A negative trigger pulse causes a positive output pulse with a duration T_1. The negative trailing edge of this pulse triggers a second monostable multivibrator and the output is a square pulse of duration T_2. Thus, this combination generates one standard-shaped output pulse delayed by a specified interval following each input trigger pulse.

Figure 9-34 Cascaded monostable multivibrators.

A simple binary triggered by a repeating waveform produces an output square wave with half-cycles as precisely regular as the frequency of the input signal. This is so because the binary changes state on every negative swing of the input wave and the precision of the square-wave output depends only upon the precision of the successive negative swings of the input signal.

Binary triggering may be accomplished by pulses fed directly to only one or the other transistor base. In this case a negative pulse induces a transition only when the transistor is conducting. Such *unsymmetrical* triggering is useful when a binary is triggered from two separate sources, one applied to each base. The output of this binary is a square pulse with a duration equal to the time interval between input pulses.

SUGGESTIONS FOR FURTHER READING

John Markus: "Electronic Circuits Manual," McGraw-Hill Book Company, New York, 1971.
Joseph A. Walston and John R. Miller (eds.): "Transistor Circuit Design," McGraw-Hill Book Company, New York, 1963.
"The Radio Amateur's Handbook" (published annually by the American Radio Relay League, West Hartford, Conn.).

EXERCISES

9-1 Derive Eq. (9-3) for the feedback network of a phase-shift oscillator. *Hint:* Use the technique of loop currents and solve for the current in the last resistor by determinants.

9-2 Draw the h-parameter equivalent circuit of the phase-shift oscillator, Fig. 9-1, and derive an expression for the feedback ratio in terms of the h parameters and the input impedance of the feedback network, z_i.

$Ans.:$ $\beta = h_{re} - [(\mathbf{Z}_i + h_{ie})/h_{fe}](h_{oe} + 1/R_L + 1/\mathbf{Z}_i)$

9-3 Using the results of Exercise 9-2, show that the circuit of Fig. 9-1 does oscillate. The h parameters of the 2N1414 are $h_{ie} = 1260 \ \Omega$, $h_{re} = 3 \times 10^{-4}$, $h_{fe} = 60$, and $h_{oe} = 2.7 \times 10^{-5}$ mho. What is the frequency of oscillation?

$Ans.:$ 1.28×10^4 Hz

9-4 Plot the signal voltage at terminal 2 and the reciprocal of the feedback ratio as a function of the output signal for the integrated-circuit Wien bridge oscillator in Fig. 9-2. What is the stabilized amplitude of the output signal?

Ans.: 1.7 V peak-to-peak

9-5 By inspection, determine the voltage output of the power-oscillator circuit in Fig. 9-7.

Ans.: 19.8 V

9-6 Develop an expression analogous to Eq. (9-10) for the phase of the feedback voltage as a function of frequency in a tickler *LC* oscillator. Show that a resonant circuit with a large *Q* is desirable for maximum frequency stability.

Ans.: tan $\theta = Q_o(\omega_o/\omega - \omega/\omega_o)$

9-7 Obtain an expression for the phase angle of the impedance of a quartz crystal near the parallel resonant frequency. Plot the phase angle as a function of frequency and compare with a plot of an *LC* resonant circuit having a *Q* of 200. Use the results of Exercise 9-6.

9-8 Find the period and amplitude of the output signal from the SCR relaxation oscillator, Fig. 9-19, if $h_{fe} = 20$ and the gate voltage required for turn-on is 3 V. Plot the output voltage waveform.

Ans.: 0.3 V, 17 V, 1.18×10^4 Hz, 6.67×10^5 Hz

9-9 Astable multivibrators are often used as light flashers to mark road barricades, construction work, etc. Qualitatively describe the operation of a typical circuit, Fig. 9-35, starting from the time when V_1 and V_2 are off and V_1 is starting to turn on.

Figure 9-35 Light-flasher circuit analyzed in Exercise 9-9.

9-10 Sketch the waveform at the output of the circuit in Fig. 9-36 with the 50-Ω potentiometer at the center. Repeat for the potentiometer set one-fourth of the way from the $+20$-V terminal. Assume the 2N491 characteristics are the same as those of the 2N2646 in Fig. 9-16.

Figure 9-36

TEN

ANALOG MEASUREMENTS

Electrical techniques are used to measure a wide variety of physical phenomena in both laboratory research and industrial control applications. Indeed, it is modern measurement practice to develop an electrical signal analogous to the phenomena of interest at the earliest opportunity so that subsequent processing can be carried out electronically. The great versatility of electronic circuits to amplify, modify, record, and detect electric signals is the principal reason they enjoy such widespread use. Circuits used to measure and process such analog signals and circuits for control purposes are investigated in this chapter. In addition, techniques for transporting signals from one point to another and recording them for subsequent analysis are considered.

CONTROL CIRCUITS

Voltage Regulators

Special circuits are widely used to stabilize the output voltage of a rectifier power supply to a greater extent than is possible with the simple zener diode circuit described in Chap. 4. Such a *voltage regulator* is an important and useful example of measurement and control with electronic circuits. The circuit measures the output voltage, compares it with the desired value, and adjusts conditions so that the difference between the two is zero. In this way control over the dc output voltage is maintained in the face of changes in load and also in ac line voltage.

The output voltage of a power supply is conveniently controlled by introducing a power transistor in series with the rectifier-filter and the load, Fig. 10-1. It is useful to analyze this circuit as a feedback amplifier in which the reference voltage V is the input voltage and V_o is the output. Since the feedback ratio is $-R_2/(R_1 + R_2)$, the output voltage is controlled by the amplifier and power transistor to be, from Eq. (8-3),

$$V_o = -\frac{R_1 + R_2}{R_2} V$$

In effect, the voltage regulator compares a fraction of the dc output voltage with the reference voltage and adjusts the control transistor to maintain the difference equal to zero; hence V_o is constant.

The stabilizing effect of the regulator is best determined by analyzing the control transistor separately from the dc amplifier. Variations in the supply voltage are assumed to arise from an ac generator v_i in Fig. 10-1 and result in a variation v_o in the output voltage. The control transistor is essentially an emitter-follower amplifier and the appropriate equivalent circuit is shown in Fig. 10-2.

Figure 10-1 Series voltage regulator.

Figure 10-2 Equivalent circuit of control transistor in Fig. 10-1.

Analysis begins by writing an expression for the output current. Considering the currents at point e and neglecting the minor current drain in the feedback divider $R_1 R_2$,

$$h_{fc}i_b + i_L + \frac{v_o - v_i}{1/h_{oc}} = 0 \tag{10-1}$$

where the third term is the current in $1/h_{oc}$. The base current is found by considering the voltages in the input circuit,

$$i_b = \frac{v_{bc} - h_{rc}v_{ec}}{h_{ic}} = -\frac{a\beta v_o + v_i + h_{rc}v_{ec}}{h_{ic}} \tag{10-2}$$

where the input voltage v_{bc} is v_i less the amplified feedback voltage $-a\beta v_o$, as may be best seen from Fig. 10-1. Finally, v_{ec} is determined by writing Kirchhoff's rule around the output circuit,

$$v_{ec} - v_o + v_i = 0 \tag{10-3}$$

Equations (10-2) and (10-3) are substituted into Eq. (10-1),

$$i_L = v_i \left[h_{oc} + \frac{h_{fc}(1 - h_{rc})}{h_{ic}} \right] - v_o \left[h_{oc} - \frac{h_{fc}(h_{rc} + a)}{h_{ic}} \right]$$

Solving for the output voltage,

$$v_o = v_i \frac{h_{ic}h_{oc} + h_{fc}(1 - h_{rc})}{h_{ic}h_{oc} - h_{fc}(h_{rc} + a\beta)} - i_L \frac{h_{ic}}{h_{ic}h_{oc} - h_{fc}(h_{rc} + a\beta)} \tag{10-4}$$

This expression has the form of the Thévenin equivalent of the regulated power supply in that the first term is the internal voltage generator and the second term is the voltage drop resulting from the load current in the equivalent internal resistance.

Before considering Eq. (10-4) further, it is useful to simplify the expression by considering the relative magnitudes of the h parameters. When this is done, Eq. (10-4) reduces to

$$v_o = v_i \frac{h_{ic}}{r_c(1 + a\beta)} - i_L \frac{h_{ic}(1 - \alpha)}{1 + a\beta} \tag{10-5}$$

which is sufficiently accurate for all practical purposes. According to Eq. (10-5) input voltage variations are reduced by a large factor through the feedback action of the regulator. Typical values of the power-transistor parameters in Eq. (10-5) are $h_{ic} = 1000\ \Omega$, $r_c = 10,000\ \Omega$, $a = 50$, and $\beta = 0.3$, so that input voltage changes are reduced by a factor of 160 at the output terminals.

Additionally, the effective internal impedance of the regulated power supply, which is the coefficient of i_L in Eq. (10-5), is very small. Taking $1 - \alpha = 0.05$, the internal resistance is only 3.1 Ω. This means that changes in load current cause only minor changes in the regulated supply voltage. It is apparent that this simple regulator is effective against both input-voltage and output-current variations.

Note that the factor by which input voltage variations are reduced at the output terminals applies to ripple voltages as well as to other voltage changes. Therefore output voltage ripple is very small in a regulated power supply. Both the reduction factor and the effective internal resistance can be improved by increasing the gain of the dc amplifier, according to Eq. (10-5). In many critical applications the gain is made large enough that residual variations in the regulated voltage reflect the stability of the reference potential.

The series control transistor must be capable of dissipating the heat generated by the entire load current without overheating. Power transistors are often used, and it is feasible to employ two or more identical units connected in parallel, if this is necessary to carry the maximum load current.

Although the series regulator is most popular, it is also possible to connect the control transistor in parallel with the load, much as in the case of a simple zener-diode regulator, Fig. 10-3. The current through the transistor is controlled so that the voltage drop across the series resistor is the same independent of load-current changes. Effective voltage stabilization is obtained by such a *shunt regulator*, but the circuit suffers from the additional power loss in the series resistor. On the other hand, it is only necessary for the control transistor to carry a fraction of the full-load current, which is a considerable advantage in high-current power supplies.

A complete regulated power supply is illustrated in Fig. 10-4. The dc amplifier is a single grounded-emitter transistor stage and the reference voltage is provided by an 18-V zener diode. The zener diode is placed in the emitter circuit of the amplifier, rather than in series with the feedback signal, so that one terminal of the

Figure 10-3 Shunt regulator.

Figure 10-4 Complete 40-V 500-mA regulated power supply.

diode can be grounded. The 33-kΩ resistor provides reverse potential for the diode. Note that the feedback ratio is adjustable. This enables the output voltage to be set at any desired value, within limits dictated by the operating conditions of the transistors. In this circuit, the output voltage can be adjusted over the range 40 to 50 V, while maintaining good regulation. The 1500-μF capacitor across the output terminals provides a very low effective internal impedance for ac signals.

Very precisely regulated voltages are obtained with more elaborate versions of this circuit. The gain is increased by using a multistage dc amplifier. In particular, operational amplifiers prove convenient because of their large gain and dual inputs, which facilitate comparing the output voltage with the reference voltage. Such circuits can hold the output voltage to within 0.1 percent over the entire range of load current. In fact, stabilities as great as one part in 10^5 are achieved in special applications.

The separate components of an entire voltage regulator can be conveniently incorporated into a single integrated circuit, Fig. 10-5a. In this circuit the voltage drop across R_s, the so-called *shut-down resistor*, acts to open the circuit when the current capability of the regulator is exceeded, as, for example, by an inadvertent short circuit at the output terminals. Thus the regulator is self-protected against current overloads. A separate control transistor, Fig. 10-5b, may be included to increase the current capacity. In this fashion, the power transistor is effectively placed in the Darlington connection with the control transistor internal to the integrated circuit to achieve enhanced current gain. It proves convenient to mount the external control transistor on a heat sink to increase power dissipation.

Figure 10-5 (a) Integrated-circuit voltage regulator and, (b) with power transistor to increase current capability.

Servos

It is often useful to control mechanical position or motion in accordance with an electric signal. To assure that the desired motion takes place, a voltage feedback signal corresponding to the mechanical motion is developed. This voltage is compared with the input signal. When the two are equal, mechanical motion ceases since the mechanism has responded to the actuating signal. Feedback systems incorporating a mechanical link in the feedback network are called *servomechanisms*, or *servos*, from the Latin word for "slave," since the mechanical

Figure 10-6 Block diagram of servo system.

motion is the slave of the input signal. Servos are widely used in control applications, from laboratory chart recorders to automatic navigation systems.

Consider the servo system sketched in block-diagram form in Fig. 10-6. The amplifier drives a mechanical actuator as, for example, a dc motor. Suppose the motor moves a mechanical arm to a position x and that the feedback voltage is proportional to the position of the arm and equal to kx. So long as the *error signal*, which is the difference between the input and feedback voltages, is not equal to zero, the amplified error signal continues to drive the motor. When the feedback signal equals the actuating signal, the error is zero, and motion ceases. Thus

$$v_i - v_f = v_i - kx_0 = 0 \tag{10-6}$$

so that

$$x_0 = \frac{v_i}{k} \tag{10-7}$$

According to Eq. (10-7), the equilibrium position of the mechanical motion x_0 is directly proportional to the input signal.

The servo cannot respond instantaneously to rapid changes in the input signal because of mechanical inertia and friction. According to the above discussion, the motive force from the motor is proportional to the error signal. This force is opposed by the inertial and frictional forces, and if the frictional force is small compared with the driving force, a damped oscillation results, Fig. 10-7. On the other hand, if the frictional force is large, the system is *overdamped* and requires an excessively long time to reach the equilibrium position. Most servo systems are designed so they are slightly underdamped because this minimizes the response time. This results in a small *overshoot*, Fig. 10-7. If desirable, additional damping may be introduced by including an *RC* filter in the input circuit. In this way the servo is never subjected to a signal change faster than the time constant of the filter.

A common laboratory instrument based on servo principles is the *recording potentiometer*. This is a self-balancing dc potentiometer in which the balance

Figure 10-7 Characteristic responses of servo to sudden input signal.

position is marked on a moving paper strip or chart producing a continuous record of the input voltage. The heart of the system, Fig. 10-8, is the slide wire R_s, in which there is a calibrated current supplied by the battery V_s and variable resistor R_A. The current is standardized by comparison with a standard battery, as described in Chap. 1. If less accuracy is satisfactory, V_s may be a mercury battery which supplies a relatively constant current.

The potentiometer in parallel with the slide-wire resistance forms a bridge circuit which is balanced when, from Eq. (1-46), $R_{C_1}/R_{C_2} = R_{S_1}/R_{S_2}$. Under this condition the feedback voltage v_f is zero. In the absence of an input signal, the

Figure 10-8 Recording-potentiometer servo.

adjustable contact on the slide wire is moved by the servomotor until balance is achieved. The recording pen mechanically attached to the slide-wire contact continuously records the balance position on the chart. This means that the zero position ($v_i = 0$) may be placed anywhere on the chart by adjusting R_C.

When an input voltage is applied, an initial signal $v_i - v_f$ is fed to the amplifier. The servomotor drives the slide-wire contact to a new equilibrium position where the unbalanced bridge voltage equals the input voltage, and $v_i - v_f = 0$. The displacement of the sliding contact is a linear function of the input voltage, as can be shown in the following way. Using Eq. (1-54), the unbalance voltage from the bridge is

$$v_f = V\left(\frac{R_{C_1}}{R_{C_1} + R_{C_2}} - \frac{R_{S_1}}{R_{S_1} + R_{S_2}}\right) \tag{10-8}$$

where V is the voltage across the bridge. If the slide wire is uniform, and the zero position, $x = 0$, is taken to be at the center,

$$R_{S_1} = \frac{R_S}{2} - mx \quad \text{and} \quad R_{S_2} = \frac{R_S}{2} + mx \tag{10-9}$$

where m is a constant of the slide-wire. Introducing Eq. (10-9) into Eq. (10-8) and noting that $R_{C_1} = R_{C_2}$ since the zero position is at the center,

$$v_f = V\left(\frac{1}{2} - \frac{R_S/2 - mx}{R_S/2 - mx + R_S/2 + mx}\right) = \frac{V}{R_S}mx \tag{10-10}$$

Therefore, according to Eq. (10-7),

$$x_0 = \frac{1}{mI_S}v_i \tag{10-11}$$

where $I_S = V/R_S$ is the standardized current in the slide wire. Equation (10-11) shows that the deflection of the slider, and hence of the recording pen, is proportional to the input signal. It is common practice to select m and the standard current I_S such that the full chart width is an integral voltage such as 100 mV.

Resistors R_1 through R_4 in Fig. 10-8 constitute a voltage divider for changing the sensitivity of the potentiometer. When the divider is in the circuit, only a fraction of the unbalance voltage is available as the feedback signal and the sensitivity is increased correspondingly. The combination of R_d and C_d is a damping filter at the input to optimize the response time of the recorder.

The recording potentiometer is a very versatile instrument in the laboratory. It combines high sensitivity for dc signals and the inherently high input impedance of a potentiometer circuit with automatic operation and a permanent record. It is widely used as part of other instruments, such as optical spectrometers, x-ray diffractometers, and gas chromatographs, or as a separate unit.

TRANSDUCERS

Devices that produce electric signals in accordance with changes in some physical effect are called *transducers*. A typical example is a microphone which generates a voltage waveform corresponding to the variations of sound pressure impinging on a diaphragm. Other transducers, such as the strain gauge, operate by changing some circuit parameter (e.g., resistance) rather than by generating a voltage. A great variety of transducers have been developed so that it is often possible to generate a signal in more than one way. The deciding factor in any given situation is the accuracy with which the transducer output represents the physical parameter being measured.

Mechanical Transducers

A useful transducer for sensing mechanical motion or position is the *differential transformer*, Fig. 10-9. This device is a transformer having two secondaries and a movable magnetic core. A sinusoidal current in the primary produces an equal voltage in each secondary when the core is positioned midway between them. Since the windings are connected in series opposition, the output voltage is zero. Displacing the core to the left causes more magnetic flux to link the left-hand secondary, and the output voltage rises. Similarly, a displacement to the right produces an output signal with opposite phase, because the voltage of the right-hand secondary exceeds that of the left.

    With proper mechanical design the output voltage can be a linear function of the displacement over a reasonable distance, say, ~ 0.1 in. A phase-sensitive detector is commonly used to sense the direction of the displacement, where this is necessary. Alternatively, if motion in only one direction need be measured, a simple ac voltmeter can be used.

Figure 10-9 Cross-sectional sketch of differential transformer.

Diaphram

Sound

Pressure

Figure 10-10 Sketch of condenser microphone.

Another position-sensitive transducer measures the separation between a movable and a fixed plate by their mutual capacitance. Variations in capacitance accompanying a change in the plate separation are detected with a capacitance bridge. Alternatively, the capacitor can be made part of the resonant circuit of an oscillator. Displacements of one plate relative to the other are reflected in frequency changes of the oscillator output. If it is only necessary to detect departures from an equilibrium position, simply connecting the capacitor in series with a resistor and a dc voltage is sufficient, as in the *condenser microphone*, Fig. 10-10. Variations in sound pressure cause the diaphragm to deflect, thus changing the capacitance C, and generating a current in the external circuit,

$$i = \frac{dQ}{dt} = V\frac{dC}{dt}$$

The output signal is

$$v_o = RV\frac{dC}{dt} \tag{10-12}$$

The capacitance transducer can be adapted to measure mechanical force, as in Fig. 10-11. Here the position of an arm to which the force is applied is detected by

Figure 10-11 Capacitance transducer used in feedback system to measure mechanical force.

Figure 10-12 Resistance strain gauge.

the capacitance between the arm and the pole face of an electromagnet. Any applied force displaces the arm, and the change in capacitance is detected by a capacitance bridge. The unbalance voltage of the bridge provides a signal for an amplifier feeding the coils of the electromagnet. The magnetic force pulls the arm back to the equilibrium position against the external force. Thus, the magnet current is a direct measure of the applied force.

A system that strives to return to the equilibrium, or *null*, position is very commonly used in measurement and control circuits. Two comments are worthy of note. First, the system is basically a feedback network and may be analyzed as such. Secondly, the system always returns to its equilibrium position so it is not necessary for the transducer to have a linear relation between input action and output signal. As a matter of fact, it is not even essential that the transducer calibration be known, since it is not deflected in the equilibrium position.

Mechanical stress in structural members and materials is conventionally measured by means of a resistance *strain gauge*. One common strain gauge is a 1-mil metal wire bent back and forth and fastened to a thin paper backing, as in Fig. 10-12. The gauge is cemented directly to the surface of the structural member under test so that changes in length are detected by changes in the resistance of the gauge. The magnitude of the change in resistance can be determined from

$$R = \rho \frac{L}{A} \tag{10-13}$$

where L is the total length, A is the cross-sectional area, and ρ is the resistivity of the wire.

A measure of the gauge sensitivity is the relative change in resistance per relative change in length, or the *gauge factor*, $K = (\Delta R/R)/(\Delta L/L)$. The gauge factor is given by

$$K = 1 + 2\sigma + L\frac{d\rho}{dL} \tag{10-14}$$

where σ is Poisson's ratio for the wire material. According to Eq. (10-14) the change in resistance arises from changes in the wire dimensions because of mechanical strain plus the possibility of changes in resistivity.

In metal-wire gauges, the third term in Eq. (10-14) is negligible and $K \cong 2$. The gauge resistance may be approximately 100 Ω and the strains of interest are of the order of 10^{-3}, so that the total resistance change is about 0.2 Ω. *Silicon strain gauges* have much larger gauge factors, of the order of 150, as a result of change in resistivity with strain. The reason for this is that the number of current carriers in a semiconductor is sensitive to mechanical strain. Therefore, the conductivity changes significantly in response to elastic deformation. Because of their increased sensitivity, silicon strain gauges are becoming ever more popular.

Semiconductor strain gauges are sensitive to temperature changes, however, because of the exponential change in resistivity with temperature characteristic of semiconductors. Actually, temperature effects are important in wire strain gauges as well. Although the change in resistivity of a metal with temperature is quite small, the changes caused by mechanical strains are small as well. Bridge circuits are universally used to measure the change in resistance of strain gauges because small resistance changes can be measured and temperature compensation is easily effected. Most often, two identical strain gauges are used in the bridge, Fig. 10-13. One is subjected to the mechanical strain and the other is isolated from the strain but positioned so that its temperature environment is the same as that of the strain detector. Thus, changes in temperature influence each gauge equally, and the bridge remains balanced.

For maximum bridge sensitivity $R_1 = R_2 = R$, the output voltage is zero, and the voltage across resistance R_1 is $V/2$ in the unstrained condition. The gauge resistance becomes $R + \Delta R$ in the presence of strain and the output voltage signal resulting from strain is

$$v_o = \frac{V(R + \Delta R)}{2R + \Delta R} - \frac{V}{2} = \frac{\Delta R}{4R + 2\,\Delta R}\,V \qquad (10\text{-}15)$$

Figure 10-13 Strain-gauge bridge circuit uses second gauge for temperature compensation.

Neglecting ΔR in comparison with $2R$, and introducing the gauge factor,

$$v_o = \frac{KV}{4}\frac{\Delta L}{L} \qquad (10\text{-}16)$$

which shows that the output signal is a direct measure of the mechanical strain.

Since $\Delta L/L \cong 10^{-3}$, the output signal is of the order of 10 mV or less, so that usually some amplification is necessary. The bridge can be excited with an oscillator so that an ac-coupled amplifier can be used to avoid the drift problems of a dc amplifier. Synchronous detection provides low-noise operation and also detects both positive and negative strains.

Mechanical vibrations in structures and machines are determined with transducers called *accelerometers*. One form of accelerometer measures the forces on a small mass caused by accelerations of the object to which the transducer is fastened. The inertial force $F = ma$ is counterbalanced by the restoring force of a spring supporting the mass and the deflections are detected by the voltage generated in a piezoelectric crystalline material such as rochelle salts or barium titanate. As discussed in Chap. 9, piezoelectric materials generate a voltage proportional to the mechanical strain on the crystal. Actually, as illustrated in Fig. 10-14, the piezoelectric element is formed from two parallel slabs such that, for example, a vertical deflection of the mass puts the upper slab in compression and the lower slab in tension. The output terminals on the flat faces of the two slabs are connected in parallel. The result of this *bimorph* construction is a more sensitive transducer than can be obtained with only a single element.

The useful frequency range of an accelerometer ranges up to the mechanical resonant frequency of the mass and the supporting spring combination. An upper frequency limit of 2 kHz is typical. Extended performance can be obtained, at a sacrifice in sensitivity, by eliminating the mass and spring altogether. In this case the mass and elastic constant of the bimorph piezoelectric element alone are used and the upper frequency limit is set by the mechanical resonance of the cantilever beam geometry of the element itself.

The output voltage of an accelerometer can be integrated electronically to develop a signal proportional to velocity. A second integration yields the displacement. Such accelerometers form an important part of inertial guidance navigation

Figure 10-14 Elementary piezoelectric accelerometer.

systems in aircraft. Three mutually perpendicular accelerometers are arranged to measure the displacement of the aircraft from a known starting point.

The pH Meter

The acidity or alkalinity of a solution is determined by measuring the concentration of hydrogen ions in the liquid with a *pH meter*. This instrument is based on the potential developed by the solution with respect to a standard electrode, which is given by

$$V = -\frac{RT}{F_y} \ln a_H \qquad (10\text{-}17)$$

where R is the universal gas constant, T is the absolute temperature, F_y is a quantity of electric charge called the faraday, and a_H is the activity of hydrogen ions which is, in effect, equal to their concentration. Actually, the quantity defined by the expression

$$pH = -\log a_H \qquad (10\text{-}18)$$

is used in practice because the numerical values so obtained are more convenient to handle. The pH of a neutral solution, for example, is equal to 7, whereas $a_H = 10^{-7}$, according to Eq. (10-18).

Standard electrode conditions pertaining to Eq. (10-17) are inconvenient in practice, so the usual pH meter employs the electrode scheme depicted in Fig. 10-15. The potential of an unknown liquid is compared with that of a solution of known and constant pH. Electric contact to both liquids is achieved with

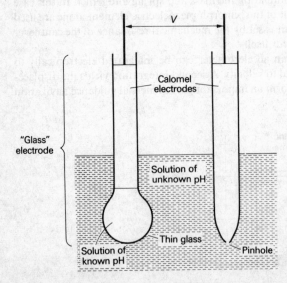

Figure 10-15 Electrode system of a pH meter. Potential V is direct measure of hydrogen-ion concentration in unknown liquid.

standard electrodes called *calomel* electrodes. The electrochemical properties of a calomel electrode (a mixture of Hg, HgCl, and KCl) are very stable and easily reproduced in the laboratory. The solution of known pH is commonly a mixture of Ag, AgCl, and HCl contained in a thin-walled glass vessel. The thin glass wall permits only hydrogen ions in the unknown solution to migrate through the system. Therefore, the potential developed results only from the hydrogen-ion concentration in the unknown even though other ions may be present as well. The combination of a calomel electrode with the standard solution is known commercially as a "glass electrode." Electric connection between the other calomel electrode and the unknown solution is provided by a tiny pinhole, as noted in Fig. 10-15.

In terms of the voltage developed by this standard arrangement, the pH of the unknown solution is, from Eqs. (10-17) and (10-18),

$$pH = \frac{V + 0.106}{0.0592} \tag{10-19}$$

where the constant 0.106 V refers to the calomel-standard-solution potential difference, and the denominator is the quantity RT/F_y evaluated at room temperature. Note that the pH is a linear function of the potential difference between electrodes, and that the potential of a neutral solution is $7 \times 0.0592 - 0.106 \cong 0.3$ V. This potential difference must be measured with a meter having an extremely large input impedance to avoid loading the cell and thereby reducing the terminal voltage. Even the small hydrogen ions migrate through the glass wall only with considerable difficulty and this produces a very large internal resistance. Voltmeters having a suitably large input impedance, called *electrometers*, are described in a later section.

Photocells

Light energy from infrared to ultraviolet wavelengths is detected and measured with various forms of *photocells*. A simple, though extensively used, photocell is a small piece of semiconductor provided with two electric contacts. The resistance of the semiconductor between contacts depends upon the intensity of light energy impinging upon the surface. Absorbed photons produce electrons and holes that act as current carriers in the semiconductor. The response of a semiconductor photocell is zero for photon energies less than the width of the forbidden energy gap because such photons have insufficient energy to promote electrons from the valence band to the conduction band across the forbidden energy gap. The photoresponse is also small at short wavelengths (high photon energies) because photons are so heavily absorbed in the surface layers that most of the semiconductor is inactive. Therefore, the response is a maximum at photon energies corresponding to the width of the forbidden energy gap, as illustrated for the common photoconductor materials CdS, CdSe, and CdTe in Fig. 10-16. Note that CdS photocells are useful in the visible region of the spectrum, CdSe is most sensitive to red light, and CdTe responds to near-infrared radiation. Other semiconductors

Figure 10-16 Wavelength response of Cds, CdSe, and CdTe photocells. Maximum response is obtained at photon energies equal to width of forbidden energy gap.

commonly used as infrared detectors are PbS and PbSe, which are useful at 3.0 microns and 4.5 microns, respectively, corresponding to forbidden energy gaps of 0.37 and 0.27 eV. In addition, InSb, with a forbidden energy gap of only 0.18 eV, responds to radiation at wavelengths as long as 7 microns.

A useful measure of the sensitivity of a photocell at the peak of the wavelength response curve is the *gain G*. The gain is the ratio of the number of current carriers generated by illumination to the number of absorbed photons per second. It turns out that the gain may be expressed as the ratio of the transit time, T, of the current carriers between electrodes to the average time, τ, a photogenerated electron spends in the conduction band before combining with a hole in the valence band, or,

$$G = \frac{\tau}{T} \tag{10-20}$$

According to Eq. (10-20), high sensitivity implies a sluggish response. That is, a large value of gain is achieved by increasing the time constant τ, and τ also determines the speed of response of the photoconductor to sudden changes in light intensity. Where speed is of secondary importance simple photoconductors are widely used because they are quite sensitive. Most often, the change in resistance of the photocell accompanying incident illumination is measured by connecting the photocell in series with a dc voltage source and a load resistor. The voltage drop across the load resistor corresponds to the resistance changes in the semiconductor resulting from photon absorption.

Figure 10-17 Internal structure and typical circuit of photomultiplier tube.

High sensitivity together with rapid response time is achieved in a vacuum-tube photocell called a *photomultiplier* or PM tube which consists of a photosensitive cathode and a number of additional electrodes called *dynodes*, Fig. 10-17. The dynodes are at successively higher voltages so that electrons emitted from the photocathode, or from a preceding dynode, are attracted to the next dynode with sufficient energy to dislodge electrons from the surface. The surfaces of the dynodes are especially processed so that more than one electron is emitted for each incident electron and the number of electrons is multiplied accordingly as they proceed from dynode to dynode. The electrons are finally collected as an output current at the anode. Gains of the order of 10^6 are achieved in PM tubes so that it is possible to detect the emission of only a single electron from the photocathode. Furthermore, the response time depends upon the transit time of the electrons and is correspondingly rapid. Useful response at frequencies of the order of hundreds of megacycles is possible using a PM photocell.

Both photoconductors and PM tubes suffer from the disadvantage of nonuniform wavelength response since photon energies must be sufficient to create electron-hole pairs in the one case and to cause photoelectric emission in the other. This means, in particular, that the response to far-infrared radiation, where photon energies are small, is negligible. *Thermistor bolometer* detectors, which measure the temperature change produced by the incident radiation, have a uniform, though comparatively small, response at all wavelengths. Conventional thermistors (the name comes from a contraction of "thermally sensitive resistors") are composed of an oxidic semiconductor material and have the rapid exponential variation of resistance with temperature that is characteristic of semiconductors. The detector consists of a small thermistor thermally insulated from its surroundings so that incident radiation increases the temperature of the sensitive element alone. Since the temperature increase depends only upon the amount of energy absorbed, independent of the individual photon energy, response is obtained at all wavelengths. Indeed, thermistor bolometers are used even at microwave frequencies.

The change in resistance of a thermistor bolometer is most often measured with a dc bridge similar to the one illustrated for strain gauges in Fig. 10-13. It is common practice to include a second thermistor to compensate for changes in ambient temperature, again as in the case of the strain gauge. Usually this compensating thermistor is mounted in the same housing as the detector element but is shielded from the incident radiation. Although not as sensitive as the other photocells previously discussed, a thermistor bolometer can detect a change in infrared radiation from an object corresponding to a variation in the object's temperature of as little as one degree centigrade. In addition, thermistors are often used as electric thermometers and are capable of measuring temperature differences as small as 10^{-6} °C.

Many semiconductor photocells having rapid response and good sensitivity are based on a reverse-biased *pn* junction. Photons absorbed in the region of the junction create electrons and holes that are swept to the opposite sides of the junction, as in Fig. 10-18. The collected carriers are observed as an increase in reverse current. Since the width of a *pn* junction is very small, the transit time T in Eq. (10-20) is also small, and a reasonable value of G can be attained even for fast response times. The *pn*-junction structure has the additional advantage that the reverse current in the dark is very small compared with the current in a simple piece of semiconductor at the same applied voltage. This means that the additional carriers produced by photon absorption can be detected with greater sensitivity. For this reason InSb infrared detectors are universally of the *pn*-junction type. In addition, it is common practice to cool such detectors in order to reduce the reverse current in the dark and obtain ultimate sensitivity.

Figure 10-18 Energy-band model of *pn*-junction photocell.

Note that electrons and holes created in the region of the junction are collected even in the absence of an applied voltage because of the internal potential rise at a *pn* junction, Fig. 10-18. This means that an illuminated *pn* junction can act as a battery. An open-circuit voltage equal to the internal potential step exists between terminals applied to the *n*-type and *p*-type regions. The short-circuit current of the battery depends upon the illumination intensity at the junction. This is the principle of the *solar battery* used as an energy source on space satellites. Silicon solar batteries are most often used because the width of the forbidden energy gap in silicon corresponds favorably with the wavelength distribution of radiant energy from the sun. Similar photocells are also used in automatic-iris cameras and exposure meters.

Highly ionizing radiation such as β rays and α rays are detected by a *pn* junction. If an α particle strikes the junction, its energy is consumed in creating electrons and holes which are subsequently collected by the field at the junction. The output current pulse corresponding to this event is a measure of the energy of the incident particle. Therefore, a *pn-junction radiation detector* not only signals the presence of nuclear radiation but also measures the energy of the individual particles. Most often, it is desirable to employ detectors having a wide *pn* junction to assure that the particle is completely absorbed in the junction region. Furthermore, the junction must be very near the surface of the semiconductor so that little energy is lost in creating carriers far from the junction where they cannot be collected. These requirements have led to special fabrication procedures for *pn*-junction radiation detectors which are different from those normally used in transistor production.

ANALOG INSTRUMENTS

Oscilloscopes

We have touched several times previously upon the circuits and applications of that most versatile instrument, the cathode-ray oscilloscope. The complete block diagram of a typical unit, Fig. 10-19, includes vertical and horizontal amplifiers

Figure 10-19 Block diagram of oscilloscope.

coupled to the vertical and horizontal deflection plates of the cathode-ray tube, a triggered sawtooth, or ramp generator to provide a linear horizontal sweep, and suitable power-supply circuits. In the most common mode of operation, the waveform of interest is applied to the vertical input terminals. The vertical amplifier increases the amplitude sufficiently to cause an appreciable vertical deflection of the electron beam. A portion of the amplified signal is used to derive a trigger pulse for the sweep generator in order to synchronize the sweep signal with the input waveform. The output of the sweep generator is amplified and applied to the horizontal deflection plates. The sweep generator also provides a pulse to the electron gun of the cathode-ray tube (CRT) which blanks off the electron beam during retrace so that the spot can return to the starting point without leaving a spurious trace.

Sometimes it is advantageous to trigger the sweep generator with an external signal as, for example, if it is necessary to start the sweep before the vertical signal amplitude is large. For this purpose, an external trigger input is provided to which the generator input can be connected. When the linear sweep is not desired, as in the case of Lissajous figures (Chap. 2), the input of the horizontal amplifier is connected to the horizontal input terminals and external waveforms can be placed on both the horizontal and vertical deflection plates. It is also possible to modulate the intensity of the electron beam by altering the grid potential in the electron gun; this external terminal is conventionally termed the z-axis input to distinguish it from the horizontal (x-axis) and vertical (y-axis) inputs.

The vertical and horizontal amplifiers are quite often of the dc-coupled differential type. Voltage-divider gain controls are included to adjust the amplitude of the deflection on the screen. Many different sweep circuits are in common use; a typical version is described in the previous chapter. Sweep-speed and trigger-sensitivity controls are necessary here in order to set the sweep for best waveform display. The power supply for the CRT is designed so that dc potentials applied to the electron-gun elements can be adjusted to achieve the optimum focus and intensity of the trace. Similarly, the quiescent position of the spot on the tube face is adjusted by altering the dc voltages on both the horizontal and vertical deflection plates.

The ability of a conventional oscilloscope to display very-high-frequency signals is limited by the high-frequency cutoff of the vertical amplifier and by the luminous efficiency of the cathode-ray-tube screen at high spot speeds. As one example of the versatility of CRT instruments, consider the so-called *sampling oscilloscope*, which can display repetitive waveforms at frequencies far in excess of a conventional instrument. The input waveform amplitude is sampled at successively later points in its cycle and then reassembled into a replica of the input waveform on the CRT screen. Since the input signal is repetitive, the sampling can be done every tenth cycle, as illustrated in Fig. 10-20, which means the sampling and display circuits operate at only one-tenth of the input frequency.

The basic circuit diagram of a sampling oscilloscope, Fig. 10-21, employs a pulse generator to produce a series of successively delayed pulses and a gate circuit which is active only during the pulse duration. The gate output is passed through a low-pass filter and applied to the CRT. Note that although most of the circuit need not be capable of handling the signal frequency, the pulse generator must produce very sharp pulses. Several different circuits are available to perform this function. It proves easier to generate extremely narrow pulses than to design an amplifier with significant gain at corresponding frequencies.

Figure 10-20 By taking sample of repetitive input waveform every tenth cycle, waveshape can be reconstructed by circuits with maximum frequency response only one-tenth of the input frequency.

Figure 10-21 Block diagram of sampling oscilloscope.

Waveform Analyzer

Complex waveforms may be thought of as combinations of harmonically related sine waves, according to the method of Fourier analysis discussed in Chap. 2. Instruments capable of determining the amplitude of the frequency components in a waveform, called *waveform analyzers*, in effect carry out Fourier analysis experimentally. As such, they are frequency-measuring instruments.

A waveform analyzer is basically a sharply tuned ac amplifier-voltmeter which has a significant response at only one frequency. The input signal is heterodyned with the output of a variable-frequency sinusoidal oscillator to produce a signal at the response frequency of the amplifier, Fig. 10-22. That is, according to Eq. (4-17), two sine waves suitably mixed produce a sine wave at the sum (and difference) frequency. If the oscillator frequency is f_o while the input waveform has a frequency component f_i, the combination will result in an output signal if

$$f_i + f_o = f_a \tag{10-21}$$

where f_a is the center frequency of the tuned amplifier. Thus, by tuning the oscillator frequency over a given interval, the output meter deflects each time Eq. (10-21) is satisfied for a frequency component of the input waveform. The magnitude of the meter deflection is proportional to the amplitude of each input-signal component. In this way, the entire frequency complement of the input waveform is determined.

Figure 10-22 Block diagram of waveform analyzer for determining frequency components in complex waveform.

The heterodyne technique is used in preference to simply tuning the amplifier itself in order to obtain the greatest possible frequency coverage. An oscillator circuit capable of output frequencies over a wide band is simpler than a tunable amplifier covering the same frequency interval. Furthermore a fixed-tuned amplifier has a constant peak response and bandpass characteristic, both difficult to achieve in a tunable system. The frequency dial of the waveform analyzer is calibrated in terms of $f_a - f_o$, so that the unknown frequency is indicated directly.

Magnetic Recorder

One of the most useful signal recording schemes is based on the *magnetic-recorder* principle in which an electric waveform is recorded in the form of magnetization on a ferromagnetic chart, or tape. After recording, the tape is passed back through the instrument and a voltage signal corresponding to the original input is obtained. This makes it possible to observe the signal again and again, if necessary, and to analyze the waveform by as many different techniques as may be required.

The magnetic tape, most often in the form of a ferromagnetic oxide powder coated onto a plastic tape, is magnetized in accordance with the signal by the magnetic field of the recording head, Fig. 10-23a. Signal current produces a magnetic field at a sharply defined gap in the core of the recording head, and this field permanently magnetizes the tape as it passes the gap. A high-frequency ac bias signal is also applied to the recording head to improve the linearity of the recording process. In playback, Fig. 10-23b, the tape is moved past the head again and the magnetic field of the tape induces a voltage in the coil of the head. Since the magnetization of the tape corresponds to variations in the original signal, voltages induced in the head replicate the input waveform. Playback does not change the tape magnetization, so that the signal can be reproduced as many times as desired.

The action of the ac bias frequency can be understood by considering the typical nonlinear properties of the magnetic tape as illustrated by the hysteresis loop in Fig. 10-24. In the absence of ac bias, the recorded magnetization is not a linear function of the signal field and the playback signal is strongly distorted. The ac bias causes the magnetization of the material to traverse minor hysteresis loops

(a)   (b)

Figure 10-23 (a) Recording and (b) playback of electric signals using magnetic recording techniques.

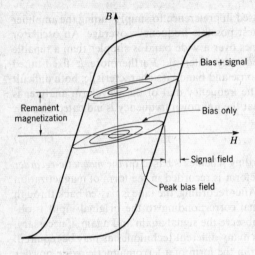

Figure 10-24 Action of ac bias is to cause magnetic tape material to traverse minor hysteresis loops as it passes under recording head. Remanent magnetization is then a linear function of signal field.

as shown. As an element of tape moves away from the recording gap, the size of the loops decreases to zero. Thus, in the absence of signal the resulting magnetization is equal to zero. The signal field displaces the minor loop so that the remanent magnetization is a finite value. Because of the straight sides of the major hysteresis loop, the relation between the remanent magnetization and the signal field is linear. The ac bias signal is of the order of five times the maximum signal frequency and the peak amplitude is approximately equal to the coercive force of the tape.

The highest frequency that can be recorded depends upon the tape speed and width of the gap in the record-playback head. The rms playback voltage, V, resulting from a sinusoidal recording of frequency ω is

$$V = MN\omega \frac{\sin(\omega l/2v)}{\omega l/2v} \qquad (10\text{-}22)$$

where M is the rms magnetic signal flux in the tape, N is the number of turns on the playback head coil, l is the gap in the head, and v is the tape velocity. According to Eq. (10-22), the playback voltage is small at low frequencies, rises to a maximum, and falls sharply to zero where

$$\frac{\omega_m l}{2v} = \pi \qquad \text{or} \qquad f_m = \frac{v}{l} \qquad (10\text{-}23)$$

Since $f = v/\lambda$, Eq. (10-23) can also be written

$$\lambda_m = l \qquad (10\text{-}24)$$

The meaning of Eq. (10-24) is that the minimum wavelength that can be played

Figure 10-25 Experimental frequency-response characteristic of magnetic recorder playback head. Fall at high frequencies is determined by size of the gap in head.

back is equal to the gap length. That is, the maximum frequency depends upon the tape speed, according to Eq. (10-23). Experimental results, Fig. 10-25, are in general agreement with Eq. (10-22).

The falloff in playback signal at low frequencies is a result of the small rate of change of flux at long tape wavelengths. Uniform frequency response is achieved by a *compensated* playback amplifier that has large gain at low frequencies and decreasing gain at high frequencies. The high-frequency response depends upon the recording-head gap length and tape speed. With suitable choice of parameters, frequencies as high as 10 MHz can be accommodated.

Although major use of magnetic recorders is in reproducing the amplitude and frequency characteristics of the input signal exactly, many other possibilities also exist. For example, according to Eq. (10-23) the playback frequency depends upon the tape speed. Therefore a very-long-duration, low-frequency recording can be played back at high tape speed. The resulting high-frequency signal can be analyzed in a much shorter time than required for the original recording. Conversely, a high-frequency recording can be played back at a slow tape speed with a corresponding reduction in the frequency components in the signal. In this way, the signal can be analyzed with instruments having only a low-frequency capability. Time compression or expansion ratios greater than factors of 1000 are possible.

Electrometer

The *vibrating-reed electrometer* uses an electromagnetically driven vibrating capacitor to measure feeble electric currents. The vibrating capacitor develops an ac signal which is amplified, rectified, and fed back to the input circuit. The system is exactly analogous to the lock-in amplifier discussed in Chap. 7 and has the stability and sensitivity advantages of this circuit. Most often, the vibrating capacitor C_e, Fig. 10-26, is a small metal reed electromagnetically vibrated by a sinusoidal voltage derived from an oscillator. The oscillator output also provides a signal for the synchronous detector to develop a dc output voltage in accordance with the input signal. The mechanical vibration of the reed results in a sinusoidal variation

Figure 10-26 Block diagram of vibrating-reed electrometer.

in the capacitance of the reed in combination with an adjacent stationary electrode. A charge Q_e placed on the capacitance produces an ac voltage

$$v_e = R_i Q_e \frac{dC_e}{dt} \tag{10-25}$$

which is presented to the amplifier input.

The operation of the vibrating-reed electrometer is analyzed in the following way, with reference to Fig. 10-26. A charge Q applied to the input terminals divides between the reed capacitance and the coupling capacitance C_c so that the charge on C_e is

$$Q_1 = \frac{QC_e}{C_e + C_c} \tag{10-26}$$

The ac signal from the vibrating capacitor is amplified and rectified and appears as an output voltage v_o across the feedback resistor R_f. This voltage also produces a charge on C_e,

$$Q_2 = \frac{v_o}{1/C_c + 1/C_e} = \frac{v_o C_e C_c}{C_e + C_c} \tag{10-27}$$

Therefore, the total charge on C_e is the sum of Eqs. (10-26) and (10-27), or

$$Q_e = Q_1 + Q_2 = \frac{C_e}{C_e + C_c}(Q + v_o C_c) \tag{10-28}$$

The overall gain of the system A is the ratio of the output voltage to the dc voltage across the vibrating capacitor, so that

$$v_o = -A\frac{Q_e}{C_e} \tag{10-29}$$

The gain involves the effective change in capacitance of the vibrating reed, Eq. (10-25), together with the amplifier gain and the rectifier efficiency. Introducing Q_e from Eq. (10-29) into Eq. (10-28) and solving for the output voltage, the result is

$$v_o = -\frac{Q}{C_c + (C_e + C_c)/A} \cong -\frac{Q}{C_c} \qquad (10\text{-}30)$$

which is true when A is large, as in all feedback amplifiers.

According to Eq. (10-30), the output voltage is a direct measure of the input charge. The output voltage is measured by a conventional d'Arsonval meter. Carefully designed vibrating-reed electrometers can measure charges as small as 10^{-16} C, equivalent to a few hundred electrons. Minute currents can be measured by timing the rate at which C_c charges or by shunting the input terminals with a large known resistance. Currents of the order of 10^{-17} A, a few hundred electrons per second, can be detected.

The input resistor R_i is present to isolate the instrument from the capacitance of the source. This is necessary because any fixed capacitance in parallel with C_e effectively reduces the ac signal developed by the vibrating reed since the net change in capacitance is reduced. The time constant $R_i C_e$ is made larger than the vibration period of the reed to accomplish this isolation. Note also that the feedback voltage is essentially equal to the input voltage. Therefore, the *guard* terminal, Fig. 10-26, can be used to reduce unwanted leakage currents between the upper input terminal and other points in the circuit. When these points are connected to the guard terminal, the potential difference is negligible and stray leakage currents are eliminated.

NOISE SIGNALS

Any spurious currents or voltages extraneous to the signal of interest are termed *noise* since they interfere with the signal. Noise voltages arise in the basic operation of electronic devices or are the result of improper circuit design and use. It is important to minimize noise effects in order to characterize the signals with the greatest possible precision and to permit the weakest signals to be amplified. A convenient measure of the influence of noise on any signal is the *signal-to-noise* ratio, the ratio of the signal power to the noise power at any point in a circuit.

Thermal Noise

When a number of amplifier stages are cascaded a random noise voltage appears at the output terminals, even in the absence of an input signal. This output voltage is caused by a random voltage generated in the input resistor. The noise voltages that appear across the terminals of any resistor are attributed to the random motion of the free electrons in the material of the resistance. Electrons in a conductor are free to roam about by virtue of their thermal energy, and at any given instant more electrons may be directed toward one terminal of the resistor than toward the other. The result is a small instantaneous potential difference between the terminals. The magnitude of the potential fluctuates rapidly as the number of electrons moving in a given direction changes from instant to instant.

Since the noise voltage across a resistor fluctuates randomly, it has Fourier components covering a wide range of frequencies. It is convenient, therefore, to specify the noise voltage in terms of the mean square noise voltage per unit cycle of bandwidth. For a resistor R this quantity is

$$\langle \Delta v^2 \rangle = 4kTR \tag{10-31}$$

where k is Boltzmann's constant and T is the absolute temperature. The noise voltage given by Eq. (10-31) is variously called *Nyquist noise*, after the physicist who derived this equation, or *thermal noise*, since its origin is a result of the thermal agitation of free electrons, or *Johnson noise*, after the engineer who first measured it.

The meaning of Eq. (10-31) is as follows. A noise voltage appears between the terminals of any resistance. The magnitude of the noise voltage actually measured with any instrument depends upon the frequency response of the instrument. For example, the rms noise voltage of a 1000-Ω resistor at room temperature as measured by a voltmeter with a bandwidth of 10,000 Hz is, using Eq. (10-31),

$$v = (1.65 \times 10^{-20} \times 10^3 \times 10^4)^{1/2} = 4.1 \times 10^{-7} = 0.41 \ \mu V$$

This rather small voltage is not inconsequential. The output voltage of an amplifier with a 10-kHz bandpass and a gain of 10^6 is nearly $\frac{1}{2}$ V if the input resistor is 1000 Ω. This output voltage is present even when no input signal is applied.

The noise voltage of resistances is independent of frequency. Accordingly, thermal noise is called "white" noise by analogy with the uniform spectral distribution of white light energy. The presence of thermal noise in any circuit is accounted for by including a noise generator given by Eq. (10-31) in series with a noiseless resistor. In practice it is usually necessary to consider the thermal noise of only those resistors in the input circuit of an amplifier. The gain of the first stage makes the amplified noise of the input resistor larger than the noise of resistors in succeeding stages.

Thermal noise is a fundamental and unavoidable property of any resistance. An amplifier should have a bandwidth only as wide as is necessary to amplify adequately all signal components in order to minimize the ever-present thermal noise voltages. If it is desired to amplify a single-frequency signal, for example, the frequency response of the amplifier should be sharply peaked at that frequency. The total noise voltage at the output is therefore reduced since only noise components having frequencies in the amplifier passband are amplified. The signal-to-noise ratio is enhanced and weak signals can be amplified usefully.

Current Noise

Noise voltages in excess of thermal noise are observed experimentally in certain resistances when a direct current is present. Although the physical origins of this additional noise are not clear, many experiments have shown that the noise is largest at low frequencies and that it increases with the square of the current. An empirical expression for this effect is

$$\langle \Delta v^2 \rangle = K \frac{I^2}{f} \qquad (10\text{-}32)$$

where K is an empirical constant involving the geometry of the resistor, the type of resistance material, and other factors; I is the dc current; and f is the frequency. According to Eq. (10-32) the mean square noise voltage per unit bandwidth depends inversely upon frequency and the phenomenon is therefore called $1/f$ noise. Since the noise also depends on I, it is also referred to as *current noise*.

The magnitude of $1/f$ noise varies markedly with the material of the conductor and its physical form. It is absent entirely in bulk metals, so that only thermal noise is observed in wire-wound resistors. Composition resistors, on the other hand, generate a large $1/f$ noise level, Fig. 10-27, which is associated with the intergranular contacts in such resistors. Although such contacts are known to be important, $1/f$ noise is also observed in single-crystal semiconductors where contact effects are negligible.

To minimize the low-frequency noise level, resistor types in which the $1/f$ noise is small, such as wire-wound units, are selected. Fortunately, at high frequencies where wire-wound resistors are unsuitable because of their inductance, the $1/f$ noise of composition resistors is usually negligible compared with thermal noise. In other situations the direct current in noisy components is minimized to reduce the current noise generated. The total current noise voltage in

Figure 10-27 Experimental noise voltage of 2.2-MΩ composition resistor. Note spectrum is $1/f$ when dc current is present and white Nyquist noise in absence of current.

any given circuit is found by integrating Eq. (10-32) over the frequency-response characteristic of the amplifier.

Low-frequency $1/f$ noise is also present in vacuum tubes and transistors. In the former it is often known as *flicker noise* and originates in the semiconducting cathode material, particularly at the emitting surface. The noise in transistors results from semiconductor properties, in which surface conditions are very important.

Noise in Transistors

Other noise effects are also present in transistors. *Shot noise* is a result of the random emission of electrons across a *pn* junction. Since each electron represents an increment of current, the collector current fluctuates slightly about the dc value. This effect is analogous to the noise of raindrops on a tin roof. That is, the basic reason for shot noise is that the electron is a discrete unit of electrical charge.

Current fluctuations due to shot noise are

$$\langle \Delta i^2 \rangle = 2eI \tag{10-33}$$

The quantity $\langle \Delta i^2 \rangle$ is analogous to $\langle \Delta v^2 \rangle$ in the thermal noise expression, Eq. (10-31), except that the noise is expressed here in terms of current fluctuations. Note that shot noise is a white noise since the right side of Eq. (10-33) is independent of frequency.

Noise voltages are treated by including noise voltage generators in the equivalent circuits. The effect of the amplifier passband, which may be determined by stray shunt capacitance or by tuned circuits, is included in assessing the magnitude of the noise voltages. The noise resulting from each generator may be treated separately since random-noise voltages are independent. The total output noise voltage is therefore simply the sum of all noise effects.

The input signal to any circuit has associated with it a given signal-to-noise ratio since thermal noise corresponding to the source resistance is present, at least. An ideal amplifier amplifies the incoming signal and the incoming noise equally and introduces no additional noise. Therefore, the original signal-to-noise ratio is preserved at the output. Practical amplifiers are not ideal because of thermal noise of the input circuit and current noise in the first stage. A useful figure of merit for any circuit is the *noise figure*, NF, which is defined as the input signal-to-noise ratio divided by the output signal-to-noise ratio. An ideal amplifier has a noise figure of unity and many practical circuits approach this value fairly closely.

The noise of transistors is a result of thermal noise of the semiconductor resistance, $1/f$ noise caused by current in the semiconductor crystal, and shot noise of carriers crossing the junctions. In addition, still another noise phenomenon has been observed in semiconductors. Electrons are promoted randomly from the valence band to the conduction band and also return randomly to the valence band, keeping the proper average number of carriers in each band. Random generation and recombination of carriers is caused by thermal energies and produces a conductivity fluctuation of the semiconductor. This generates a

Figure 10-28 Experimental internal noise level of a type 2N929 transistor with a source resistance of 10,000Ω.

noise voltage when a direct current is present; this second type of current noise in semiconductors is termed *generation-recombination*, or *g-r noise*.

Analysis of noise phenomena in transistors is complicated by these many factors and by the inherent input-output coupling. It turns out that at intermediate frequencies the noise figure of a transistor can be expressed as

$$\text{NF} = 1 + \frac{r_b}{R_s} + \frac{r_e}{2R_s} + \frac{(R_s + r_b + r_e)^2}{2h_{fe}r_e R_s} \tag{10-34}$$

where R_s is the source resistance and the other symbols have their usual meaning. According to Eq. (10-34) the noise figure approaches unity if the emitter and base resistances are small with respect to the source resistance and if the forward current gain h_{fe} is large. Note that the internal noise of a transistor depends upon the operating point.

Equation (10-34) ignores $1/f$ noise, which increases the noise level at low frequencies, Fig. 10-28. Additionally, an increase in noise at frequencies of the order of the α-cutoff frequency is observed. The latter effect, also apparent in Fig. 10-28, can be quite well explained on the basis of the influence of current gain and the α-cutoff frequency upon g-r noise.

The internal noise of field-effect transistors is a result of thermal noise in the channel. Therefore, at intermediate frequencies the drain noise is represented quite well by

$$\langle \Delta i_d^2 \rangle = 4kTg_m \tag{10-35}$$

where g_m is the transconductance of the FET. The analogy between Eqs. (10-35) and (10-31) is evident; note that the equivalent noise resistance in the drain circuit is just $1/g_m$ and, therefore, depends upon the operating point. The noise increases at low frequencies because of $1/f$ noise and also increases at high frequencies as a result of capacitive coupling between channel noise and the gate electrode. With careful design the FET may have the lowest noise figure of all active devices, particularly at high impedance levels.

Figure 10-29 (a) Stray capacitive coupling introduces noise pickup signals into sensitive circuits. (b) Grounded conductor shields the circuit from surroundings.

Shielding and Grounding

Noise is often introduced in practical circuits by extraneous voltage signals coupled into the circuit from the surroundings. The most common source of this noise comes from the 60-Hz electric and magnetic fields produced by power mains. The 60-Hz signal induced by these fields is called *hum* because it is audible as a low-frequency tone in amplifiers connected to a loudspeaker. Other *stray pickup* may result from electric fields generated by nearby electronic equipment, electric motors, lightning discharges, etc.

It is useful to *shield* those portions of a circuit where the signal level is small and, consequently, where noise voltages are most troublesome. Electric fields induce noise voltages capacitively, so it is only necessary to surround the circuit with a grounded conducting shield in order to reduce stray pickup. This is illustrated schematically in Fig. 10-29, where the capacitive coupling to external sources in Fig. 10-29a is interrupted by interposing a grounded conductor, Fig. 10-29b. Such shielding is also effective in reducing so-called *crosstalk* between different stages of the same circuit as, for example, between the input stage and the power output stage of a complete amplifier.

Additionally, it is useful to shield a circuit to minimize induced currents resulting from stray magnetic fields. This is accomplished with high-permeability ferromagnetic enclosures which reduce the intensity of the magnetic field inside. Such shielding is never complete because of the properties of ferromagnetic materials, and it is always advantageous to minimize the area of the circuit by using the shortest possible signal leads, since the induced voltage in any circuit resulting from changing magnetic fields decreases if the enclosed area of the circuit is reduced. Transformers are particularly troublesome with respect to inductive pickup because of their many turns of wire. They are kept well removed from all power transformers because of the strong magnetic fields generated by such units.

In general it is good practice to keep all circuits physically small in order to minimize stray pickup, crosstalk, and stray capacitance. All grounded compon-

Figure 10-30 When several electronic devices are connected, system must be grounded only at one point, preferably at input. Multiple grounds can lead to large ground-loop currents.

ents, such as bypass capacitors, pertaining to a given stage are returned to a single point. This reduces so-called *ground loops*, which are current paths through the metal chassis on which electronic circuits are often mounted. If all components of one stage are not grounded at the same point, the currents may cause undesirable signal coupling between stages.

When a number of individual electronic units are interconnected, shielded cable is used for all signal leads between units. The shield is used as the ground lead, as illustrated in Fig. 10-30. It is important that the entire system be grounded at only one point, usually the input terminal. If each unit is grounded separately, as shown by dashed lines in Fig. 10-30, large 60-Hz currents can be induced in the circuit because of the large area encompassed. The current induced in loop A by stray 60-Hz magnetic fields introduces a large stray pickup signal into the amplifier.

Note that the capacitance between the central wire and its shield tends to shunt the signal at high frequencies. For this reason such cables are kept as short as possible. The circuit output impedance is also made small since this reduces the effect of shunt capacitance. Therefore, the output stage of many electronic circuits is a source or emitter follower.

TRANSMISSION LINES

Usually the components of a measuring system are physically separate from one another. For example, several distinct electronic instruments may be connected together in cascade to make a measurement. The signal must then be transported from the output of one unit to the input of the next. This is done by connecting the various circuits with *transmission lines*. In the simplest case a transmission line is just two wires, as has been implicitly assumed in previous chapters. A transmission line must, however, faithfully transmit the signal between instruments with a minimum of waveform or amplitude distortion. If this is not the case, the ultimate output of the system may be erroneous.

In low-frequency and dc systems it is only necessary to consider the resistivity of the conductors in the transmission line, as well as, perhaps, shunt resistance between conductors. These effects are examined by straightforward circuit analysis as studied in Chap. 1. Even a straight piece of wire has associated with it a

Figure 10-31 Representation of transmission line in terms of series and shunt impedances per incremental length.

small inductance however, and there is also a small capacitance between the two conductors. At high frequencies the reactances of these unavoidable inductances and capacitances become significant and influence signal propagation along the transmission line. This is particularly important in pulse circuits because of the high frequencies associated with pulse waveforms. In many practical high-frequency circuits transmission lines only a few centimeters long introduce objectionable waveform distortions if they are improperly used.

Characteristic Impedance

It is useful to represent the incremental impedances of a short section of transmission line by a series resistance r_1 and inductance l per unit length together with a shunt resistance r_2 and capacitance c per unit length, as in Fig. 10-31. Thus the series and shunt complex impedances of a small section Δx of the line are

$$\mathbf{z}_1 \, \Delta x = \left(r_1 + j\omega l\right) \Delta x$$

$$\frac{1}{\mathbf{z}_2} \, \Delta x = \left(\frac{1}{r_2} + j\omega c\right) \Delta x \tag{10-36}$$

Variations of voltage and current along the transmission line are produced in response to an input signal. Consider one section Δx of the line for which the input voltage is v and the output voltage is incrementally different, $v + \Delta v$. Similarly, the input and output currents of the line element are i and $i + \Delta i$ respectively. The voltage equation around the outside loop of this circuit is then

$$v - i \frac{\mathbf{z}_1}{2} \, \Delta x - (i + \Delta i)\frac{\mathbf{z}_1}{2} \, \Delta x = v + \Delta v \tag{10-37}$$

Dividing through by Δx,

$$\frac{\Delta v}{\Delta x} = -\mathbf{z}_1 i - \frac{\mathbf{z}_1}{2} \Delta_i \tag{10-38}$$

In the limit as $\Delta x \to 0$, $\Delta v / \Delta x$ becomes the derivative dv/dx and the second term on the right side of Eq. (10-38) vanishes,

$$\frac{dv}{dx} = -\mathbf{z}_1 i \tag{10-39}$$

Similarly, the voltage equation including the shunt path is

$$v - i\frac{\mathbf{z}_1}{2}\Delta x - [i - (i + \Delta i)]\frac{\mathbf{z}_2}{\Delta x} = 0 \tag{10-40}$$

$$v = i\frac{\mathbf{z}_1}{2}\Delta x + \frac{\Delta i}{\Delta x}\mathbf{z}_2 = 0 \tag{10-41}$$

In the limit, $\Delta x \to 0$, Eq. (10-41) reduces to

$$\frac{di}{dx} = -\frac{1}{\mathbf{z}_2} v \tag{10-42}$$

Differentiating Eq. (10-39) with respect to x and using Eq. (10-42), the result is

$$\frac{d^2 v}{dx^2} = -\mathbf{z}_1 \frac{di}{dx}$$

$$\frac{d^2 v}{dx_2} = \frac{\mathbf{z}_1}{\mathbf{z}_2} v \tag{10-43}$$

Similarly, differentiating Eq. (10-42) with respect to x and using Eq. (10-39) yields an equation in i alone,

$$\frac{d^2 i}{dx^2} = \frac{\mathbf{z}_1}{\mathbf{z}_2} i \tag{10-44}$$

Equations (10-43) and (10-44) are called the transmission-line equations. Their solutions yield the current and voltage at any point along the line.

The general solution of Eq. (10-43) is

$$v = Ae^{-\gamma x} + Be^{\gamma x} \tag{10-45}$$

where A and B are constants that depend upon the input and output conditions of the transmission line, and γ is called the *propagation constant*. The propagation constant is a complex number with a real part α and an imaginary part β given by

$$\gamma = \alpha + j\beta = \sqrt{\frac{\mathbf{z}_1}{\mathbf{z}_2}} \tag{10-46}$$

as can be verified by substituting Eqs. (10-45) and (10-46) into the transmission-line equation, Eq. (10-43).

Since the differential equation for i, Eq. (10-44), is identical in form to the voltage equation, Eq. (10-43), the solution also has the same form. It is possible to

Figure 10-32 Transmission line terminated in its characteristic impedance.

evaluate the arbitrary constants in terms of A and B by using Eq. (10-42). The result is

$$i = \frac{A}{\mathbf{Z}_c} e^{-\gamma x} - \frac{B}{\mathbf{Z}_c} e^{\gamma x} \tag{10-47}$$

where

$$\mathbf{Z}_c = \sqrt{\mathbf{z}_1 \, \mathbf{z}_2} \tag{10-48}$$

is called the *characteristic impedance* of the line. Equations (10-45) and (10-47) give the variation of voltage and current with distance along the line. We postpone examination of their significance to consider the meaning of the characteristic impedance.

Suppose a transmission line d meters long is terminated in its characteristic impedance, as in Fig. 10-32. Using Eqs. (10-45) and (10-47), the output voltage v_o can be written

$$v_o = i_o \mathbf{Z}_c$$

$$Ae^{-\gamma d} + Be^{\gamma d} = \mathbf{Z}_c \left(\frac{A}{\mathbf{Z}_c} e^{-\gamma d} - \frac{B}{\mathbf{Z}_c} e^{\gamma d} \right) \tag{10-49}$$

which means $B = -B = 0$. At the input end $x = 0$, so that $v_i = A$. Therefore, the voltage and current equations become

$$v = v_i e^{-\gamma x} \quad \text{and} \quad i = \frac{v_i}{\mathbf{Z}_c} e^{-\gamma x} \tag{10-50}$$

In particular, the input impedance of the line is

$$\mathbf{Z}_i = \frac{v_i}{i_i} = \frac{v_i}{v_i / \mathbf{Z}_c} = \mathbf{Z}_c \tag{10-51}$$

This means that the input impedance of a transmission line terminated in its characteristic impedance is simply equal to the characteristic impedance, independent of the length of the line. For this reason, as well as an equally significant one discussed subsequently, a transmission line is most often terminated in its characteristic impedance.

In most applications the series and shunt resistances of the transmission line can be neglected in comparison with the reactances. Thus, if $r_1 \ll \omega l$ and

$1/r_2 \ll \omega c$, the characteristic impedance, Eq. (10-48), reduces to

$$\mathbf{Z}_c = \sqrt{\mathbf{z}_1\,\mathbf{z}_2} = \sqrt{\frac{r_1 + j\omega l}{1/r_2 + j\omega c}} = \sqrt{\frac{l}{c}} \tag{10-52}$$

which means that the characteristic impedance is a pure resistance.

The magnitude of $\mathbf{Z}_c$ depends upon the geometry of the line since both l and c are functions of the wire size, shape, separation, etc. In particular, the characteristic impedance of a transmission line comprising two parallel wires is given by

$$\mathbf{Z}_c = 276 \log \frac{D}{a} \tag{10-53}$$

where D is the separation between the wires and a is the radius of each conductor. A transmission line commonly used to connect television antennas to television receivers has $D/a = 10$, so that $\mathbf{Z}_c = 276\ \Omega$. Actually, such "twinlead" has a characteristic impedance closer to 300 Ω. The difference is attributable to the dielectric constant of the insulation separating the two conductors. Equation (10-53) applies to the case of air (dielectric constant of unity) between the wires.

Coaxial transmission lines are also commonly employed. The characteristic impedance of a coaxial line is given by

$$\mathbf{Z}_c = 138 \log \frac{b}{a} \tag{10-54}$$

where b is the radius of the outer conductor and a is the radius of the central wire. A common value for the characteristic impedance of a coaxial line is 72 Ω, although other values are also commercially available.

Delay Time

According to Eq. (10-50) the voltage along a transmission line terminated in its characteristic impedance is given by

$$v = v_i e^{-j\beta x} \tag{10-55}$$

if the series and shunt resistances can be neglected. Suppose that the input signal is sinusoidal. Then, Eq. (10-55) can be written

$$v = V_i \sin \omega t(\cos \beta x - j \sin \beta x) = V_i \sin (\omega t - \beta x) \tag{10-56}$$

Equation (10-56) represents a wave that is periodic both in time and in space.

A plot of Eq. (10-56) for one cycle when $t = 0$ is illustrated in Fig. 10-33. Similarly, plots of Eq. (10-56) when $t = T/4$ and $t = T/2$ (where T is the period of the input signal) show that the input signal moves away from the input end of the line. The *wavelength* λ of the signal at any given time is the distance along the line at which the voltage has gone through one complete cycle of 2π radians. Therefore, from Eq. (10-56), $\lambda = 2\pi/\beta$. Note that, according to Fig. 10-33, the input signal moves a distance $\lambda/4$ in a time $T/4$, a distance $\lambda/2$ in a time $T/2$, etc.

Figure 10-33 Motion of sinusoidal signal along transmission line.

Therefore, the speed with which the input signal is transported along the line is

$$v = \frac{\lambda/4}{T/4} = \frac{\lambda}{T} \quad \text{or} \quad v = \lambda f \quad (10\text{-}57)$$

where f is the frequency of the input signal. Equation (10-57) is a fundamental relation between the wavelength, frequency, and velocity of any propagating phenomenon.

The time required for an input signal to travel from the input of a transmission line to the output end is given by, using Eq. (10-57),

$$t_d = \frac{d}{v} = \frac{d}{\lambda f} = \frac{d}{\lambda} T \quad (10\text{-}58)$$

where d is the length of the line. According to (10-58), the time delay in terms of the period of the input signal depends upon the length of the line in terms of the wavelength. The time delay is a function of the parameters of the line since $\lambda = 2\pi/\beta$ depends upon the line geometry.

The propagation characteristics of transmission lines are often used to introduce a given time delay between input and output signals. Such *delay lines* are also constructed with actual inductances and capacitances connected according to Fig. 10-31. This is necessary if long delay times are required, since the delay time increases when l and c are large.

Reflections and Resonance

A transmission line need not be terminated by an impedance equal to its characteristic impedance. As a useful illustration, consider the case when the output end is short-circuited so that $v_o = 0$. Thus at $x = d$, where d is the length of the line,

$$0 = Ae^{-j\beta d} + Be^{j\beta d} \tag{10-59}$$

Similarly, from Eq. (10-47),

$$i_o = \frac{A}{Z_c} e^{-j\beta d} - \frac{B}{Z_c} e^{j\beta d} \tag{10-60}$$

Solving (10-59) and (10-60) for A and B yields

$$A = \frac{i_o Z_c}{2} e^{j\beta d} \quad \text{and} \quad B = -\frac{i_o Z_c}{2} e^{-j\beta d} \tag{10-61}$$

Using Eq. (10-45) the voltage on the line is

$$v = \frac{i_o Z_c}{2} e^{j\beta d} e^{-j\beta x} - \frac{i_o Z_c}{2} e^{-j\beta d} e^{j\beta x}$$

$$v = \frac{i_o Z_c}{2} \left(e^{-j\beta(x-d)} - e^{j\beta(x-d)} \right) \tag{10-62}$$

Similarly, the current is, from Eq. (10-47),

$$i = \frac{i_o}{2} \left(e^{-j\beta(x-d)} + e^{j\beta(x-d)} \right) \tag{10-63}$$

Comparing Eqs. (10-62) and (10-63) with Eq. (10-50) we interpret the second term in (10-62) and (10-63) as a wave traveling from the receiving end to the input end. That is, the incident wave is *reflected* at the output of the line because of the improper termination at the receiving end. Reflections are avoided in pulse applications particularly since pulses that are reflected back and forth between the input and output ends of the line completely obscure the true input signal. This is the second major reason why a transmission line is normally terminated in its characteristic impedance.

A short-circuited transmission line does have useful properties, however, as can be illustrated by computing the input impedance. Using Eqs. (10-62) and (10-63) the complex input impedance is

$$Z_i = \frac{v_i}{i_i} = Z_c \frac{e^{j\beta d} - e^{-j\beta d}}{e^{j\beta d} + e^{-j\beta d}} = Z_c \frac{\cos \beta d + j \sin \beta d - \cos \beta d + j \sin \beta d}{\cos \beta d + j \sin \beta d + \cos \beta d - j \sin \beta d} \tag{10-64}$$

$$= jZ_c \tan \beta d = jZ_c \tan 2\pi \frac{d}{\lambda} \tag{10-65}$$

According to Eq. (10-65) the input impedance of a short-circuited transmission line is equal to zero when the length of the line is such that $d = \lambda/2$, because $\tan \pi = 0$. Of particular importance is the case when $d = \lambda/4$, for which $Z_i \to \infty$. That is, the line acts as a parallel resonant circuit at the frequency corresponding to $d = \lambda/4$. Such quarter-wavelength lines are often used as resonant circuits at very high frequencies. This is useful since normal parallel resonant circuits consisting of a single inductance and capacitance are not possible at extreme frequencies because of the very small inductance and capacitance values required. Analogous results can be achieved with transmission lines that are open-circuited at the receiving end.

Waveguides

At very high frequencies, where the signal wavelength is of the order of the spacing between conductors, parallel-wire and coaxial transmission lines do not transport signals efficiently from the input to the output end. Such signal frequencies are propagated in hollow conductors, called *waveguides*. Analysis of the transmission properties of waveguides is accomplished by focusing attention on the electric and magnetic fields in the interior of the hollow conductor, rather than on the currents in the conductor itself. In effect, the signal is carried by waves of electric and magnetic fields and the function of the hollow conductor is to guide these waves from the transmitting to the receiving end.

Some appreciation of the electric-field configuration in a circular waveguide is attained by considering the situation in a coaxial line as the central conductor is made smaller and smaller until it vanishes altogether, Fig. 10-34. When the central conductor is large, the electric field in the space between conductors is essentially radial, Fig. 10-34a, and currents are present in both conductors corresponding to the electric-field pattern. A small central conductor results in an electric-field configuration that originates and terminates on the same conductor, Fig. 10-34b. This suggests that the central wire may be eliminated altogether; the electric-field pattern that results is shown in Fig. 10-34c. One reason for the increased efficiency of the waveguide over that of the two-conductor line is the fact that resistance losses in the center wire are eliminated.

Waveguides with a rectangular cross section are more commonly employed than are circular waveguides because the electric-field configurations are somewhat less complicated. This means that input and output terminations are simpler. In both types the electric-field configuration is such that only signals of frequencies higher than a certain *cutoff frequency* can be propagated; that is, a waveguide acts as a high-pass filter. For example, the longest wavelength that can be transmitted through a rectangular waveguide having the longer cross-sectional dimension a is given by $\lambda_c = 2a$. Minimum losses are achieved if the waveguide dimension is only just sufficient in cross section to pass the lowest signal frequency of interest. For this reason, the size of a waveguide is a good indication of the signal frequency it is designed to transmit.

If the receiving end of a waveguide is short-circuited, a resonance effect identical in principle to that of a short-circuited transmission line is observed. Most often, both ends of the waveguide are closed, which produces a *resonant cavity* useful as a resonant circuit at microwave frequencies.

Figure 10-34 Electric-field configurations (*a* and *b*) in coaxial transmission lines and (*c*) in hollow circular waveguide.

SUGGESTIONS FOR FURTHER READING

C. F. G. Delaney: "Electronics for the Physicist," Penguin Books, Inc., Baltimore, 1969.
A. James Diefenderfer: "Principles of Electronic Instrumentation," W. B. Saunders Company, Philadelphia, 1972.
Franklin Offner: "Electronics for Biologists," McGraw-Hill Book Company, New York, 1967.
A. van der Ziel: "Noise," Prentice-Hall, Inc., Englewood Cliffs, N.J., 1954.

EXERCISES

10-1 Determine the regulation factor and internal resistance of the transistor regulated power supply, Fig. 10-4. Assume properties of the control transistor given in the text and that the *h* parameters of the 2N2049 are the same as those given in Exercise 6-6.

Ans.: 4.5×10^{-4}, 2.3 Ω

10-2 Analyze the simple shunt regulator in Fig. 10-3 to determine the effective internal resistance and regulation factor. The *h* parameters are $h_{ie} = 100$ Ω, $h_{re} = 10^{-4}$, $h_{fe} = 60$, $h_{oe} = 175 \times 10^{-6}$ mho.

Ans.: 5.9×10^{-2}; 1.6 Ω

10-3 Determine the percentage change in output voltage of the voltage regulator in Fig. 10-35 from no-load to a full-load current of 10 A. Repeat for a change in input voltage from 40 to 50 V at a load current of 5 A. Estimate the range of input voltages over which the regulator is effective. Take the small-signal parameters of the 2N1613 to be the same as for the 2N1415 in Fig. 7-12, and the value of β for the 2N1479 to be 50. The characteristics of the 2N1489 are similar to the 2N2016, Fig. 7-13.

Ans.: 0.46 percent, 0.35 percent, 18 to 36 V.

Figure 10-35 Voltage regulator.

10-4 Given that a typical thermistor has a 6 percent change in resistance per degree centigrade at room temperature, what is the minimum temperature change detectable by a 1000-Ω thermistor if the limiting factor is thermal noise in the thermistor resistance. Assume that 10 V is applied to the thermistor and that the bandwidth is 10 Hz.

Ans.: 2.2×10^{-8} °C

10-5 What is the minimum tape speed necessary if a magnetic recorder is used to record signals up to 1 MHz? Assume that the recording gap is 1 μm long.

Ans.: 1 m/s

10-6 Suppose an amplifier with a bandpass from 10 Hz to 100 kHz and a gain of 10^5 uses a 1-MΩ input resistor. What is the rms noise output voltage?

Ans.: 4.1 V

10-7 Repeat Exercise 10-6 if the input resistor is the carbon resistor of Fig. 10-27 and the direct current indicated is present.

Ans.: 7.3 V

10-8 What is the minimum detectable input signal voltage of the chopper amplifier of Fig. 7-32, assuming chopper noise is negligible?

Ans.: 1.6×10^{-7} V

10-9 Calculate the time delay of a 100-MHz signal transmitted along a transmission line 10 m long. Assume the properties of the line are such that the wavelength on the line is 1 m.

Ans.: 10^{-7} s

10-10 What is the input impedance of a transmission line if the output end is an open circuit? Compare the condition for parallel resonance with that corresponding to the short-circuited line, as given by Eq. (10-65).

Ans.: $-j\mathbf{Z}_c \cot (2\pi d/\lambda)$; $\lambda/2$

ELEVEN

DIGITAL ELECTRONICS

Numerical digits can be represented by electrical signals which have only two possible magnitudes, say, zero and some finite value. In this situation only the existence of one signal state or the other is significant, and the actual magnitude is relatively unimportant. Therefore, digital circuits need have only two stable conditions represented by a transistor fully conducting, or completely cut off. Such circuits are inherently more reliable than conventional analog circuits, which must handle a continuous range of signal levels. Any given number can be represented by a digital waveform, so accuracy is not limited by the stability of circuit parameters. Digital signals are manipulated by circuits according to specific logic statements which make possible exceedingly flexible and powerful processing of information.

DIGITAL LOGIC

Binary Numbers

Voltage signals are used to represent the digits of numbers in order that mathematical logic operations such as addition and subtraction can be carried out by electronic circuits. The stability advantages of digital circuits are best realized if the signal waveforms have only two amplitudes, called state 0 and state 1, or *off* and *on*. This is so because a transistor, for example, need then only be either completely cut off or fully conducting. Since only two digits are available, the *binary* number system is used, rather than the more familiar *decimal* system based on the use of 10 digits. It is, of course, not particularly surprising that a counting system based on the number of fingers we possess is not the most efficient one for electronic circuits. Although the binary system is less familiar, there is no difference in principle between the two number systems.

Consider, for example, the meaning of the decimal number 528. This array of decimal digits is a shorthand notation for the increasing powers of ten in the number. Thus,

$$528 = (5 \times 10^2) + (2 \times 10^1) + (8 \times 10^0)$$
$$= \quad 500 \quad + \quad 20 \quad + \quad 8 \quad = 528$$

Similarly, the two digits in the binary system, 0 and 1, tell the number of increasing powers of two in the number. The binary number 10110, for example, means

$$10110 = (1 \times 2^4) + (0 \times 2^3) + (1 \times 2^2) + (1 \times 2^1) + (0 \times 2^0)$$
$$= \quad 16 \quad + \quad 0 \quad + \quad 4 \quad + \quad 2 \quad + \quad 0$$
$$= \quad 22$$

That is, the binary number 10110 represents the same quantity as the decimal number 22. Table 11-1 contains the decimal-binary equivalents for the numbers from 0 through 16.

Arithmetic manipulations with binary numbers involve mathematical logic quite familiar from decimal numbers. Addition and subtraction operations involve digits that are carried and borrowed in an analogous fashion. It is useful to state the mathematical logic contained in the addition of two binary numbers A and B by means of a *truth table*, Table 11-2, which accounts for all of the possible combinations of A and B. A digit is carried, as in the fourth entry in Table 11-2, by shifting it to the next higher position, that is, to the left. For example, the sum of 0110 and 0101 is written

$$
\begin{array}{rr}
0110 & 6 \\
+0101 & +\ 5 \\
\hline
1011 & 11 \\
\end{array}
$$

Table 11-1 Binary and decimal numbers

Decimal number	Binary number
0	0000
1	0001
2	0010
3	0011
4	0100
5	0101
6	0110
7	0111
8	1000
9	1001
10	1010
11	1011
12	1100
13	1101
14	1110
15	1111
16	10000

Binary numbers can be represented by voltage waveforms with regularly spaced pulses of uniform amplitude. Conventionally, the pulses corresponding to increasing powers of 2 appear in time sequence beginning with 2^0. The waveforms of the binary numbers 1011 and 0011 are illustrated in Fig. 11-1, together with their sum, 1110. Note that the pulse lengths of both 0 and 1 digits are equal. One digit, be it a 0 or a 1, is termed a *bit*, which is a contraction for *binary digit*.

It should be noted in passing that the waveform of the sum bears no direct relationship to the waveforms of the two numbers. That is, the waveform of the sum is not obtained simply by adding the voltages of the waveforms representing the individual binary numbers. This points up the fact that waveforms in digital circuits are a coded representation of numbers and the voltages of the waveforms have no significance in themselves.

Table 11-2 Truth table for addition

A	B	Sum	Carry
0	0	0	0
0	1	1	0
1	0	1	0
1	1	0	1

Decimal
equivalents
= 11

= 3

= 14

Figure 11-1 Waveforms of the binary numbers 1011 and 0011 and their sum, 1110.

Logic Gates

The various operations performed on digital waveforms are accomplished using circuits termed *logic gates* which have two or more inputs and one output. The output waveform of a logic gate depends upon the input waveforms and the input-output characteristic of the circuit described in terms of mathematical logic.

Consider the logic statement "if A is true and B is true, then T is true," which is written symbolically, as

$$A \cdot B = T \qquad (11\text{-}1)$$

This is called the AND concept and is signified by the dot between the two quantities A and B in Eq. (11-1). Conventionally the condition *true* is identified with state 1 and the opposite condition, *false*, is identified with state 0 in digital circuits. In this fashion a truth table for the AND operation may be devised as in Table 11-3. Extending the truth table for more than two input quantities is easily accomplished by applying the logic in Table 11-3 several times to each pair of inputs. According to Table 11-3 an output is obtained (that is, T is true) only when A AND B are true.

Table 11-3 Truth table for AND

A	B	T
0	0	0
0	1	0
1	0	0
1	1	1

Figure 11-2 (a) Diode AND gate, and (b) circuit symbol.

A simple three-input diode AND gate that accomplishes the logic in Table 11-3 is illustrated in Fig. 11-2a. The diodes are biased in the low-resistance forward direction and the output voltage is zero, which means that the output condition is state 0. If positive input voltage signals somewhat greater than V_c are applied simultaneously to all three inputs, the diodes become reverse-biased and the output voltage rises to V_c, or state 1. Note, however, that if even one input remains at state 0, the corresponding diode is under forward bias and the output signal remains at state 0. This action is logically described by saying that an output signal is obtained only when there are A AND B AND C input signals.

The AND gate is also called a *coincidence circuit* in nuclear radiation measurements since, in effect, it detects the coincidence of pulses presented to the inputs. The operation of an AND gate should not be confused with the mathematical operation of addition. That is, the output of an AND gate is not the sum of the input signals, as may be verified by noting the differences between the corresponding truth tables. AND gates are so extensively used in digital electronic circuits that it is convenient to use the special symbol in Fig. 11-2b.

The logical-OR statement is "if A or B is true, then T is true," written as

$$A + B = T \qquad (11-2)$$

where the + symbol indicates the OR concept. The corresponding truth table, Table 11-4, shows that an output is obtained whenever any input is present.

A diode OR gate and its circuit symbol are shown in Fig. 11-3a and b, respectively. Note that a positive signal at any input biases the corresponding diode in

Table 11-4 Truth table for OR

A	B	T
0	0	0
0	1	1
1	0	1
1	1	1

Figure 11-3 (*a*) Diode OR gate, and (*b*) circuit symbol.

the forward direction and appears at the output. At the same time the other diodes are reverse-biased so that signals fed back to the other inputs are negligible. In effect the OR gate is a simple mixing circuit that brings several input signals to a common output with minimum interaction between signal sources.

In binary logic the NOT operation inverts the polarity of an input signal, as indicated schematically in Fig. 11-4. Since a digital waveform has only two states, 0 and 1, if at any given instant the signal is in state 1 it may equally be said to be in state NOT 0. The symbol for NOT is a bar over the quantity,

$$\text{NOT } A = \bar{A} \tag{11-3}$$

and the truth table is quite simple, Table 11-5. A simple grounded-emitter amplifier produces the logic in Table 11-5 as a result of the 180° phase shift between input and output. Accordingly, the appropriate circuit symbol for a NOT or *inverter* gate is similar to that previously used for an amplifier, Fig. 11-5. The small circle at the apex of the triangle (sometimes placed at the input) specifically indicates inversion.

It is common practice to combine the NOT inversion with other logic-gate functions. This results in amplification of the signals in the logic gate itself and consequently preserves the signal amplitude throughout a network of multiple logic gates. Consider the single-transistor amplifier in Fig. 11-6a with three equivalent inputs so that the waveform at each input may appear at the output

Figure 11-4 Waveforms of binary numbers *A* and NOT *A*.

Figure 11-5 Inverter circuit symbol for NOT operation.

Table 11-5 Truth table for NOT

A	$\bar{A}$
0	1
1	0

terminal. This is the same logic as in Table 11-4 and therefore this circuit is a NOT-OR, or NOR, gate. The circuit symbol, Fig. 11-6b, includes the small circle to indicate inversion. A NOT-AND, or NAND, gate is equally possible and a typical circuit is illustrated in a later section. The circuit symbol for a NAND gate is shown in Fig. 11-7. Truth tables for NOR and NAND gates are simply obtained by applying the NOT operation to the last columns of the truth tables for OR, Table 11-4, and for AND, Table 11-3.

A logic gate related to the NAND gate is a combination of an AND gate and a NOR gate, Fig. 11-8. The operation of this circuit is best examined by developing the appropriate truth table, as in Table 11-6. The output of the NOR gate $\bar{B}$ is found with the aid of Table 11-5 and the output $A \cdot \bar{B}$ uses columns A and $\bar{B}$ in

Table 11-6 Truth table for anti-coincidence circuit

A	B	$\bar{B}$	$A \cdot \bar{B}$
0	0	1	0
0	1	0	0
1	0	1	1
1	1	0	0

Figure 11-6 (a) Simple NOR gate performs the combined operation NOT-OR, and (b) circuit symbol.

$A \cdot B \cdot C$

Figure 11-7 NAND gate.

Figure 11-8 Combination of an AND gate and a NOR gate to produce an anticoincidence circuit.

connection with Table 11-3. According to the truth table, no output is ever achieved when signal B is present. That is, the B input inhibits the A output and in pulse applications the circuit is termed an *anticoincidence* circuit by analogy with the coincidence circuit.

One further simple logic statement is the exclusive-OR function. The everyday usage of *or* is ambiguous for it can mean "one or the other or both" or "one or the other and not both." The logical-OR function previously defined is the "one or the other or both" situation according to the truth table, Table 11-4. It could correctly be called the inclusive-OR function. The exclusive-OR statement is

$$A \oplus B = A \cdot \bar{B} + B \cdot \bar{A} \qquad (11\text{-}4)$$

which can be implemented by the array of logic gates in Fig. 11-9. Here again, it is useful to examine the truth table, Table 11-7, which is constructed with the aid of Tables 11-3, 11-4, and 11-5. According to the last column in Table 11-7, an output from the exclusive-OR circuit is produced only when either A or B is present, but not both, as expected.

Table 11-7 Truth table for exclusive-OR

A	B	$\bar{A}$	$\bar{B}$	$A \cdot \bar{B}$	$B \cdot \bar{A}$	$A \cdot \bar{B} + B \cdot \bar{A}$
0	0	1	1	0	0	0
0	1	1	0	0	1	1
1	0	0	1	1	0	1
1	1	0	0	0	0	0

Figure 11-9 Exclusive-OR logic.

Boolean Algebra

Analysis of logical networks is greatly aided by the logical algebra developed in the last century by George Boole, an English mathematician. The theorems of Boolean algebra are used to simplify digital logic networks in much the same fashion that mathematical logic is used to manipulate ordinary algebraic expressions. One major difference, of course, is that the variables in Boolean expressions can assume only one of two possible values.

A list of theorems in Boolean algebra is given in Table 11-8. Theorems 1 through 4 may be recognized as OR logic and are proved using the OR truth table, Table 11-4. Similarly, Theorems 5 through 8 are based on the AND concept in Table 11-3. Theorem 9 is a formal definition of the NOT function. The commutation, association, distribution, and absorption theorems are relatively straightforward and may be proved by means of truth tables.

De Morgan's theorems, numbers 18 and 19, are particularly interesting as they show a useful relationship between and AND and OR functions. Theorem 18 is proved in Table 11-9 by first finding $\overline{(A + B)}$ and then separately finding $\bar{A} \cdot \bar{B}$. The result is that the two truth tables are equal and the theorem is proved. A similar process may be used to establish Theorem 19. The basic duality of Boolean algebra is expressed by these theorems and can also be noted in pairs of other expressions in Table 11-8. Compare, for example the two association theorems, and also Theorem 1 with Theorem 6. In each case the OR and AND functions are dually related.

Table 11-8 Boolean algebra theorems

OR function	1	$0 + A = A$
	2	$1 + A = 1$
	3	$A + A = A$
	4	$A + \bar{A} = 1$
AND function	5	$0 \cdot A = 0$
	6	$1 \cdot A = A$
	7	$A \cdot A = A$
	8	$A \cdot \bar{A} = 0$
NOT function	9	$(\bar{\bar{A}}) = A$
Commutation	10	$A + B = B + A$
	11	$A \cdot B = B \cdot A$
Association	12	$A + (B + C) = (A + B) + C$
	13	$A \cdot (B \cdot C) = (A \cdot B) \cdot C$
Distribution	14	$A \cdot (B + C) = A \cdot B + A \cdot C$
	15	$(A + B) \cdot (A + C) = A + B \cdot C$
Absorption	16	$A + A \cdot B = A$
	17	$A \cdot (A + B) = A$
De Morgan's Theorems	18	$\overline{(A + B)} = \bar{A} \cdot \bar{B}$
	19	$\overline{A \cdot B} = \bar{A} + \bar{B}$

Table 11-9 Proof of De Morgan theorem

A	B	$A + B$	$(\overline{A + B})$	$\overline{A}$	$\overline{B}$	$\overline{A} \cdot \overline{B}$
0	0	0	1	1	1	1
0	1	1	0	1	0	0
1	0	1	0	0	1	0
1	1	1	0	0	0	0

An elementary example of the use of Boolean algebra to simplify logic networks is provided by the circuit in Fig. 11-10 which contains three NOR gates. The outputs of the two input gates are $\overline{A}$ and $\overline{B}$, respectively, so that the output of the total circuit may be written down and subsequently simplified by means of De Morgan's Theorem 18

$$(\overline{\overline{A} + \overline{B}}) = \overline{\overline{A}} \cdot \overline{\overline{B}} = A \cdot B \tag{11-5}$$

In arriving at Eq. (11-5), Theorem 9 is also used. According to this result, the combination of three NOR gates may be simplified to one simple AND gate.

A second example of logic-network reduction is the exclusive-OR logic network illustrated in Fig. 11-9 and repeated in Fig. 11-11a. Observe that the output logic may be written in the form

$$A \cdot \overline{B} + B \cdot \overline{A} = (\overline{\overline{A \cdot \overline{B}}) + (\overline{B \cdot \overline{A}})} = (\overline{A \cdot \overline{B}}) \cdot (\overline{B \cdot \overline{A}}) \tag{11-6}$$

where Theorems 9 and 18 have been used. The last logic statement may be accomplished by replacing the output OR gate with a NAND gate, if the inputs are $(\overline{A \cdot \overline{B}})$ and $(\overline{B \cdot \overline{A}})$. These logic functions are easily produced by replacing the two AND gates with NAND gates. The result is the logic network in Fig. 11-11b. This network may be further simplified to Fig. 11-11c by noting that a NAND gate can replace the two inverters. This is proved by determining the output of the upper NAND gate,

$$\overline{A \cdot (\overline{A \cdot B})} = \overline{A \cdot (\overline{A} + \overline{B})} = \overline{(A \cdot \overline{A}) + (A \cdot \overline{B})} \qquad \text{Theorems 19 and 14}$$

$$= \overline{0 + (A \cdot \overline{B})} = \overline{0 \cdot (A \cdot \overline{B})} \qquad \text{Theorems 8 and 18}$$

$$= \overline{1 \cdot (A \cdot \overline{B})} = (\overline{A \cdot \overline{B}}) \qquad \text{Theorem 6}$$

$$\tag{11-7}$$

Figure 11-10 Three NOR gates produce an AND gate.

Figure 11-11 Reduction of exclusive-OR logic.

A similar analysis shows that the output of the lower NAND gate is $(\overline{B \cdot \bar{A}})$. Therefore the network in Fig. 11-11c is exactly logically equivalent to the exclusive-OR logic in Fig. 11-11a.

The results of the two previous paragraphs illustrate an important feature of Boolean logic networks. This is that any logic function may be accomplished with NOR gates alone, or with NAND gates alone. Although it may seem wasteful to employ three NOR gates to achieve AND logic as in Fig. 11-10, it is an appreciable convenience to be able to assemble large logic networks of thousands of gates from the same simple subunit. This greatly simplifies design, production, and maintenance considerations. Furthermore, it turns out that, as discussed in a later section, the operating characteristics of either the NAND or NOR gate of a particular design are more favorable than the other type of gate. Therefore, considerable advantage is realized in using only the more favorable form. The choice of which to use is entirely a matter of convenience and the logic statements may be converted from one form to the other by means of the theorems of Boolean algebra in Table 11-8.

LOGIC CIRCUITS

Logic Signals

In the preceding discussions it has been assumed that signal waveforms corresponding to state 1 are at a small positive potential while state 0 is represented by ground potential. This is in keeping with current practice in which the voltage of a state 1 bit is approximately +5 V and that of a state 0 bit is near 0 V. This choice is completely arbitrary, of course, since Boolean algebra does not depend on the specific choice of signals used but only upon the presence of two distinct states.

It is interesting to note the effect of interchanging the designation of logic levels upon actual gate circuits. That is, suppose state 1 is represented by 0^- V and state 0 by $+5$ V. This is called *negative logic* because state 1 is more negative than state 0. Consider the action of the diode AND gate in Fig. 11-2a in negative logic. The diodes are biased in the forward direction so that the output is state 1 when any one of the inputs is at ground potential, or state 1. The output is positive, state 0, only when all inputs are positive, or state 0. The truth table for this logic is written in Table 11-10 for the case of only two inputs. Table 11-10 is identical to Table 11-4 for the OR function. According to this analysis, an AND gate in positive logic becomes an OR gate in negative logic. This also means that a positive logic OR gate is a negative logic AND gate.

It is useful to think of an inverter gate as a logic level exchanger since a state 1 input yields a state 0 output, according to Table 11-5. Thus, for example, a logical AND function can be produced using a positive logic OR gate by first inverting signals to negative logic. The negative logic signals are acted upon by the positive logic OR gate to give negative AND logic. Subsequently, the signals may be converted back to positive logic by means of another inverter gate. The steps in this process can be symbolized in the following fashion:

1. Positive logic signals A, B
2. Negative logic signals $\bar{A}, \bar{B}$
3. Positive OR gate $\bar{A} + \bar{B}$
4. Theorem 19 $\overline{A \cdot B}$
5. Positive logic signal $A \cdot B$ $\qquad$ (11-8)

This illustrates how the combination of positive and negative logic may be analyzed by means of the standard theorems of Boolean algebra.

There are two fundamentally different ways of transmitting signals representing the several bits of a complete binary number. The first, already encountered in connection with Fig. 11-1, employs a time sequence of pulses corresponding to increasing powers of 2 in the number. This is called *serial* representation and is

Table 11-10 Truth table for negative logic AND gate

A	B	T
0	0	0
0	1	1
1	0	1
1	1	1

(a) Serial   (b) Parallel

Figure 11-12 (a) Serial, and (b) parallel representation of binary numbers.

indicated schematically in Fig. 11-12a. An alternate method, *parallel* representation, illustrated in Fig. 11-12b, is equally possible. In parallel transmission, pulses representing the increasing powers of 2 in the number all appear simultaneously.

Both serial and parallel representation of numbers are commonly used in the same digital circuitry, as is examined in subsequent sections. Clearly, serial transmission has some advantages since only one signal path is required, while a path for each bit is necessary in the case of parallel transmission. On the other hand, a time interval of only one bit length is needed to transmit an entire digital number when the parallel representation is used, whereas the time required is much longer for the serial representation and the time interval also depends upon the number of digits in the number. The method selected in any given instance depends upon which feature is more advantageous.

DTL, TTL, and ECL Logic

Several different designs of electronic circuits are used as logic gates. The simplest, illustrated in Figs. 11-2a and 11-3a, use diodes and, accordingly, are known as *diode logic*, or DL, gates. These simple circuits perform the desired logic function satisfactorily but are not easily cascaded to carry out complicated logical functions. The reason for this is the inevitable signal loss through the circuit and the impedance mismatch between the output of one gate and the input of the next.

Much more satisfactory in these respects is *resistor-transistor logic*, RTL, already shown in Fig. 11-6a as a NOR gate. Actually this design is used only in the form of a NOR gate, but this is sufficient to carry out all logic functions, as pointed out previously. The transistor provides gain so that logic signals are not degraded in traversing the gate, and the transistor also significantly reduces the effect of load impedance upon the input circuit. One useful measure of this effect is the *fan-out* characteristic. Fan-out refers to the number of identical logic gates that

Figure 11-13 DTL NAND gate.

can be connected to the output without seriously impairing circuit operation. Clearly a large fan-out is advantageous since complex logic networks are then possible. A fan-out of about 5 is characteristic of RTL gates. Better fan-out and somewhat faster circuit operation is achieved by shunting the summing resistors by capacitors, very much like the commutating capacitors in a binary. This configuration is called RCTL, *resistor-capacitor-transistor logic*.

The combination of a diode AND gate and an inverter, Fig. 11-13, yields a very useful NAND gate. Diodes D_1, D_2, and D_3 perform the logic function, and the transistor inverts and amplifies the signal. This circuit is capable of higher operating speeds than the RTL gate and has even further reduced output-input interaction because of coupling diode D_4 (which, however, can be omitted at some degradation in circuit performance). Fan-out in DTL gates can be as large as 10. Although an analogous NOR gate is possible, DTL logic normally employs only NAND gates.

All the foregoing logic gates are available as integrated circuits, but obviously discrete component circuits can be used as well. The versatility of integrated-circuit fabrication permits *transistor-transistor logic*, TTL, in which a multiple-emitter transistor is the logic element, Fig. 11-14. The operation of the circuit may be looked upon as a DTL gate with the diode anodes in common. When any one

Figure 11-14 Integrated-circuit TTL NAND gate.

Figure 11-15 Integrated-circuit ECL NOR gate.

of the emitters is at ground potential, state 0, the collector is essentially at ground and the output transistor is cut off since no current flows in its emitter-base junction. With the output transistor cut off the output is at state 1. If all three emitters are at state 1, both transistors conduct and the output goes to state 0. This is logically equivalent to a NAND gate.

TTL logic has a fan-out as great as 15, and eight or more input emitters are not unusual. Furthermore, operating speed is faster than the other gate types and the so-called noise-immunity properties are better. Logic speed is usually stated in terms of the *propagation delay* of the gate, that is, the time between introducing a change of logic state at the input and achieving a change at the output. The propagation delay is about 10×10^{-9} s (called nanoseconds, or ns) for TTL gates compared to values of 50 ns for DTL and RTL gates.

The noise immunity of logic gates is an important parameter that specifies how well a gate remains in a given logic state in the face of spurious signals due, for example, to changes in supply voltages, etc. Noise immunity is the difference between the minimum effective input signal in state 1 and the maximum state 0 output signal. A large value implies the gate is most likely to remain in a given logic state. Noise immunity exceeds 1 V for TTL gates, is about 0.75 V for DTL gates, and may be as small as 0.2 V for RTL gates.

Exceedingly small propagation delays are achieved in *emitter-coupled logic*, or *ECL* gates, Fig. 11-15, in which the input transistors Q_1, Q_2, and Q_3 are coupled together through a common-emitter resistor. When the input transistors are all cut off (state 0), the base of Q_5 is at 0 V and the NOR output signal is just the built-in emitter-base voltage, or state 1. If one or more inputs are raised to state 1, the corresponding input transistor conducts, causing Q_5 to conduct and put the output at state 0.

Table 11-11 Characteristics of logic gates

Gate type	Propagation delay ns/gate	Power dissipation mW/gate
DL	500	0
FET	200	0.25
DTL	50	10
TTL	10	20
ECL	1	60

The bias on Q_4 is set so that the gate transistors are not saturated when in state 1, which materially reduces the time required for them to be subsequently turned off again. This feature, together with the relatively small voltage swing between state 0 and state 1, leads to the small propagation delay of ECL gates. It should be remarked that in *ECL* logic state 0 is at -1.5 V, while state 1 is -0.75 V. While this is positive logic, the voltage levels are not compatible with the logic levels of other gate circuits.

Gate circuits using FETs are particularly appropriate for integrated-circuit logic arrays, as discussed in subsequent sections. Actually, practical digital circuits based on all logic gates rely on integrated-circuit technology in order to achieve useful operational characteristics in reasonably sized units. In this connection, it is useful to note the propagation delay and power dissipation of various logic gates because either the total time delay or the total power dissipation prove to be the limiting factor in extensive logic arrays. According to Table 11-11, small propagation delays imply high power dissipation, so that ECL gates are useful when speed is important and FET gates are most appropriate when power dissipation is to be minimized.

ADD Gates

It is pertinent to examine the addition of binary numbers since addition is an arithmetic operation basic to numerical computation. Multiplication, for example, can then be accomplished simply by repeated additions. Similarly, division can be carried out by the inverse of addition, that is, subtraction. Logic gates are used to achieve binary addition in the following two-step way. First, the digits in each column are added and then the carry digits are added to the column representing the next higher power of 2. This is identical to conventional addition as illustrated in a previous section.

The truth table for the addition of bits, Table 11-2, shows that the sum S of two bits is just exclusive-OR logic, Table 11-7, while the carry C is AND logic, Table 11-3. Therefore, the addition of bit A to bit B is stated by the following logic

$$S = A \oplus B \tag{11-9}$$

$$C = A \cdot B \tag{11-10}$$

Figure 11-16 Half-adder circuit.

A network to accomplish this logic is called a *half-adder*, and one form is illustrated in Fig. 11-16. Note that Eq. (11-10) is immediately evident, while the sum output is

$$S = \overline{(\overline{A + B}) + (A \cdot B)} = \overline{(\overline{A + B})} \cdot \overline{(A \cdot B)} \qquad \text{Theorem 18}$$

$$= (A + B) \cdot (\bar{A} + \bar{B}) \qquad \text{Theorems 9}$$

$$\text{and 19}$$

$$= A \cdot \bar{A} + A \cdot \bar{B} + B \cdot \bar{A} + B \cdot \bar{B} \qquad \text{Theorem 14}$$

$$= A \cdot \bar{B} + B \cdot \bar{A} \qquad \text{Theorem 8}$$

which establishes Eq. (11-9).

The half-adder accomplishes the first step in the addition of numbers, that is, the addition of bits. Two half-adders are combined into a *full adder*, Fig. 11-17, in order to add carry bits to the bit sum. The OR gate is needed to include the possibility that adding the carry bit to the sum generates a new carry bit.

The way in which a full adder is used to add binary numbers depends upon whether the numbers to be added are available in serial or parallel representation. In serial representation, Fig. 11-18a, the waveforms of the two numbers are introduced to a full-adder circuit, and each bit sum is added in sequence. Any carry bit produced is returned to the carry input, delayed by a time interval equal to the time between successive bits. This delay has the effect of adding the carry bit to the next column of bits in the binary number. A convenient 1-bit delay might be the combination of two cascaded binaries discussed in Chap. 9.

Figure 11-17 Full-adder.

Figure 11-18 (a) Serial, and (b) parallel addition

Addition in parallel representation is similar but a full adder is required for each digit, Fig. 11-18b. In this case the carry bit from the least significant bit addition is introduced into the full adder of the next most significant bit, etc. Although parallel addition may appear to require very elaborate circuitry since some 28 logic gates are required to add the two 4-digit numbers in Fig. 11-18b, it should be recalled that fabrication processes in integrated circuits are readily capable of producing elaborate networks. Thus a complete integrated-circuit full-adder gate is physically no larger than the container needed to provide the necessary terminals.

INFORMATION REGISTERS

Flip-flops

If two inverting gates are cross coupled, as in the case of the two NAND gates in Fig. 11-19, the combination has two stable states. Suppose, for example, the S input (for *set*) is state 1 and the R input (for *reset*, or *clear*) is also state 1. If the upper NAND gate has a state 1 at the other input, its output, Q, must be state 0. The state 0 and state 1 input at the lower NAND gate makes the $\bar{Q}$ output state 1, as originally assumed at the upper gate input. That is, Q and $\bar{Q}$ always have opposite logic states. Note that, if S is state 1 and R state 0, the output Q becomes state 0 and $\bar{Q}$ becomes state 1. Actually, this cross-coupled logic gate is quite

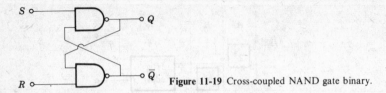

Figure 11-19 Cross-coupled NAND gate binary.

equivalent to the binary discussed in Chap. 9, as may be verified by comparing the logic diagram, Fig. 11-19, with the circuit of a binary, Fig. 9-23, in the case of RTL logic, Fig. 11-6b.

The truth table for a NAND binary is given in Table 11-12. When S is state zero the flip-flop is said to be *set* since $Q = 1$. Similarly, a state 0 input to R *clears* the flip-flop by putting Q at state 0. Note that the binary is, in effect, a memory circuit since it indicates which input was at state 0 last. Simultaneous state 1 inputs to both R and S leave the circuit in its original state, while two state 0 inputs gives an indeterminate result, since the logic of the circuit is not satisfied and the final state after these input signals are removed is only a matter of chance. A similar binary results from cross-coupled NOR gates; the corresponding truth table is related but not identical to Table 11-12.

If input gates are added to the NAND flip-flop, Fig. 11-20, the circuit responds to input logic signals only when the T, or *clock* input is in state 1. The reason for this is that when the clock input, T, is at state 0, the input gates must both have state 1 outputs (independent of the signals at R and S). According to Table 11-12, the binary stays in its original state under this condition. When the T input is in state 1, logic signals at R and S can cause transitions in the circuit, as described by Table 11-13. Note that this logic is reversed compared to the simple NAND flip-flop because of inversion in the input gates. In effect, this circuit is controlled, or gated, by signals at the T input since logic signals at R and S are only effective when T is at state 1. Note the similarity of this configuration with the RS/T binary discussed in Chap. 9. The present gated flip-flop is completely dc coupled, however, and responds to logic signals even if they are not pulses.

It is not possible to connect the output terminals of a gated flip-flop to the input terminals as in the RS/T binary in order to make the circuit toggle. The

Table 11-12 Truth table for NAND binary

R	S	Q
0	0	Indeterminate
0	1	0
1	0	1
1	1	Q

Figure 11-20 Gated NAND flip-flop.

reason for this is that the output cannot change state while connected to the input signals. Two cascaded flip-flops can be used, however, to achieve toggling while retaining the advantages of dc coupling. The logic network of a *RS master-slave* flip-flop in Fig. 11-21 illustrates this technique.

Note that the circuit is two gated flip-flops controlled by the same clock input. The slave flip-flop is gated by an inverted clock signal, however, and because of the 220-Ω resistor in the clock line is activated at a slightly different time than the master flip-flop. When the clock is at state zero, logic signals at the R and S inputs are ineffective, as in a regular gated flip-flop. The inverted clock signal is state 1 at the slave flip-flop, however, and the slave assumes the same state as the master. As the clock signal goes towards state 1, the input of the clock inverter gate reaches state 0 before the master input gates reach state 1. Therefore, the slave input gates isolate the slave flip-flop from the master, and the state of the slave flip-flop stores that of the master.

Next, the master input gates receive the state 1 clock signal and the logic signals at R and S determine the state of the master flip-flop according to Table 11-13. At the end of the clock pulse, the master input gates isolate the master flip-flop from the R and S inputs. With the clock back at state 0, the inverted clock signal at the slave flip-flop causes the slave to take up the state of the master flip-flop. Thus the output state reflects the logic signals at R and S, but the output state is not attained until the end of the clock pulse.

Note that multiple terminals are available at both the R and S inputs. According to Table 11-13, a state 0 at any S input puts Q at state 0, and a state 0 at any R input puts Q at state 1. Direct set and clear inputs are also provided. A state 0 signal at the set input overrides all other signals and produces $Q = 1$. Similarly, a state 0 signal at the reset input makes $Q = 0$ and the flip-flop is cleared.

Table 11-13 Truth table for gated NAND flip-flop

R	S	Q
0	0	Q
0	1	1
1	0	0
1	1	Indeterminate

Figure 11-21 RS master-slave flip-flop.

Toggle action is achieved by connecting $\bar{Q}$ in Fig. 11-21 to one of the S inputs and Q to one of the R inputs. The remaining S terminals are renamed J inputs and the free R terminals are called K inputs. The configuration is known as a JK *master-slave* flip-flop. The new truth table, given in Table 11-14, is similar to that for a gated NAND flip-flop except that the circuit toggles, as indicated by the last entry. Stated in words, the logic of a JK flip-flop is: a state 0 at any J input prevents a state 1 at Q, a state 0 at any K input prevents a state 0 at Q, and the circuit toggles unless inhibited by a state 0 at either J or K inputs. Note also that the toggling action is in response to the signals at T whether or not they are pulses.

Such elaborate flip-flops are only practical in integrated-circuit form. A typical JK master-slave flip-flop in TTL logic uses a total 18 transistors yet is contained with a conveniently sized package, Fig. 11-22.

A useful modification of a gated flip-flop has a single input terminal connected to the S input and uses an inverter to connect $\bar{S}$ to the R input, Fig. 11-23. The appropriate truth table, Table 11-15, shows that the D flip-flop (for *data latch*) stores the logic signal at the input terminal during the last interval that the clock was in state 1. An application of this circuit is considered in the next section.

Table 11-14 Truth table for NAND
JK **master-slave flip-flop**

$J_1 \cdot J_2$	$K_1 \cdot K_2$	Q
0	0	Q
0	1	0
1	0	1
1	1	$\bar{Q}$

Figure 11-22 Integrated-circuit JK master-slave flip-flop. (*Motorola Semiconductor Products, Inc.*)

Figure 11-23 Data latch flip-flop.

Table 11-15 Truth table for D flip-flop

D	Q
0	0
1	1

Counters

A single JK flip-flop produces one output pulse for every two pulses applied to the clock input. This is so because two pulses cause the circuit to shift from one stable state to the other and then back to the first state again, as in the case of the RS/T toggle binary discussed in Chap. 9. If the Q output of one binary is connected to the T input of a second binary, a total of four input pulses produces a single output pulse. That is, a string of cascaded binaries is an electronic *counter* which denumerates the number of input pulses. The state of each binary tells the total number of input pulses at any time.

Consider the four-binary counter in Fig. 11-24. Suppose that initially all four flip-flops are cleared so that $Q = 0$ in each stage. The waveforms at the output of each flip-flop as a result of a regular series of input pulses are shown in Fig. 11-25. The first input pulse causes FFA to change state while the others remain in state 0.

At the second input pulse FFA returns from state 1 back to state 0 and its output signal triggers FFB into state 1. The net result of two input pulses is that

Figure 11-24 Cascaded flip-flops form binary counter.

Figure 11-25 Waveforms in scale-of-16 counter.

FFB is in state 1 while the other three are in state 0. This process continues with further input pulses, as illustrated in Fig. 11-25, until at the sixteenth input pulse a single negative-going output signal is produced. The cascade combination of four binaries is, accordingly, called a *scale-of-16* counter.

A major application of cascaded RS/T binaries is to reduce or scale down the pulse rate produced, for example, by nuclear radiation detectors, to the point where the output pulses may be recorded by an electromechanical register. The scale factor of such a *scaler* may be increased simply by adding additional binary stages. Cascaded flip-flops, called *registers*, are also widely used to count, store, and manipulate digital logic signals and signals representing binary numbers.

In the register application the simple cascade configuration in Fig. 11-24 suffers from the propagation delay associated with the transition time of each stage. On the eighth and sixteenth input pulses, for example, all four flip-flops change state in sequence, so that a total time equal to the transition time of each stage times the number of stages is required for the last transition. At fast input pulse rates the first flip-flop may already have responded to succeeding input pulses before the last flip-flop reaches its final state. Thus the state of the counter at any given instant does not represent an exact number.

This difficulty is eliminated in the so-called *synchronous* binary counter, Fig. 11-26, in which all four flip-flops change state simultaneously. Note also that in this figure the input terminal is placed to the right so that a digital number stored in the register is presented with its most significant digit on the left as it is normally read. Each flip-flop is toggled by input pulses applied to all clock inputs. Interconnections between the flip-flops inhibit transitions until the appropriate count is reached. Note, for example, that FFA toggles on every input pulse, while

Figure 11-26 Synchronous binary counter.

FFD is inhibited until the eighth (and sixteenth) input pulse when all its J and K inputs are at state 1. The waveforms are identical to those for the nonsynchronous binary counter, Fig. 11-25, since the final state of each JK flip-flop is attained on the negative transition of the clock pulse. This waveform pattern may be verified by using the truth table for the JK flip-flop, Table 11-14.

It is usually desirable to display information in decimal form, and therefore conversion from the binary number system to the decimal system is necessary. Decoding logic to accomplish this change becomes very complicated when more than 4 or 5 bits are involved. Instead, it is often more convenient to store each digit of a decimal number in binary form separately. Thus each digit of the decimal number is represented by a four-flip-flop register counting from 0 to 9, and, for example, five such registers can store decimal numbers from 0 to 99,999. The decoding logic need therefore handle only decimal numbers from 0 to 9 and large numbers are readily accommodated by adding an additional register for each decade.

A four-flip-flop register is capable of counting to 16, but a total count of 10 is desired in such a *binary-coded decimal*, or BCD, counter. This can easily be accomplished by interconnections between flip-flops, as in Fig. 11-27. Input pulses cause *FFA* to toggle and *FFA*, in turn, toggles *FFB* until the eighth count when the Q output from *FFD* is inhibiting. Similarly, *FFD* is inhibited by the Q outputs of *FFB* and *FFC* at its J inputs, one or the other of which is at state 0 until the

Figure 11-27 Asynchronous binary-coded decimal counter.

Table 11-16 BCD counter states

Count	FFD	FFC	FFB	FFA
0	0	0	0	0
1	0	0	0	1
2	0	0	1	0
3	0	0	1	1
4	0	1	0	0
5	0	1	0	1
6	0	1	1	0
7	0	1	1	1
8	1	0	0	0
9	1	0	0	1

seventh count. FFD flips on the eighth input pulse. The ninth pulse toggles FFA only and on the tenth count both FFA and FFD flip so that all Q's are at state zero and the cycle is complete. This is best illustrated by a table showing the state of each flip-flop as a function of the input count, Table 11-16. Note that this pattern is identical to the waveforms of the binary counter, Fig. 11-25. That is, the count-ing pattern of this BCD counter is the same as the first nine states of a scale-of-16 counter. This is called the natural 8421 code and is the one most often used.

The cycle, or *modulus*, of a BCD counter is 10 since all flip-flops are cleared every 10 counts. Similarly, the modulus of a four-flip-flop binary counter is 16. Through appropriate logic connections, counters of any modulus are possible. Note, for example, that the combination of FFB, FFC, and FFD of the BCD counter is a *modulo*-5 counter if the input is at T of FFB. Alternatively, a simple binary counter may be converted to a counter of any desired modulus by using the direct clear inputs and a logic circuit to clear all flip-flops at the proper count. This technique is illustrated in Fig. 11-28 for a direct-cleared modulo-7 counter. The circuit counts as a normal binary until the seventh input pulse when the Q outputs

Figure 11-28 Modulo-7 counter using direct clearing.

Figure 11-29 BCD counter and storage register.

are all state 1 (refer to Table 11-16) for the first time. The output of the NAND gate goes to state 0, all the flip-flops are cleared, and the count cycle repeats.

It is often desirable to store the total count in a register, as for display purposes, while the counter proceeds to count a new signal waveform. One way to accomplish this is by means of a D flip-flop connected to each counter flip-flop, Fig. 11-29. A state 1 pulse at the memory transfer terminal causes each D flip-flop to assume the state of its respective counter flip-flop, according to Table 11-15. Therefore, the number in the BCD counter at the time of the transfer pulse is stored in the memory register while the counter is cleared to count again. The output of the memory register may be used to activate a visual display that remains constant until the next transfer pulse. Note that this configuration also converts a serial representation into the parallel representation of the binary number.

Registers

A combination of flip-flops arranged to store logic signals is called a register, and the four D flip-flops so grouped in Fig. 11-29 is a typical example. Actually, of course, a set of cascaded binaries is also a register since it stores information. It is often useful to move the information stored in a register along the chain of flip-flops without changing the relative waveform of the logic signal, and such a device is called a *shift register*.

Consider the 4-bit shift register of JK flip-flops in Fig. 11-30a together with typical corresponding waveforms in Fig. 11-30b. The input signal, I, and $\bar{I}$ are applied to the J and K inputs, respectively. Whenever I is state 1, the same logic

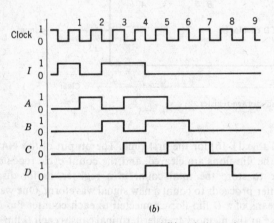

(b)

Figure 11-30 (a) Four-bit shift register, and (b) typical waveforms.

appears at Q of FFA at the trailing edge of the clock pulse. This, in turn results in state 1 at Q of FFB at the cessation of the next clock pulse, and so on. That is, information presented to the input appears at each succeeding flip-flop output delayed by one additional clock pulse. Therefore, the information pattern is shifted intact along the register, as indicated by the typical waveforms in Fig. 11-30b.

If the output of a shift register is returned to the input, the logic signals circulate continuously. A useful arrangement of such a *circulating register* is shown in Fig. 11-31. With the control input at state 1, logic signals at the output of the register are reintroduced at the input. The signals reappear at the output every m pulses, where m is the number of flip-flops in the register. New logic signals can be introduced by a state 0 signal at the control terminal. Integrated-circuit circulating registers many hundreds of bits long are available.

A related register with useful properties is obtained by connecting the outputs of the shift register in Fig. 11-30 to the input terminals and regarding the clock terminal as the input signal. First, a state 1 signal is placed in FFA and state 0 in

Figure 11-31 Circulating register.

all other flip-flops using the direct-set and direct-clear inputs on each flip-flop. According to the waveforms in Fig. 11-30 this single 1 circulates around the register in response to input pulses at the clock terminal. The successive states of this 4-bit *ring counter* are shown in Table 11-17. Note that this operation is, in effect, a modulo-4 counter, and that the total decimal count is readily discernible by, say, connecting an indicator to each flip-flop output. That is, the ring counter is self-decoding. This useful feature compensates for the fact that only one bit per flip-flop may be accommodated.

VISUAL DISPLAYS

Often it proves useful to show the information stored in a register by means of a visual display. Simple indicators which tell the logic state of each flip-flop in the register may be sufficient, but more elaborate displays which present the stored information in terms of decimal digits are usually more convenient.

Single-element Displays

The logic state of a flip-flop can be indicated directly by a small incandescent lamp, as illustrated in Fig. 11-32. When $Q = 1$ the transistor base is biased in the forward direction and the lamp lights. Conversely, the transistor is cut off and the lamp is dark when the flip-flop is in state $Q = 0$. The transistor switch is preferable, rather than, say, simply putting the lamp in series with one collector load of

Table 11-17 States of 4-bit ring counter

Input pulse	FFD	FFC	FFB	FFA
1	0	0	0	1
2	0	0	1	0
3	0	1	0	0
4	1	0	0	0

Figure 11-32 Incandescent-lamp circuit indicates flip-flop logic state.

the flip-flop itself, because the current required to light the lamp, approximately 50 mA, could easily disrupt normal operation of the circuit.

The state of each binary in a counter can be indicated by a separate lamp and the total count determined at any time by noting which lamps are lit. The lamps connected to each binary in the scale-of-16 counter in Fig. 11-24 are labeled 1, 2, 4, and 8, respectively, according to the waveforms in Fig. 11-25. The total count is found by adding the numbers associated with those lamps which are lit. This simple display obviously presents information in the binary number system.

In order to provide visual indicators in the decimal number system it is first necessary to convert binary signals into signals corresponding to decimal digits. Consider, for example the BCD counter in Fig. 11-26. The logic statement (that is, the state of the counter) corresponding to each decimal number for this counter is shown in Table 11-18. This is just a restatement of the presentation in Table 11-16 which shows the state of each flip-flop resulting from successive input pulses.

The binary statements in Table 11-18 are all AND logic which can be accomplished by the array of diode AND gates in Fig. 11-33. This is illustrated most easily by isolating the circuit associated with the 0 lamp, Fig. 11-34. Note that if $\bar{B}$ and $\bar{C}$ and $\bar{D}$ are present, the diodes are all reverse-biased and the transistor base becomes forward-biased. If now, in addition, the transistor emitter is grounded (that is, $\bar{A}$), the lamp lights. A similar analysis can be made for each of the other BCD logic input statements. This circuit, in effect, takes the input information in a BCD code and converts it to output signals in decimal code and, accordingly, the array is called a *decoder*. The driver transistors are needed to provide the necessary lamp currents, as discussed in connection with Fig. 11-32.

In arriving at the configuration of this decoder, advantage has been taken of certain simplifications in BCD logic. Note, for example, that D and $\bar{D}$ only distinguish states 0 and 1 from 8 and 9 in Table 11-18; that is, $\bar{D}$ is really not needed to describe states 2 through 7. Correspondingly, no connection to the D and $\bar{D}$ lines is made for these digits in the decoder array. Also, A and $\bar{A}$ can be introduced in connection with the driver transistors rather than as separate vertical logic lines, which, though equally possible, would result in a more complicated circuit.

+5 V

Decoder array

Driver transistors

2.2kΩ

2.2 kΩ

2.2 kΩ

2.2 kΩ

2.2 kΩ

Decimal output

0
1
2
3
4
5
6
7
8
9

D   D̄   C   C̄   B   B̄   A   Ā

BCD input

Figure 11-33 BCD-to-decimal decoder and driver.

Once the decoder logic is established by the logic statements, the operation of converting from binary to decimal numbers may be looked upon in the following way. Each 4-bit binary word which describes the state of the counter is decoded into a specific decimal number that is indicated by the lamp associated with that number. This display is obviously more convenient and informative than one which simply indicates the state of each flip-flop in the counter.

Several indicator schemes have been developed that show the actual number corresponding to the count in a decade register. One of these, called a *Nixie* tube, is basically a 10-cathode neon tube in which each cathode is a wire shaped into the form of one of the digits from 0 to 9. These are connected to the decoder-driver in a fashion similar to that used for the 10 individual lamps, so that only one cathode

Figure 11-34 Isolated AND gate of BCD decoder for 0 lamp.

is lit at a time. The characteristic red neon glow covers the entire cathode wire and this numeral appears at the face of the tube, Fig. 11-35. One Nixie tube is associated with each decade register and when the tubes corresponding to cascaded decades are arranged in a row, the total count is indicated directly as a decimal number.

Seven-Segment Displays

A very popular numerical indicator uses the seven-segment display illustrated in Fig. 11-36. Each segment is lit by a small lamp and the appropriate combination of lit segments forms the numbers from 0 to 9. A decoder-driver logic circuit for the seven-segment display analogous to Fig. 11-33 is discussed in the next section.

A second type of seven-segment display is based upon the property of some *pn* junctions to emit visible light when placed under forward bias. In *pn* junctions of

Table 11-18 Binary to decimal code

Decimal number	Binary logic statement
0	$\bar{A} \cdot \bar{B} \cdot \bar{C} \cdot \bar{D}$
1	$A \cdot \bar{B} \cdot \bar{C} \cdot \bar{D}$
2	$\bar{A} \cdot B \cdot \bar{C} \cdot \bar{D}$
3	$A \cdot B \cdot \bar{C} \cdot \bar{D}$
4	$\bar{A} \cdot \bar{B} \cdot C \cdot \bar{D}$
5	$A \cdot \bar{B} \cdot C \cdot \bar{D}$
6	$\bar{A} \cdot B \cdot C \cdot \bar{D}$
7	$A \cdot B \cdot C \cdot \bar{D}$
8	$\bar{A} \cdot \bar{B} \cdot \bar{C} \cdot D$
9	$A \cdot \bar{B} \cdot \bar{C} \cdot D$

Figure 11-35 Nixie tube numerical indicator. (*Burroughs Corporation.*)

certain semiconductors, most notably combinations of gallium arsenide and gallium phosphide, injected minority carriers recombine quickly in the junction region. Each recombination is accompanied by the emission of light energy of a wavelength corresponding to the width of the forbidden energy gap, as illustrated in the energy-band model, Fig. 11-37. This effect is just the inverse of photoconductivity discussed in connection with the energy-band model for *pn*-junction photocells, Fig. 10-18.

Most often, such a *light-emitting diode*, or *LED*, emits red light, although green and yellow LEDs are also available. A seven-segment display is produced by fabricating a linear junction shape for each segment, Fig. 11-38, which can be made quite compact for multidigit applications. In addition to this useful attribute, the LED display is characterized by an exceptionally long operating life.

Other seven-segment displays include fluorescent-target vacuum tubes and multiple-filament incandescent lamps. The first type is a vacuum tube containing seven fluorescent targets arranged according to the pattern in Fig. 11-36. Electron bombardment of those target segments made positive with respect to the cathode causes them to fluoresce with a blue-green color. This form of display has a fast response time and requires very little current per segment. The several incandescent filaments in the multiple-filament lamp display are made linear and are placed in the seven-segment pattern. This indicator may be looked upon as a more compact and convenient version of the individual-lamp segment type in that all visible elements of the digit are contained within one vacuum enclosure.

Figure 11-36 Seven-segment decimal display.

Figure 11-37 Energy-band model of light-emitting diode, or LED.

Decoder Logic

Each of these seven-segment display types requires a decoder logic circuit to activate the proper segments of the display corresponding to the BCD input word describing each decimal number. A truth table specifying the lit segments for each numeral is easily established, Table 11-19. This means, for example, that the logic

Figure 11-38 LED seven-segment display. (*Dialight Corporation.*)

Table 11-19 Truth table for seven-segment decimal display

Decimal number	BCD word	Segment						
		a	b	c	d	e	f	g
0	$\bar{A} \cdot \bar{B} \cdot \bar{C} \cdot \bar{D}$	1	1	1	1	1	0	1
1	$A \cdot \bar{B} \cdot \bar{C} \cdot \bar{D}$	0	0	1	1	0	0	0
2	$\bar{A} \cdot B \cdot \bar{C} \cdot \bar{D}$	1	0	1	0	1	1	1
3	$A \cdot B \cdot \bar{C} \cdot \bar{D}$	0	0	1	1	1	1	1
4	$\bar{A} \cdot \bar{B} \cdot C \cdot \bar{D}$	0	1	1	1	0	1	0
5	$A \cdot \bar{B} \cdot C \cdot \bar{D}$	0	1	0	1	1	1	1
6	$\bar{A} \cdot B \cdot C \cdot \bar{D}$	1	1	0	1	1	1	1
7	$A \cdot B \cdot C \cdot \bar{D}$	0	0	1	1	1	0	0
8	$\bar{A} \cdot \bar{B} \cdot \bar{C} \cdot D$	1	1	1	1	1	1	1
9	$A \cdot \bar{B} \cdot \bar{C} \cdot D$	0	1	1	1	1	1	1

necessary to light segment a, which is activated for decimal numbers 0, 2, 6, and 8, is

$$a = (\bar{A} \cdot \bar{B} \cdot \bar{C} \cdot \bar{D}) + (\bar{A} \cdot B \cdot \bar{C} \cdot \bar{D}) + (\bar{A} \cdot B \cdot C \cdot \bar{D}) + (\bar{A} \cdot \bar{B} \cdot \bar{C} \cdot D)$$
$$(11\text{-}11)$$

This expression may be simplified using Theorem 14 of Table 11-8 so that

$$a = (\bar{A} \cdot \bar{B}) \cdot (\bar{C} \cdot \bar{D} + \bar{C} \cdot D) + (\bar{A} \cdot B) \cdot (\bar{C} \cdot \bar{D} + C \cdot \bar{D})$$

$$= (\bar{A} \cdot \bar{B}) \cdot (\bar{C} \cdot \bar{D} + D) + (\bar{A} \cdot B) \cdot (\bar{D} \cdot \bar{C} + C)$$

$$= \bar{A} \cdot \bar{B} \cdot \bar{C} + \bar{A} \cdot B \cdot \bar{D} \qquad (11\text{-}12)$$

A similar logic expression can be written for the other six segments. For certain segments such as d and e the logic statement for NOT lit is shorter since these segments are activated on all but one or two of the ten numerals. In each case the result is similar to Eq. (11-12) and can be implemented by AND and OR logic arrays analogous to Fig. 11-33. In this connection it should be noted that OR logic is achieved in such a DL array simply by reversing the forward direction of the diodes.

Most often, decoder logic arrays are best implemented by integrated-circuit techniques because of the repetitive complexity involved. Thus a single integrated-circuit decoder accepts the four input signals corresponding to a BCD word and provides the proper output signals at each of its seven output terminals.

Certain digital instruments may employ displays involving 6-, 8-, or even 12-decimal numbers. This means that an equivalent number of decoder-driver arrays are needed, which implies extensive circuitry. Alternatively, it proves possible to use a single decoder-driver in a multidigit display by rapidly switching to each digit in succession. In this case, persistence of vision makes it appear that the entire display is lit continuously.

Figure 11-39 Multiplexing circuit for digital display.

The rapid-switching technique, called *multiplexing*, is illustrated in Fig. 11-39. All common segments in the display are connected together and to the appropriate output terminal of the decoder. The decoder input word is delivered from each of the BCD registers through a combination of NOR and NAND gates such that only one BCD register is connected at any instant. At the same instant only the corresponding decimal number is activated by the associated digit-enable gate. Each set of NAND gates and its corresponding digit-enable gate is activated sequentially by the logic 1 state circulating in the ring counter (compare Table 11-17).

The net result of this multiplexer is that each number is supplied with its proper seven-segment code and is lit once per cycle of the ring counter. The frequency of the oscillator driving the ring counter is made great enough to eliminate flicker in the display. In addition to a significant reduction in circuit complexity, multiplexing also lowers the overall power required to illuminate the display since, in effect, only one digit is lit at any time.

MEMORY CIRCUITS

Read-Only Memories

The display logic decoder arrays discussed in the previous section are each a simple example of what has come to be called a *read-only memory*, or *ROM*. A ROM associates a specific output signal with each input signal according to its fixed internal logic. The fixed relationship between input and output distinguishes the ROM from other memory circuits described subsequently.

Consider the circuit configuration of a 256-bit ROM arranged in a 32-word, 8-bit format, Fig. 11-40. As fabricated by integrated-circuit techniques, each word line is the collector of a transistor having an emitter connected to all eight bit lines. A state 1 at the base terminal of a selected word transistor turns on output driver transistors, that is, results in a state 0 at the bit outputs. During final stages of

Figure 11-40 Circuit configuration of 256-bit ROM.

fabrication, certain emitter connections between word transistors and bit lines are etched away to correspond to the desired logic pattern. Accordingly, those bit lines which are disconnected from a given word transistor result in a state 1 output when that word transistor is activated. The pattern of connected and absent transistors at the intersections of word lines and bit lines is the fashion by which ROM permanently remembers the relationship between input words and output signals.

The binary numbers from 00000 to 11111 specify 32 different words in Fig. 11-40, and an address decoder array is used to select one of the 32 word transistors in accordance with each 5-bit input word. Thus the overall operation of this ROM is to store permanently 256 bits of information which are accessed by a 5-bit binary word and which are presented as an 8-bit binary word.

Note that the presence or absence of a transistor at the intersection between word lines and bit lines specifies the logic at that location. This is exactly analogous to the diodes used in the decoder-driver ROM, Fig. 11-33. Similarly, a MOSFET ROM is fabricated such that the presence or absence of a gate electrode at the intersection determines the logic. All three types of logic connections are used in practice. ROMs capable of storing as many as 2^{14} (16,384) bits are available.

A major application of ROMs in digital circuits is to look-up tables for mathematical functions, such as trigonometric, exponential, square root, and logarithmic functions. In the case of a sine table, for example, the input signal corresponds to a given angle and the output represents the sine of that angle. A single 1024-bit ROM can store 128 angular increments corresponding to an angular resolution of 0.7 degree and can generate an 8-bit output signal that errs less than 0.1 percent from the handbook value of the sine. Obviously, accuracy is improved by increasing the bit storage capacity of the memory.

Many times it proves more efficient to use ROM multiplication tables rather than to carry out an actual multiplication calculation, which involves a sequence of bit additions and shifts. As in the case of other mathematical functions, the memory capacity needed for ROM multiplication tables increases exponentially with the accuracy required. Whereas, for example, a 2048-bit ROM is adequate to contain the multiplication tables for two 4-bit binary numbers, a capacity exceeding 10^6 bits is required in the case of two 8-bit numbers.

This extreme can be circumvented by representing each of the numbers to be multiplied together as the sum of two smaller binary numbers,

$$M_1 = m_1 + n_1 \qquad \text{and} \qquad M_2 = m_2 + n_2 \qquad (11\text{-}13)$$

where m_1 and m_2 are the most significant bits and n_1 and n_2 are the least significant bits in each number. The product is the sum of four terms,

$$M_1 M_2 = m_1 m_2 + m_1 n_2 + n_1 m_2 + n_1 n_2 \qquad (11\text{-}14)$$

Each term in Eq. (11-14) can be accommodated by a ROM of much smaller capacity than that required for $M_1 M_2$. If, for example, M_1 and M_2 are each 8-digit numbers, then m_1, n_1, m_2, and n_2 are all 4-digit numbers, which means that each term in Eq. (11-14) requires a 2048-bit ROM. Thus a total memory capacity of only $4 \times 2048 = 8.2 \times 10^3$ bits is needed. Even considering the several full adders used to implement Eq. (11-14), this results in a significant reduction in circuit complexity. A similar approach is equally applicable to ROM look-up tables for the other mathematical functions.

MOS Shift-Register Memories

In many digital-circuit applications it proves useful to store digital signals temporarily for recall at a later time. This is, in effect, a memory into which information can be written and changed, as well as read out, as distinguished from the case of the ROM. Shift registers are convenient and effective memory circuits for this purpose. They are universally fabricated by integrated-circuit techniques in order to achieve a usefully large memory capacity. In this connection MOSFET circuits are particularly advantageous because of the small power dissipation per gate, as already noted in Table 11-11.

A simple MOSFET binary, Fig. 11-41, has cross-connected inverter stages which use transistors for collector load resistors, as discussed in Chap. 6. Input driver transistors Q_5 and Q_6 are used to isolate the binary from the previous stage. A negative input signal at Q_5 turns Q_5 on and Q_2 off, which, in turn, turns Q_1 on because of the feedback connection from the collector of Q_2 to the gate of Q_1. A subsequent negative input pulse to Q_6 can turn Q_1 off and place the binary in its other stable state. It is noteworthy that the MOSFET binary consists entirely of FETs which can be easily fabricated by integrated-circuit technology. This, together with the minimal power-dissipation characteristic of FET gates, proves particularly advantageous for memory circuits where a great many gates make up a single unit.

Figure 11-41 *p*-channel MOSFET binary.

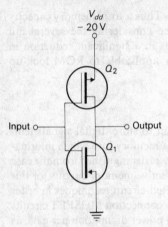

Figure 11-42 Complementary MOS inverter.

The lowest possible power dissipation per gate is achieved in *complementary MOS*, or *CMOS*, circuits which employ both *n*-channel and *p*-channel MOSFETs. Consider the CMOS inverter gate, Fig. 11-42, which is simply an *n*-channel MOSFET Q_1 in series with a *p*-channel MOSFET Q_2. When the input terminal is at ground potential, state 0, the gate of Q_1 is unbiased and so Q_1 is cut off while Q_2 is conducting because its gate is positive with respect to the source. Since Q_1 is off and Q_2 is on, the output terminal is at V_{dd}, or state 1, as expected for an inverter gate. Conversely, if the input terminal goes to state 1, (that is, V_{dd}), Q_1 is biased on and conducts while the Q_2 gate is unbiased and Q_2 is cut off. Therefore, the output terminal is at ground, or state 0. Note that in the two states either Q_1 or Q_2 is cut off and no current is present. Therefore, the power dissipation of the CMOS gate is zero.

Cross-coupled CMOS inverter gates analogous to Fig. 11-41 yield a binary that remains in either state without consuming power. Thus extensive logic arrays which entail minimal power dissipation become possible. This very real advantage compensates for the considerable complexity involved in fabricating integrated circuits comprised of *n*-channel and *p*-channel devices in the same unit. It is noteworthy that power is required to switch CMOS logic gates from one state to the other, however. This means that the advantage of zero power dissipation under static conditions is somewhat annulled in actual operation because power is dissipated as the logic gates switch states in response to logic signals. This is particularly so for fast logic circuits with very rapid clock pulse rates.

The operation of many digital logic circuits is, in fact, dynamic in nature such that logic gates in the network are activated repetitively by clock pulses, as, for example, is true for the circulating register, Fig. 11-31. In this situation it is not necessary for a bit storage stage in the register to have the two static stable states characteristic of a flip-flop. Consider the bit storage stage of such a *dynamic shift register* in Fig. 11-43, which is comprised of two inverter gates. In this circuit the Q_1, Q_2 inverter is coupled to the Q_4, Q_5 inverter through Q_6, while Q_3 serves to

Figure 11-43 Two MOS inverters comprise one bit of a dynamic shift register.

isolate the stage from the one preceding. Capacitors C_1, C_2, and C_3 are stray capacitances, which, however, are important to the dynamic operation of the circuit. Note that two clock pulses, v_1 and v_2, are required, but that these replace the usual dc drain supply potential.

The dynamic operation of this circuit is a series of sequential steps as follows. When v_1 is at state 1, Q_2 and Q_3 conduct so that C_1 charges through Q_3 to the input state level (either state 1 or state 0) while C_2 charges through Q_2 to state 1. When v_1 returns to state 0, Q_1 is either on or cut off, depending upon the charge on C_1, that is, upon the input state level during the time v_1 was state 1. If the input level was state 1, C_1 is charged, Q_1 conducts and this discharges C_2. Conversely, if the input level was state 0, C_1 is uncharged, Q_2 is cut off, and the state 1 charge on C_2 remains. Note that the result of v_1 cycling from state 1 to state 0 is to put on C_2 the inverse of the input state. That is, Q_1 and Q_2 (together with Q_3) comprise a *dynamic inverter gate*.

During the time v_1 is state 0, v_2 goes to state 1 and then returns to state 0. The dynamic inverter gate consisting of Q_4, Q_5, and Q_6 performs in a similar manner to the input gate. The net result is that C_3 is charged to the same state as C_2 and Q_4 is either conducting or cut off depending upon the charge on C_2. Thus, the pair of clock pulses v_1 and v_2 have caused the circuit to store the input bit dynamically for the duration of the pulse pair and to present the input bit level at the output terminal.

A cascade of such stages causes the input signal to be successively shifted from stage to stage as in a conventional shift register comprised of flip-flops. It is true, of course, that the clock pulses must persist in order to preserve the information stored in the register. This also implies that dynamic registers are most often used

Figure 11-44 *n*-word, *b*-bit shift register memory.

as circulating registers because data reaching the output terminals is preserved by reintroducing it at the input. The advantage of dynamic operation is circuit simplicity (compare Fig. 11-43 with Fig. 11-41 as the basis of a master-slave flip-flop, Fig. 11-21) and minimal power dissipation.

The organization of a shift-register memory based on circulating-register principles is shown in Fig. 11-44. Each register stores one bit of each digital word and the words are shifted along the registers in synchronism since the registers are all activated by the same clock signals. The read, write, and output gates, analogous to those in Fig. 11-31, determine whether the output signals are recirculated or lost, read as output and recirculated, or replaced by new input signals. These gates are activated in unison by mode control circuitry so that all bits of a word are written and read synchronously. In this connection the mode control circuit also performs address logic by keeping track of the location of the words as they circulate through the memory.

Memory capacity may be increased by paralleling additional shift registers to increase the number of bits in each word. Alternatively, the number of words may be increased by cascading additional shift registers. This has the disadvantage that the *access time*, that is, the time required to read out any given word, increases proportionally since the total time required for information to circulate through the memory depends upon the number of bits in each register.

Random-Access Memories

The access time in a shift-register memory depends upon the word address and upon the word storage capacity of the memory since information is only available sequentially at the shift-register outputs. In a *random-access memory*, or *RAM*, the access time is independent of the location of information in the memory and addressing logic permits immediate access to any information stored in the memory. A RAM is organized into word lines and bit lines much like a ROM except that the information is stored at each intersection by the state of a flip-flop memory cell.

Consider a single MOS memory cell in Fig. 11-45, which is similar to the MOSFET binary, Fig. 11-41, except that a total of four control FETs are used. When both the x-address line and the y-address line are at state 0, the four control

Figure 11-45 MOS memory cell for RAM.

transistors Q_3, Q_4, Q_7, and Q_8 are all cut off and the memory cell is electrically isolated from its surroundings. The cell is accessed by putting both the x-address line and the y-address line at state 1. This turns on all four control transistors so that logic levels at the bit and NOT bit lines correspond to the state of the binary. Alternatively, the state of the binary can be set by the logic levels presented externally to the bit and NOT bit lines.

This static memory cell is wasteful of space, since eight transistors are required per bit, and is also wasteful of power, since current is present in one or the other transistors in the binary. Power dissipation can be reduced by clocking the load transistor gates, V_{gg}, so that Q_2 and Q_6 are cut off except when the cell is addressed. This causes the current to be zero and memory information is retained by the charges on the stray capacitances at the gates of Q_1 and Q_5, much as in the case of the dynamic inverters in Fig. 11-43. In this case it is necessary to *refresh* the charges on the stray capacitances by periodically turning on Q_2 and Q_6 so that unavoidable leakage currents do not cause information to be lost. Dynamic memory cells can also be designed using as few as three transistors per bit, which considerably increases the memory capacity of the RAM.

A 16-bit RAM is shown in Fig. 11-46. Address decoders select the appropriate x and y lines corresponding to the address logic input and the state of the binary at the selected intersection appears at the output terminals. The information at any intersection is set by introducing the desired bit states to the write input. Practical RAMs are available with storage capacities as great as 2^{12} (4096) bits and with access times of the order of 10 ns.

Figure 11-46 16-bit RAM.

SUGGESTIONS FOR FURTHER READING

Arpad Barna and Dan I. Porat: "Integrated Circuits in Digital Electronics," John Wiley & Sons, Inc., New York, 1973.

H. V. Malmstadt and C. G. Enke: "Digital Electronics for Scientists," W. A. Benjamin, Inc., New York, 1969.

Basil H. Vassos and Galen W. Ewing: "Analog and Digital Electronics for Scientists," John Wiley & Sons, Inc., New York, 1972.

EXERCISES

11-1 Carry out the binary additions for $20 + 14$, $15 + 34$, and $56 + 25$. Check by forming the binary numbers of the decimal sums. Sketch the waveforms of the individual numbers and their sums.

Ans.: 100010; 110001; 1010001

11-2 Carry out the binary subtractions for $25 - 14$, $35 - 16$, and $12 - 3$. Check by forming the binary numbers of the decimal subtractions. Sketch the waveforms of the individual numbers and their differences.

Ans.: 1011; 10011; 1001

11-3 Sketch the waveforms of the binary numbers 1101 and 1001 and their sum. Compare the waveform of the sum with the output waveform of an AND gate with 1101 and 1001 as input signals.

11-4 Using De Morgan's theorem, simplify the logic expression for NOT exclusive-OR. Why is this sometimes called an *equality* comparator?

Ans.: $A \cdot B + \bar{A} \cdot \bar{B}$

11-5 Develop the truth table for the full adder in Fig. 11-17.

11-6 Sketch the waveforms at the output of each gate of a serial full adder using Figs. 11-18a, 11-16, and 11-17, if the input waveforms are the binary numbers 1101 and 1001. Verify that the output waveform represents the sum of the inputs.

11-7 Develop the truth tables for a NOR binary and a gated NOR binary and contrast with Tables 11-12 and 11-13.

11-8 Specify the states of the 4-bit ring counter similar to Table 11-17 if the Q output of FFD is connected to J of FFA and $\bar{Q}$ of FFD to K of FFA. Assume the initial state of this so-called *switch-tail* counter is all flip-flops cleared.

11-9 Determine the output of the BCD-to-decimal decoder in Fig. 11-33 to an input state specified by $\overline{ABCD}$. Repeat for $AB\bar{C}D$.

Ans.: 4 and 8; 3 and 9

11-10 What is the maximum number of lamps that can be lit using any input state in the BCD-to-decimal decoder in Fig. 11-33? Is there an input state for which no lamp is lit?

Ans.: two; no

TWELVE

DIGITAL MEASUREMENTS

Digital instruments use the logic circuits and techniques described in previous chapters to carry out measurements and to process data. A significant advantage of digital over analog instruments is that digital data is decisive. The least significant bit must be either a 1 or a 0, so that precision is increased simply by employing additional digits. Furthermore, digital signals can be amplified indefinitely and stored accurately. Because of the off-on character of digital circuits, drift and stability problems are inherently negligible. On the other hand, digital instruments tend to be more complex than their analog counterparts. The advantages of reliability, accuracy, and flexibility of digital circuits were first realized in digital computers but now find application in a wide range of instrumentation and measuring circuits.

DIGITAL INSTRUMENTS

Time-Interval Meter

The time interval between two events can be accurately measured by counting the number of cycles of a stable oscillator during the time interval between pulses signaling the two events. This is accomplished, Fig. 12-1, by triggering a binary by the input pulses to provide a gate pulse with a duration equal to the interval between pulses. If the oscillator frequency is 1000 Hz, the cascaded decade counters display the time interval directly in milliseconds.

The timing accuracy depends upon oscillator stability and is subject to a one-cycle uncertainty, depending upon when the gate pulse is initiated relative to the oscillator pulses. One extra count can be registered if the gate pulse starts during an oscillator pulse compared with the count recorded if the gate opens between pulses, as illustrated in Fig. 12-2. Because of this 1-bit uncertainty it is desirable to have the oscillator frequency high enough to ensure a large total count.

Frequency Meter

Similar techniques are used to measure the frequency of an unknown oscillator. The number of cycles in the unknown signal is counted during an accurately known time interval. For this purpose the input signal is first converted into a

Figure 12-1 Block diagram and waveforms of time-interval meter.

Figure 12-2

series of pulses, one per cycle, by means of a Schmitt trigger followed by a differentiator and clipper, as shown in Fig. 12-3. The number of input pulses is counted during a given time interval. If, for example, the interval is 1 s, the frequency of the unknown is displayed directly in hertz.

It is convenient to derive the time interval by scaling down the output of a stable crystal oscillator. The output pulse from the last flip-flop of the scaler is used directly to control the gate. Note that the frequency range of the instrument may be changed easily by changing the scale factor of the oscillator counter. In effect, the instrument compares the unknown frequency of the input signal with the known frequency of the crystal oscillator.

Figure 12-3 Digital frequency meter.

The frequency meter is started by closing the start switch and subsequently opening it after one counting interval. In practical instruments, control circuits are included to perform this operation, as well as to transfer the count to memory registers, and to clear the frequency-count register. In this mode the instrument repetitively samples the input frequency and can, for example, detect and display slow frequency changes of the input signal. Control circuits employ the same logic gates and networks as used in the signal-handling portions of the instrument.

The frequency measurement is subject to the same 1-bit uncertainty as in the case of the time-interval meter. The effect is minimized by choosing a scaling factor for the oscillator counter that maximizes the number of input signal pulses during the gate interval. Alternatively, low-frequency input signals can be measured by deriving the gate pulse from one period of the input signal and counting the number of oscillator pulses during this time, just as in the time interval meter.

Digital Voltmeter

A basic feature of digital instruments is conversion of input signals into digital data. The instruments discussed in the previous section employ quite simple pulse circuits to accomplish this *analog-to-digital* or, *A-D*, conversion. A number of different techniques have been devised to convert other analog signals into digital signals. A typical example of interest is used in one type of *digital voltmeter*, an instrument that provides a direct-reading numerical display of the voltage being measured.

Consider the simplified block diagram of a digital voltmeter sketched in Fig. 12-4. The unknown dc voltage applied to the input terminals is compared with a staircase waveform produced by a diode pump circuit fed from a 10-kHz oscillator. When the staircase signal is equal to the input voltage, the comparator interrupts transmission of further oscillator pulses. The parameters of the diode pump are selected so that the height of each step of the staircase is equal to 1 mV.

Figure 12-4 Simplified block diagram of digital voltmeter. (*Princeton Applied Research Corp.*)

Therefore, the number of steps is directly equal to the unknown voltage in millivolts, and the steps are counted and displayed by a three-decade counter, memory register, and decimal display.

The control circuits subsequently execute a memory transfer for display and discharge the diode pump capacitor so that the cycle repeats. A new reading is attained in less than 1 s. Actually, in practical instruments the control circuit has other functions as well. It is possible, for example, to change input-multiplier resistors automatically so that voltages from 100 mV to 999 V can be measured with the same 1-digit precision. In addition, the circuit can automatically adjust to either input-voltage polarity.

A-D AND D-A CONVERSION

Conversion of analog signals into their digital counterparts is characterized by the speed and accuracy with which the equivalences can be established. The reverse process, *digital-to-analog*, or *D-A*, conversion is of equal interest, and this generally proves easier to accomplish than the A-D step. Both A-D and D-A conversion can be achieved in a number of different ways which vary in complexity and effectiveness.

The simple staircase-waveform A-D conversion discussed in the previous section is limited in accuracy by the precision with which each step can be generated. In addition, the possibility of erroneous readings is present if the input is not strictly a dc signal. Suppose, for example, the input signal contains noise pulses or a 60-Hz component from the power mains. Then it is possible for the staircase waveform to become equal to the input signal plus a noise voltage and produce a false reading. This leads to annoying fluctuations in the numerical display on successive measurement cycles.

Dual-Slope Integration

The above difficulty is overcome in the *dual-slope integrator*, Fig. 12-5, which is a simple and popular A-D converter. In this approach the input signal is integrated for a fixed time interval determined by the time required for the counter to fill up completely. At the last count the overflow pulse causes the control logic to switch the integrator input to a fixed negative reference voltage. Integration of the reference voltage causes the output voltage of the integrator to decrease linearly towards zero. At this point the zero-detector comparator activates control logic to transfer the count in the BCD counters to the output registers and to initiate the next measurement cycle.

In effect the system compares the charge placed on the integrating capacitor by the unknown input signal during a fixed time interval with the charge removed by the known reference voltage in the measured time interval. Since the same

Figure 12-5 Dual-slope integrator.

operational-amplifier integrator and oscillator is used in both measurements, circuit errors are minimized and accuracy is limited primarily by the stability of the reference voltage.

Since the input signal is integrated during the initial input period, random-noise signals and alternating components are averaged out. The dual-slope integrator is capable of fifteen 5-digit readings per second compared to only two 3-digit readings per second in the case of the staircase A-D converter. The dual-slope technique is used in digital voltmeters as well as in other A-D conversion applications.

Successive-Approximation Converter

Very rapid operation is achieved by the *successive-approximation* converter, Fig. 12-6. In this technique a sequence of voltages having a 8421 BCD weighted code are successively compared with the input signal. Each successive voltage level is stored if it is lower than the input signal or rejected if it is greater than the

Figure 12-6 Simplified block diagram of successive-approximation A-D converter.

Figure 12-7 The use of successive approximations to convert 11.55 V into its BCD equivalent.

input signal. After the most significant digit in the input signal is thus approximated, a second series of voltages in the 8421 pattern, but a factor of 10 smaller, is added to the stored value and the successive sums compared to the input signal. The process is repeated digit by digit until the desired precision is attained.

The successive-approximation process is indicated schematically in Fig. 12-7 for the specific case of converting an 11.55 V input signal into its binary equivalent. At START the control logic, Fig. 12-6, generates the binary word 1000 and places it into the BCD register. The D-A converter, which may be similar to those discussed in the next section, converts this word into its analog signal, 8.00 V. This value is compared with the input signal and is stored since it is less than the input signal. Next, the control generates the binary word 1100, which has the analog equivalent $8.00 + 4.00 = 12.00$ V. Since this value exceeds the input signal, the second approximation value, $0100 = 4.00$ V, is rejected. The third approximation is $1000 + 0010 = 1010 = 8.00 + 2.00 = 10.00$ V and this value, being smaller than the input, is stored. The final approximation, $1010 + 0001 = 10.00 + 1.00 = 11.00$ V, is also less than the input and is kept. The result of this four-step process is the first approximation to the input signal, that is, the first digit, which is 1011, or 11.

This process is repeated for the second digit, as illustrated in Fig. 12-7, except that the analog equivalents of the binary trial words are reduced to 0.80, 0.40, 0.20, and 0.10 V by the control logic. The result, 0101 = 5, is the second digit. This is now repeated for the next digit, etc., until the desired precision is attained. In the specific case considered, the BCD equivalent of 11.55 V is 1011 . 0101 0101.

The speed of the successive-approximation converter is limited only by propagation delays in the digital logic circuits and the converter is capable of as many as one thousand 4-digit readings per second. Furthermore, the conversion time is independent of the magnitude of the signal, in contrast to the staircase-waveform and dual-slope integration approaches. The successive-approximation converter is nonintegrating, however, so that the input signal is often introduced through a low-pass filter to reduce spurious noise signals.

D-A Ladder Networks

It often proves useful to convert binary output signals into their analog equivalents so that, for example, in digital control applications the output signal may be used to adjust the system being controlled. Another instance is the D-A conversion used in the successive-approximation converter discussed in the previous section. Most often, D-A converters use resistor arrays called *ladder networks*.

Consider the *weighted-resistor ladder* network and operational amplifier in Fig. 12-8. As discussed in Chap. 8, the output of the operational amplifier is just the sum of the input signals with each input multiplied by the ratio of the feedback resistance to its input resistor. That is, the analog output signal is

$$V_a = -V\left(\frac{D}{1} + \frac{C}{2} + \frac{B}{4} + \frac{A}{8}\right) \tag{12-1}$$

where V is the magnitude of the logic 1 state for any of the $ABCD$ input bits. Note that this network is weighted in the usual 8421 BCD code.

Table 12-1 lists the analog output signal calculated from Eq. (12-1) for the BCD words corresponding to decimal digits. Clearly the output signal voltage is analogous to the decimal values.

For good conversion accuracy, precision resistors must be used in the ladder network. Also, the resistance values must cover a wide range, particularly if binary

Figure 12-8 Weighted-resistor 4-bit D-A converter.

Table 12-1 Analog output from weighted resistor D-A converter

Decimal	BCD word DCBA	$V_a - V^*$
0	0000	0
1	0001	−0.5
2	0010	−1.0
3	0011	−1.5
4	0100	−2.0
5	0101	−2.5
6	0110	−3.0
7	0111	−3.5
8	1000	−4.0
9	1001	−4.5

* Assuming logic 1 state is +4.0 V.

words longer than 4 bits are converted. Both of these difficulties are overcome by the so-called *binary ladder network*, Fig. 12-9. In this circuit only two values of resistance are required, independent of the number of bits in the input binary words. Note also, that an operational amplifier is not necessary to sum the individual bit input signals.

The analog output signal corresponding to BCD word inputs can be determined from Fig. 12-9 by straightforward circuit analysis. It is assumed that all logic 1 bits are at the same potential and that all logic 0 bits are at ground. The result is that the analog output voltages are just one-third of those listed in Table 12-1 for the weighted-resistor converter.

Figure 12-9 Binary-ladder 4-bit D-A converter.

Figure 12-10 Practical 4-bit D-A converter.

In either ladder network the accuracy of the analog output depends upon the voltage magnitude of each bit in the binary input signal. Since this quantity is not often closely controlled in binary logic circuits, a practical D-A converter, Fig. 12-10, may apply diode limiting to each bit in order to set the voltage levels accurately. This has the further advantage that the total input signal into the summing amplifier is less likely to saturate the amplifier. With diode limiting, the voltage level of each logic 1 bit is 0.6 V, rather than 4.0 V or so as in the case of normal positive logic signals. Note also that a circuit similar to Fig. 12-10 is required for the BCD input word corresponding to each decimal digit. It often proves possible to minimize circuit complexity by multiplexing the D-A converter in a fashion similar to that discussed in the previous chapter in connection with ROM display decoders.

DIGITAL PROCESSORS

Digital logic circuits permit signal processing of an entirely different nature than is possible using analog instruments. A number of digital instruments which are, in effect, small digital computers have been designed to carry out highly specialized calculations. Although the input and output signals used with such instruments are most often analog in nature, the power and flexibility of digital logic circuits result in extremely versatile and flexible performance.

Digital Filter

The so-called *digital filter* performs much the same task as, for example, an inductance-capacitance filter network, but in a fashion that provides filter parameters not possible with conventional component values. The operation of a digital filter rests upon the same duality between frequency and time already noted in connection with the Fourier series representation of complex waveforms. That is, the spectrum of frequency components in a signal is an equally valid description of the signal as is the time variation of the signal given by the waveform. Correspondingly, while most often a filter network is described by means of its frequency-response characteristic, an equally useful description is possible using the transient properties of the circuit. Note that the frequency-response characteristic is given by the ratio of the output signal, $v_o(f)$, to the input signal, $v_i(f)$ at a given frequency,

$$\frac{v_o(f)}{v_i(f)} = F(f) \tag{12-2}$$

A similar input-output relation exists in the time domain, except that the input and output signals are noted at separate times

$$\frac{v_o(t)}{v_i(t')} = T(t, t') \tag{12-3}$$

Both $F(f)$ and $T(t, t')$ are satisfactory descriptions of filter network properties.

It is possible to develop an expression for $T(t, t')$ if the frequency-response characteristic is known in much the same fashion as the Fourier components of a complex waveform are found. The details of this transformation are not of direct interest here. It is important to observe, however, that the timelike representation permits a direct calculation of filter performance by digital techniques in the following way. The output of the digital filter at any instant is written in terms of the input signal at several discrete times earlier,

$$v_o(t) = a_1 v_i(t - \delta) + a_2 v_i(t - 2\delta) + a_3 v_i(t - 3\delta) + \cdots \tag{12-4}$$

Here, the a coefficients are the representation of the $T(t, t')$ characteristic of the filter. The calculations indicated by Eq. (12-4) are accomplished by the circuit shown in Fig. 12-11. First, the input signal is sampled at discrete intervals and digitized. Each sample is delayed an appropriate time interval, multiplied by a constant characteristic of the filter performance desired, and the result is summed to yield the output signal. The output may be converted to analog form, if desirable.

The digital filter, in effect, continuously calculates the filter output arising from input signals. Since filter performance is described by the a coefficients which are easily adjustable, the same digital filter may be used to synthesize a variety of frequency-response characteristics. No inductances are needed, which is often advantageous in low-frequency and integrated-circuit applications. Furthermore, because of the discreteness of digital data, precise performance, such as very large attenuation at specific frequencies, is achievable.

Figure 12-11 Block diagram of digital filter.

The frequency-response characteristic of a simple low-pass digital filter, Fig. 12-12, shows the very steep attenuation and the excellent low-frequency properties that can be achieved. Note that the a coefficients in Fig. 12-11 refer to Eq. (12-4) and that the sampling rate is $1/\delta = 100$ Hz. A simple change in the coefficients, which is accomplished by altering amplifier gains, produces an entirely different frequency-response characteristic with the same basic circuit. This aspect is explored further in the Exercises.

Signal Correlators

Recurrent signals heavily masked by noise can often be recovered by an averaging technique which takes advantage of a basic difference between recurrent or periodic signals and random-noise signals. This is that the average value of a random-noise signal is equal to zero. Instruments called *signal averagers* in effect average out the random noise so that only the signal waveform of interest remains.

Figure 12-12 Frequency-response characteristic of digital filter.

Figure 12-13 Simplified block diagram of signal averager. (*Princeton Applied Research Corp.*)

According to the block diagram of a signal averager, Fig. 12-13, the input signal plus noise is sampled successively by each of the 100 FET gates which are activated in sequence by a ring counter. The voltage level of each sample is averaged and stored by the *RC* capacitor memories. After many repetitive waveforms have been sampled, the noise is averaged toward zero and the voltage stored in each capacitor represents the signal voltage at a specific point on the waveform. The full waveform may be displayed at the output terminal by successively activating the gates in conjunction with the sweep output signal.

The trigger input signal keeps the sampling in synchronism with the signal waveform so that the same point on the waveform is sampled each time. It is usually not possible to derive the trigger signal from the input signal itself because of the disturbing influence of the noise. Note also that display of the averaged waveform may be at the same rate as the input signal, or, alternatively at a rate either faster or slower than that corresponding to the input signal. This makes it possible, for example, to preserve the output waveform on a chart recorder if desired. An example of a signal waveform recovered from noise is shown in Fig. 12-14.

Figure 12-14 (*a*) Signal masked by noise and (*b*) waveform recovered by signal averager. (*Princeton Applied Research Corp.*)

Another version of this technique is the so-called *boxcar integrator* in which a single gate and capacitor memory is used. The point on the waveform being sampled is swept slowly through the waveform. The signal is averaged at each position and read out immediately before moving on to the next point. Because of the inherently simpler circuitry, boxcar integrators can be used for waveform durations of the order of 10^{-9} s, whereas the multiple channel types are limited to signal durations longer than about 10^{-5} s. On the other hand, the boxcar integrator requires about 100 times as long to recover the entire waveform because only a single channel is used.

A more sophisticated signal-processing technique related to the signal averager is based on the correlation between two time-dependent signals. The so-called *cross-correlation function* between two signals $f_A(t)$ and $f_B(t)$ is defined as

$$C_{AB}(\tau) = \lim_{T \to \infty} \frac{1}{T} \int_T f_A(t) f_B(t + \tau) \, dt \qquad (12\text{-}5)$$

In the special case where $f_A(t) = f_B(t)$, $C_{AA}(\tau)$ is called the *autocorrelation function* and, because of the time average involved in Eq. (12-5), noise reductions similar to those achieved with signal averaging are possible. The autocorrelation function contains much additional information, however, and may be used, for example, to determine the Fourier components of a waveform. Analogously, the cross-correlation function basically measures how closely two signals correspond to each other and therefore it provides a much deeper insight into the phenomenon represented by the two signals than any analysis of the properties of either one separately.

The calculation corresponding to Eq. (12-5) is carried out by a *signal correlator*, Fig. 12-15. One input signal is digitized and presented to a 100-channel shift register. The output of each stage in the register is multiplied by the second input signal and the result averaged and stored in each RC memory. The movement of

Figure 12-15 Simplified block diagram of signal correlator. (*Princeton Applied Reasearch Corp.*)

Figure 12-16 Transient recorder based on dynamic shift register.

the first input signal along the shift register provides the delay time τ in Eq. (12-5). The correlation function is displayed by scanning the capacitor memory bank in conjunction with a sweep signal output. While the many applications of signal correlation analysis are not of direct interest here, it is clear from these examples that digital techniques provide diverse and powerful signal-processing instruments.

Transient Recorder

A dynamic shift register is a useful and convenient way of storing transient analog signals for subsequent display or analysis. A typical unit based on a 128-word, 6-bit shift register, Fig. 12-16, is capable of storing transient signals as short as 10^{-5} s or as long as 5 s and replaying either one at rates suitable for oscilloscope or chart recorder display.

The input A-D converter samples at rates adjustable from 10^{-7} s per word to 5×10^{-2} s per word while the dynamic shift register operates at a constant 10-MHz rate. The input gates are activated by the control logic to insert the digital words into the register at the proper times so that a full 128-word recording can be made of both short and long duration signals. Stored transient signals circulate through the memory until replaced by new signals or purposefully erased during subsequent read out.

During playback, output signals may be obtained at the same rate as during recording. Most often, however, the output signal is delivered at a rate selected to be consistent with the display or analysis circuitry. This flexibility is possible using a sampling technique at the output gates similar to that of the sampling oscilloscope described in Chap. 10. As the transient signal circulates through the memory, it becomes, in effect, a recurrent signal and can be sampled on successive portions of the cycle. The logic circuitry also develops a linear ramp sweep output signal which is consistent with the duration of the output signal. Ramp durations ranging from as short as 10^{-3} s, suitable for oscillographic display, to 5 s, appropriate for chart recordings, are feasible.

DIGITAL COMPUTERS

Organization

A *digital computer* is a complex array of logic gates and associated circuitry organized to perform logic computations by manipulating waveforms representing digital numbers. The great power of electronic digital processing stems from the variety of phenomena that can be represented and analyzed logically. Typical examples range from simple arithmetic calculations to scientific computations which obey some physical law and bookkeeping activities which follow the principles of accounting. Because the speed of electronic circuits is so great, digital computers can complete extremely complex and extensive calculations in practical lengths of time.

The circuits of digital computers are designed to carry out numerical calculations of all kinds. Therefore, the machine is furnished specific instructions pertaining to a desired computation, in addition to all the digital numbers involved. These instructions are represented by pulse waveforms similar to those representing numbers. A complete set of instructions, together with the numbers associated with a given problem, is called the *program*.

A digital computer comprises five major parts: input, output, memory, control, and arithmetic units, Fig. 12-17. The input and output units present the program to the machine and retrieve results. The computer *memory* stores each instruction and number of the program until needed during the course of the computation. The instructions of the program are stored separately from numbers since the two are used in fundamentally different ways in the calculation. The location of each number is specified by an *address* which is used to tell the machine where a particular number is stored in the memory.

The *arithmetic* unit contains logic circuits such as ADD gates which actually perform the calculations specified by the program. The arithmetic unit also includes *registers* for temporarily storing digital numbers. The reason for this is that, ordinarily, only one number at a time is recalled from the memory. If two numbers are to be added, the first one must be stored until the second one is available before they can be presented simultaneously to the ADD gate.

The function of the *control* unit is to interpret each instruction and set the circuits of the computer accordingly. It is the most heterogeneous part of the computer and is composed of logic gates, registers, and a *clock* circuit which

Figure 12-17 Organization of digital computer.

regulates the basic speed at which the machine operates. The clock provides a continuous train of pulses which are used in connection with the various logic gates to direct the digital waveforms through the different parts of the computer. Because the functions of the control unit and the arithmetic unit are central to the performance of the computer, the two units together are referred to as the *central processing unit*, or *CPU*.

Microcomputers

The circuits of a digital computer are exemplified by the most recent development in digital computers, the pocket-sized calculator, or *microcomputer*, Fig. 12-18. Microcomputers use digital-computer principles and circuits to carry out arithmetic calculations extending from simple addition and subtraction to evaluations of transcendental functions. This range of capability is exhibited in the keyboard nomenclature visible in Fig. 12-18.

Actually, the keyboard represents the input unit of the microcomputer, while the numerical display, which is most often of the seven-segment LED type, is the output unit. The keyboard is used to introduce numbers into the memory and to select the proper internal program to carry out the desired calculation. Numbers introduced by the keyboard are indicated by the display, as are the results of computations.

Figure 12-18 Typical Microcomputer. (*Texas Instruments, Inc.*)

DIGITAL MEASUREMENTS **395**

The operation of a microcomputer may be described with the aid of the internal block diagram, Fig. 12-19. The keyboard and the display are multiplexed by the keyboard scan circuit which sequentially pulses each key. The keys representing numbers are connected to the digit drive circuit which activates the digits of the display much as in Fig. 11-39 of the previous chapter. The display segments are activated in synchronism by the segment drive circuit.

Depressing a key causes the keyboard encoder to generate a digital word corresponding to the number or instruction represented by that key. This word is passed to the subroutine register where it may be modified by the result of a previous calculation, and thence to the address register of a 13,000-bit ROM. This ROM contains the instructions and look-up tables necessary to carry out the calculations. The ROM instruction words are interpreted by the instruction decoder and activate control logic which actually sets the digital circuits to perform calculations. Under the direction of the control logic, instruction words may be

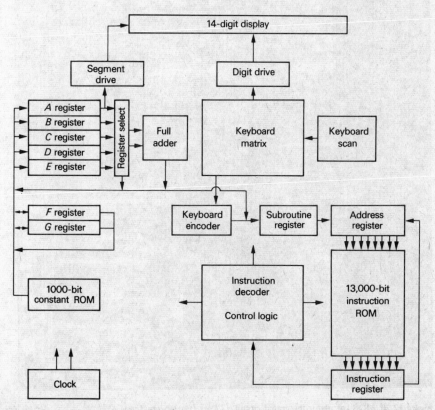

Figure 12-19 Simplified block diagram of pocket-sized microcomputer. (*Texas Instruments, Inc.*)

reintroduced into the ROM so that an entire series of instructions can be generated in order to calculate, for example, a mathematical function by a series approximation.

Actual calculations are performed by the arithmetic unit which consists of registers A through E together with a full adder. Words in selected pairs of registers can be added, exchanged or compared and the results stored in a third register. The A register is connected to the segment display decoder and its contents are displayed at the end of the computation cycle. Also part of the arithmetic unit are the F and G registers, which are used to store intermediate results, and a 1000-bit ROM which contains 16 numerical constants used in calculations. A typical example is the constant ln 10, 2.302585092994, which is used for calculating logarithmic functions.

Figure 12-20 MOS calculator integrated circuit. (*Texas Instruments, Inc.*)

Microcomputers are feasible because of large-scale integration possible through MOS technology. The microcomputer in Figs. 12-18 and 12-19, for example, has 25,000 transistors, yet it is entirely contained in a small enclosure and operates from a small internal battery. Actually, the entire circuit of a microcomputer of somewhat less capability can be fabricated as a single integrated circuit, Fig. 12-20. Coupling this integrated circuit together with a keyboard, visual display, and appropriate display driver yields a very useful hand calculator.

The microcomputer hand calculator is a specific example of a digital instrument organized around a CPU which has come to be called a *microprocessor*. Very extensive digital logic circuits capable of a wide range of measurement and control applications have been designed using microprocessors associated with specific ROMs which contain the programs and data appropriate to the assigned tasks. The design of a microprocessor may emphasize one or another particular aspect to maximize effectiveness. For example, one design may feature ease of arithmetic operations whereas another may facilitate input- and output-data transfer. While either type has all of the usual components of a CPU, the former is used for lengthy numerical analysis and the latter for control applications.

Programming Languages

Digital computers in general have much greater computational flexibility than that available in microcomputers because the program is not made an inherent part of the machine. Rather, a different program is inserted into the computer memory for each computational task so that very elaborate and sophisticated programs can be employed.

A digital computer is operated by first inserting the program into the memory, as, for example, by means of a set of punched cards. When the machine is started, the control unit reads the first instruction, prepares the circuits of the computer accordingly, and causes the appropriate number to be read from memory as specified by the address in the instruction. At the completion of the indicated operation, the result is returned to the memory at the address also specified in the program. The control unit then passes on to the next instruction. The machine proceeds sequentially through the instructions, placing final results in the output unit until the final instruction *stop* is reached. At this point the computation is finished and the desired result is located in the output unit.

It is useful to illustrate by means of a rudimentary example how the control unit prepares the circuits of the machine in accordance with the instructions. Consider the partial block diagram, Fig. 12-21, in which either of two numbers can be operated upon and returned to the memory, depending upon the instruction. Suppose the instruction 10101, as shown, is abstracted from the memory and stored in a register connected to the several AND gates. This means that the number A passes through the circuit, emerges as NOT A, and is returned to the memory. The number B is not used since it does not pass the first AND gate.

Figure 12-21 Illustrating how an instruction determines path of number through digital computer.

Other instructions cause different operations to be performed, as may be determined by following the path through the circuit in each case:

Instruction	Interpretation
10101	Take A and return NOT A to memory
10010	Take A and return A to memory
01101	Take B and return NOT B to memory
01010	Take B and return B to memory

In effect, instructions set up signal paths through the computer by activating logic gates of the control unit.

It is not feasible, or necessary, to trace the signal path for each instruction through the complete circuit of the computer when preparing a program. Rather, in the design of the machine, a tabulation of the instructions to carry out the various operations is developed, corresponding to the control logic circuits. This dictionary of *machine language* is used to write programs, and it obviates the need to refer to the circuits of the computer itself. Nevertheless, because each minute step must be detailed in the instructions, the preparation of a program in machine language is a long and arduous task and programming errors are frequent. For example, transposing a 0 for a 1 in any one instruction may result in a nonsense step, or, more seriously, a false instruction, and such errors are difficult to locate.

To help this situation, symbolic programming languages have been developed in which readily recognizable symbols are used to represent machine-language instructions or groups of machine-language instructions. The computer manufacturer supplies translating programs which use the computer itself to convert symbolic programs into machine language. The simplest symbolic language is *assembly language*, and the program for translating a program written in assembly language into binary-coded instructions is called an *assembler*.

Symbolic statements in assembly language have nearly a one-to-one correspondence to machine-language instructions. Consider, for example, the simple program to add the numbers Z, B, and C together, which is represented by the set

Table 12-2 Comparison of assembly language and Fortran statement with machine language instructions

Assembly language	Machine language	Fortran statement
LDA Z	0 110 001 100 100 001	
ADA B	0 100 001 100 100 010	$D = Z + B + C$
ADA C	0 100 001 100 100 011	
STA D	0 111 001 100 100 100	

of binary instructions in the center column of Table 12-2. The corresponding statements in assembly language in the left column of Table 12-2 are quite mnemonic. They may easily be read as follows: "Load the number at address Z into register A; add the number at address B to the number in register A; add the number at address C to the number in register A; and store the number in register A at address D." Thus, this series of assembly-language statements accomplishes the desired routine. The assembler program translates the humanly interpretable statements in assembly language into the corresponding binary-coded statements in machine language.

Because each detailed instruction must be included, programming in assembly language is very tedious. Higher-level languages have been developed in which single statements can be translated into large groups of machine-language instructions. In this case, the translating program is called a *compiler* because, in effect, the computer compiles its own program following quite general instructions.

The most widely used programming language, called *Fortran* (for *formula translator*), employs stylized algebraic symbols and formulas as program statements. The familiar algebraic expression in the third column of Table 12-2, for example, is the Fortran statement which translates into the set of machine-language instructions in the center column. Clearly, it proves much simpler to program in Fortran than in either assembly language or machine language. It often turns out, however, that a machine-language program compiled in this fashion is inefficient in the use of computer memory space and in the total time required to execute the program. This is so because of the generalities inherent in a compiler. Thus, while Fortran and similar higher-level languages are widely used, assembly language is necessary to realize the maximum capabilities of any computer.

SUGGESTIONS FOR FURTHER READING

Thomas C. Bartee: "Digital Computer Fundamentals," 2d ed., McGraw-Hill Book Company, New York, 1966.

G. G. Bell and A. Newell: "Computer Structures: Readings and Examples," McGraw-Hill Book Company, New York, 1971.

Sam P. Perone and David O. Jones: "Digital Computers in Scientific Instrumentation," McGraw-Hill Book Company, New York, 1973.

EXERCISES

12-1 Devise a suitable control circuit in block-diagram form for the digital voltmeter in Fig. 12-4. Arrange the control so that a new measurement cycle is initiated immediately after each reading is attained.

12-2 Suppose the input signal to the digital voltmeter in Exercise 12-1 is 0.5 V dc together with a 0.5 V peak-to-peak sawtooth waveform with a 2s period. Determine graphically the first five readings displayed if the voltmeter happens to start a measuring cycle when the input signal is 0.5 V.

    Ans.: 0.67, 0.89, 0.52, 0.68, 0.92 V

12-3 The dual-slope integrating A-D converter, Fig. 12-5, delivers ten 4-digit readings per second. What is the oscillator frequency? Assume the input and reference voltages are equal.

    Ans.: 200 kHz

12-4 Determine the minimum gain necessary in the zero-detector comparator of the dual-slope integrating A-D converter of Exercise 12-3.

    Ans.: 10^4

12-5 In the so-called *voltage-to-frequency* A-D converter, Fig. 12-22, the pulse generator completely discharges the integrating capacitor each time the comparator senses equality between the integrated input signal and the reference voltage. What are the output frequencies for input signals of 10, 5, and 1 V?

    Ans.: 10^4 Hz, 5×10^3 Hz, 10^3 Hz

Figure 12-22 Voltage-to-frequency A-D converter.

12-6 How many reject trials occur in the successive-approximation A-D converter, Fig. 12-6, if the input signal is 8.08 V? Repeat for 0.62 V. If each comparison takes 10^{-4} s, determine how long it takes to achieve the conversion in each case.

    Ans.: 10, 9; 1.2×10^{-3} s, 1.2×10^{-3} s

12-7 Using circuit analysis of the binary-ladder D-A converter, Fig. 12-9, develop a tabulation of decimal numbers, BCD word input signals, and analog output signals analogous to Table 12-1.

    Ans.: 1, 0001, 1/24; 2, 0011, 2/24; etc.

12-8 In the practical D-A converter of Fig. 12-10, what are the analog output signals corresponding to the input words 0010, 0100, 0110, and 1000?

  Ans.: 0.15, 0.30, 0.45, 0.60 V

12-9 Calculate the frequency-response characteristic of the digital filter corresponding to Eq. (12-4) using the coefficients listed in Table 12-3. Do this by introducing sine waves of different frequencies at the input. Sample each input signal at the 100-Hz rate for one full cycle and compute the output signals using Eq. (12-4). Use these values to determine the rms output signal at each frequency. Plot the frequency-response characteristic over the frequency interval from 0.1 to 50 Hz. What is the ratio of output signal to input signal at a frequency of 10 Hz?

  Ans.: 0.1

Table 12-3 Digital filter coefficients for exercise 12-9 ($\delta = 10^{-2}$ s)

$a_1 = -a_{10}$	$= +0.0008$
$a_2 = -a_9$	$= -0.0025$
$a_3 = -a_8$	$= +0.0094$
$a_4 = -a_7$	$= -0.0350$
$a_5 = -a_6$	$= +0.1890$

12-10 Repeat Exercise 12-9 for an input signal which is a 2-V peak-to-peak 1-Hz square wave. From the output waveform, what is the name of this filter?

  Ans.: Differentiating filter

VACUUM TUBE CIRCUITS

The basis for all electronic circuits before the invention of the transistor was the electron vacuum tube. A great many variations in tube design have permitted applications ranging from very sensitive detectors to very-high-power amplifiers. Because of their small size and minimal power requirements, transistors have largely replaced vacuum tubes in electronic circuits, but many applications for which vacuum tubes prove to be useful still remain.

THE VACUUM DIODE

Thermionic Emission

The vacuum diode comprises a hot *cathode* surrounded by a metal *anode* inside an evacuated enclosure, usually glass, as sketched in Fig. A1-1. At sufficiently high temperatures electrons are emitted from the cathode and are attracted to the positive anode. Electrons moving from the cathode to the anode constitute a current; they do so when the anode is positive with respect to the cathode. When the anode is negative with respect to the cathode, electrons are repelled by the anode and the reverse current is zero. The space between the anode and cathode is evacuated, so that electrons may move between the electrodes unimpeded by collisions with gas molecules.

Cathodes in vacuum diodes take several different forms. The free electrons in any conductor, given sufficient energy by heating, will escape from the solid. Some materials are much more satisfactory in this respect than others, however, either because it is relatively easy for electrons to escape, or because the material can safely withstand high temperatures. Tungsten, for example, is a useful cathode material, because it retains its mechanical strength at extreme temperatures. A thin layer of thorium on the surface of a tungsten cathode filament increases electron emission, and appreciable current is attained at temperatures of about 1900 K. This form of cathode is directly heated by an electric current similarly to the filament in an incandescent lamp. It is used in vacuum diodes suitable for high-voltage applications. The conventional symbol for a vacuum diode of this type is shown in Fig. A1-2a.

The modern oxide-coated cathode, which consists of a metal sleeve coated with a mixture of barium and strontium oxides, is the most efficient electron emitter developed to date. Copious electron emission is obtained at temperatures

Cathode
Anode
Heater

Figure A1-1 Sketch of vacuum diode.

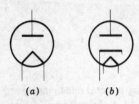

(a)   (b)

Figure A1-2 Circuit symbols for vacuum diodes of (a) filamentary-cathode and (b) heated-cathode types.

near 1000 K, which means that the power required is much less than for tungsten. Usually the oxide cathode is indirectly heated by a separate heater located inside the metal sleeve, as in Fig. A1-1. This isolates the heater current electrically from the cathode connection, which is a considerable advantage in electronic circuits. Also, an ac heater current may be used without introducing undesirable temperature variations in the cathode at the frequency of the heater current. Oxide cathodes are employed in the great majority of vacuum tubes. The heater is often omitted in the conventional circuit symbol, Fig. A1-2b, since it is not directly an active part of the rectifier.

Child's Law

When the anode is negative with respect to the cathode, the electron current is zero and the reverse characteristic is essentially that of an ideal diode. The forward characteristic is determined by the motion of electrons in the space between the cathode and the anode when the anode has a positive potential. Most often, the number of electrons emitted from the cathode is so great that the electric fields due to the electron charges drastically alter the electric field between the cathode and anode produced by the anode potential. Under these conditions the actual current is given by Child's law,

$$I_b = AV_b^{3/2} \qquad (A1-1)$$

where V_b is the voltage between anode and cathode, and A is a constant involving the tube geometry. According to Eq. (A1-1), the current increases as the three-halves power of the applied voltage; that is, the vacuum diode is nonlinear.

The current depends upon the separation between anode and cathode and upon the area of the cathode. Thus it is possible to obtain different current-voltage characteristics by altering the geometric shape of the anode and cathode. Two examples of practical diode characteristics are given in Fig. A1-3. The type 5U4 diode is a medium-power rectifier; the type 1V2 diode is designed for high-voltage low-current power supplies. The anode-cathode separation is considerable in the 1V2 in order to minimize the possibility of a discharge passing between anode and cathode on the reverse portion of the voltage cycle. Consequently, the current in the forward direction is smaller than in the case for the 5U4. The 5U4 is designed for use at lower voltages, so the separation between anode and cathode is smaller. The forward current is correspondingly greater at the same forward voltage.

Figure A1-3 Experimental forward characteristics of two practical vacuum diodes are much different because of different geometric construction.

In any practical vacuum diode the reverse current is not truly zero, because of leakage currents on the surfaces of the glass insulators and similar secondary effects. Typically, the reverse resistance is of the order of 10 MΩ. Since the forward resistance may be of the order of 100 Ω at a suitable operating potential, the ratio of the reverse resistance to the forward resistance, or the *rectification ratio*, is appreciable. The interelectrode capacitance between the cathode and anode limits the maximum frequency at which a vacuum diode is useful. The cathode-anode capacitive reactance is effectively in parallel with the electron current and tends to short out the reverse resistance at high frequencies.

VACUUM TUBES

The current in a vacuum diode is controlled by introducing a third electrode, the grid, between the anode and the cathode. Changes in grid voltage alter the anode current independently of the anode potential. This action makes the triode vacuum tube a useful amplifier.

The Grid

The grid in a vacuum triode usually consists of a wire helix surrounding the cathode, as sketched in Fig. A1-4a. If the grid potential is always negative, electrons are repelled and the grid current is negligible. This means that the power expended in the grid circuit to control the anode current is very small. It is desirable to minimize the area of the grid, so the number of electrons intercepted on their way to the anode is negligible. On the other hand, if the wires are spaced too widely, the grid's ability to control the anode current is reduced. Practical triodes are designed to strike a useful compromise between these conflicting requirements. The anode in vacuum tubes is commonly termed the *plate*, because of

Figure A1-4 (a) Vacuum triode and (b) circuit symbol.

its shape in early tube designs. The conventional circuit symbol for a triode is shown in Fig. A1-4b.

The grid potential alters the electric-field configuration in the space between the cathode and plate from that corresponding to the vacuum diode. Since the grid is much closer to the cathode, its potential is relatively more effective in controlling the current than is the plate voltage. Therefore, using Child's law, the current in the tube may be written†

$$I_b = A(\mu V_c + V_b)^{3/2} \tag{A1-2}$$

where A is a constant involving the tube geometry, V_c is the grid voltage, V_b is the plate voltage, and μ is a constant called the *amplification factor*. The amplification factor accounts for the greater effect of the grid voltage compared with the plate voltage. This equation agrees reasonably well with experimental current-voltage characteristics of practical triodes. It is difficult to evaluate A from first principles, however, and the exponent is often not exactly three-halves. Therefore, it is common practice to display the characteristics of practical tubes graphically, rather than attempt an accurate mathematical representation.

The action of V_c in Eq. (A1-2) opposes that of the plate voltage V_b since the grid potential is negative. An expression for the amplification factor is obtained by noting that these opposing actions cancel out if the electric charge induced on the cathode by the grid potential is equal and opposite to the charge produced by the plate voltage, or

$$-C_{gk}V_c + C_{pk}V_b = 0 \tag{A1-3}$$

† It has become commonly accepted to associate the subscripts p and b with the plate circuit and g and c with the grid circuit of a vacuum tube. Similarly, the subscript k refers to the cathode.

where C_{gk} is the capacitance between the grid and cathode. Solving for the amplification factor

$$\mu = \frac{V_b}{V_c} = \frac{C_{gk}}{C_{pk}} \tag{A1-4}$$

According to Eq. (A1-4) the amplification factor is increased if the grid is close to the cathode since the grid-cathode capacitance is increased. The plate is further away, and is shielded from the cathode by the grid, so that μ is always greater than unity. Actually, triodes with amplification factors ranging from 10 to 100 are commercially available.

Of the several graphical ways to represent the current-voltage characteristics of a triode, the most useful is a plot of the plate current as a function of plate voltage for fixed values of grid voltage. The curves for several grid potentials are called the *plate characteristics* of the tube. A typical set of plate characteristics, as provided by the tube manufacturer, is illustrated in Fig. A1-5. Note that each curve is similar to the current-voltage curve of a vacuum diode. Furthermore, the current at a given plate voltage is reduced as the grid is made increasingly negative. The plate current-voltage curves are displaced to the right with little change in shape for each negative increment in grid potential, in conformity with Eq. (A1-2). Comparing Eq. (A1-2) with the experimental curves shows why it is

Figure A1-5 Plate characteristics of type 6SN7 triode. Curve for zero grid voltage is in approximate agreement with Child's law.

necessary to use graphical data: although Child's law represents the general behavior of the plate characteristics, it is not sufficiently accurate to yield satisfactory quantitative results.

According to Fig. A1-5, the plate current is essentially zero at sufficiently large negative grid potentials. The negative grid voltage necessary to put the tube in this cutoff condition depends upon the plate voltage. In effect, the tube is an open circuit when cut off. If the grid is at a positive potential, there is appreciable grid current. Then the grid is the anode of a diode biased in the forward direction and represents a much lower resistance than is the case for a negative grid voltage. In most circuits this effect prevents the grid from ever becoming positive. Therefore, at zero grid voltage the tube is said to be saturated since it is in its maximum conducting condition. The range between cutoff and saturation is the normal grid-voltage range.

Pentodes

The capacitance between the grid and plate in a vacuum triode introduces serious difficulties when the tube is used as an amplifier at high frequencies. The ac plate signal introduced into the grid circuit through the grid-plate capacitance interferes with the proper operation of the circuit. Another grid is interposed in the space between the grid and plate to circumvent this difficulty. This *screen grid* is an effective electrostatic shield that reduces the grid-plate capacitance by a factor of 1000 or more. The screen grid is at a positive potential with respect to the cathode, and the current of electrons from cathode to plate is maintained. The helical grid winding is much more open than is the case for the control grid, so that the screen current is smaller than the plate current.

Two-grid four-electrode tubes called *tetrodes* are virtually obsolete except for certain special-purpose types. The reason for this is that electrons striking the plate dislodge other electrons. These may be attracted to the screen, particularly when the screen voltage is greater than the instantaneous plate potential. This introduces serious irregularities in the plate characteristics at low plate voltages. A *suppressor grid* is introduced between the screen and plate to eliminate this effect. The suppressor is held at cathode potential and effectively prevents all electrons dislodged from the plate from reaching the screen. The pitch of the suppressor grid helix is even larger than that of the screen grid. Therefore, the suppressor does not interfere with electrons passing from cathode to plate.

A tube with three grids is called a *pentode* because there are a total of five electrodes. The conventional circuit symbol of a pentode is shown in Fig. A1-6. In normal operation the screen and suppressor grids are maintained at fixed dc potentials. The influence of the plate potential on the electric field near the cathode is practically zero in pentodes, because of the shielding action of the screen and suppressor. This means that plate-voltage changes cause little or no change in the plate current. Since the plate current is almost independent of the plate voltage, the plate characteristics are nearly straight lines parallel to the voltage axis. The action of the control grid is essentially the same as in a triode, however.

Plate

Suppressor grid

Control grid

Screen grid

Cathode

Figure A1-6 Circuit symbol for pentode.

Typical plate characteristics of a pentode for a given positive-screen potential and for the suppressor connected to the cathode are illustrated in Fig. A1-7. Note that the plate current is relatively independent of plate voltage, as anticipated above. Actually, however, the plate characteristics also depend upon the fixed screen potential. Higher screen voltages shift the curves in Fig. A1-7 upward to higher current values with little change in shape. It is common practice to specify pentode characteristics at two or three specific values of screen potential.

The screen voltage is maintained below the plate potential in most circuits, and it turns out that the screen current is about 0.2 to 0.4 of the plate current. As mentioned above, it is almost universal to connect the suppressor grid to the cathode; in many tubes this connection is made internally.

The pentode is extensively used in amplifier circuits because of its very high amplification factor. It surpasses the triode as a high-frequency amplifier where small grid-plate capacitance is important. Plate-voltage excursions can be nearly as large as the plate supply voltage without introducing excessive distortion, so high-power operation is also possible.

6AU6

Grid potential, V = 0

−0.5

−1.0

−1.5

−2.0

−2.5

Plate current, mA

Plate potential, V

Figure A1-7 Plate characteristics of type 6AU6 pentode.

Other Multigrid Tubes

Many other vacuum-tube designs suitable for special circuit applications have been developed. A pentode having a control grid wound with a helix of varying pitch has an amplification factor that depends markedly upon the grid bias. This is so because electrons are controlled best in the region of the grid where the spacing is small. At sufficiently large values of negative grid bias the electron stream is cut off at this portion of the grid. Therefore, the amplification factor of the tube corresponds to that of a vacuum tube whose grid wires are widely spaced. At less negative biases the amplification factor is larger. These *variable-μ* tubes are used in automatic volume-control circuits where the amplification of the tube is automatically adjusted by controlling the dc grid voltage. The voltage is adjusted to maintain the output signal constant against variations in input signal. These designs are also called *remote-cutoff tubes*, since a very large negative grid voltage is required to reduce the plate current to a small value.

More than one electrode structure may be included within a single envelope where two tube types are often used together in circuits. An example of this is the dual vacuum diode for full-wave rectifier circuits. Other common structures are the dual triode and the dual diode-triode. Many other combinations are possible and have been constructed. The symbol for a dual triode is indicated in Fig. A1-8a. This symbol also illustrates the method commonly employed to indicate the tube base pin connections to various electrodes. The base pins are numbered clockwise when viewed from the bottom; this convention facilitates identification of the various electrodes when the tube socket is examined in an actual circuit.

Vacuum tubes with more than three grids have been designed for special purposes. An example is the *pentagrid converter*, Fig. A1-8b. This tube has two control grids, g_1 and g_3, shielded from each other by two screen grids, g_2 and g_4. The fifth grid is the suppressor, g_5. This tube is used in frequency-converter circuits where two signals of different frequencies are applied to the two control grids. Nonlinearities of the tube characteristics result in sum and difference frequencies analogous to the diode first detector described in Chap. 4. Actually,

Figure A1-8 (*a*) Base diagram of 12AX7 dual triode; (*b*) base diagram of 6BE6 pentagrid converter.

the pentagrid converter can be used also to generate one of the signals within the tube itself, so only the input signal need be supplied externally.

In certain *beam power tubes*, most notably the type 6L6, the screen-grid wires are aligned with wires of the control grid. This forms the electron stream into sheets and reduces the screen current by a factor of 5 or so below that of conventional pentodes. There is no suppressor grid as such, but beam-forming plates at cathode potential further shape the electron stream. This specific electrode design causes the electron charges in the beam to produce an effective suppressor action. The result is that the plate characteristics are straight nearer to the $V_b = 0$ axis than is the case for a standard pentode. Accordingly, the beam power tube is useful as a power amplifier since the allowable plate-voltage range is nearly as large as the plate supply voltage. In addition, such tubes are often constructed with sturdy cathodes and anodes, so high-current operation is possible.

THE TRIODE AMPLIFIER

Cathode Bias

Suitable electrode potentials to permit the vacuum triode to be used as a simple amplifier are provided by a plate supply battery, V_{bb}, and a resistor in series with the cathode, Fig. A1-9. The IR drop across the cathode resistor places a negative bias on the grid with respect to the cathode. The operating point is found by drawing the load line

$$I_b = \frac{V_{bb}}{R_L + R_k} - \frac{1}{R_L + R_k} V_b \tag{A1-5}$$

on the plate characteristics as shown in Fig. A1-10. By definition, the operating point must be located somewhere along the load line. Since the grid bias depends upon the plate current, a second line is drawn connecting the points $I_b = -V_c/R_k$ on every plate-characteristic curve. The intersection of this curve with the load line is the operating point. Capacitor C_k shunting the cathode-bias resistor R_k in

Figure A1-9 Circuit diagram of a practical triode amplifier using cathode bias.

Figure A1-10 Determination of operating point for 6C4 amplifier.

Fig. A1-9 prevents ac signals caused by the ac plate current in the cathode resistor from appearing in the grid circuit.

It is advantageous to derive the screen potential for a pentode from the plate supply voltage. This can be accomplished by means of a resistive voltage divider across the plate supply voltage. A simple series dropping resistor, Fig. A1-11, is even more convenient, however. The value of R_s is selected to yield the desired screen voltage at the given screen current and plate supply potential. Values in the range from 0.05 to 1.0 MΩ are typical. The screen resistor is bypassed with capacitor C_s to maintain the screen voltage constant independent of variations caused by signal voltages.

Determining the operating point for cathode-biased pentodes is slightly more involved than for triodes because the bias voltage depends upon the screen current as well as the plate current. It is usually satisfactory to assume the screen current is

Figure A1-11 Practical pentode amplifier circuit.

a fixed fraction of the plate current and proceed as in the case of the triode. The correct fraction to choose depends somewhat upon the tube type in question and may be estimated from the tube's screen-current characteristics. For most pentodes the total cathode current is approximately $1.3I_b$ at the operating point. This approximation usually yields a sufficiently accurate determination of the quiescent condition.

In many cases it is simpler to resort to the following "cut-and-try" procedure for determining the operating point. Choose a point on the load line corresponding to an arbitrary grid-bias voltage and plate current. The screen current is then determined from the screen characteristics for these values of V_c and V_b. The product $-(I_b + I_s)R_k$ is compared with the chosen value of V_c. If these two values are equal, the original choice is satisfactory. If they are not, the process is repeated until the desired accuracy is obtained.

Aside from the obvious economy achieved by providing grid, plate, and screen potentials from one voltage source, these bias techniques also result in more stable quiescent operation than does fixed bias. Suppose, for example, that the plate current tends to increase because of tube aging. This increases the negative grid bias and this, in turn, tends to lower the plate current. The net change in operating point is much less than is the case for fixed bias. The same situation exists with regard to the screen voltage.

Small-signal Parameters

Very frequently the signal amplitudes applied to a vacuum tube are small compared with the full range of voltages covered by the plate characteristics. It then proves possible to represent the tube characteristics using an approximation to Child's law which is quite accurate for small signals. Once the operating point is established graphically, small departures about the operating point caused by small ac signals are treated by assuming the triode is a linear device. This approach may be illustrated in the following way. According to Child's law, the change in plate current caused by a change in plate voltage is

$$\Delta I_b = \frac{3A}{2}(\mu V_c + V_b)^{1/2}\,\Delta V_b = \frac{3}{2}A^{2/3}I_b^{1/3}\,\Delta V_b \qquad (A1\text{-}6)$$

where Eq. (A1-2) has been used. The ratio $\Delta I_b/\Delta V_b$ can be identified as the reciprocal of an equivalent resistance called the *plate resistance*

$$\frac{1}{r_p} = \frac{\Delta I_b}{\Delta V_b} \qquad (A1\text{-}7)$$

The plate current also changes with a change in grid voltage,

$$\Delta I_b = \frac{3A}{2}(\mu V_c + V_b)^{1/2}\mu\,\Delta V_c = \frac{3\mu}{2}A^{2/3}I_b^{1/3}\,\Delta V_c \qquad (A1\text{-}8)$$

The ratio $\Delta I_b/\Delta V_c$ is the mutual transconductance,

$$g_m = \frac{\Delta I_b}{\Delta V_c} \qquad (A1\text{-}9)$$

The total change in plate current, from Eqs. (A1-7) and (A1-9), is written

$$\Delta I_b = \frac{1}{r_p} \Delta V_b + g_m \, \Delta V_c \qquad (A1\text{-}10)$$

A useful relation between r_p and g_m is obtained by noting that if the net change in plate current is zero, Eq. (A1-10) becomes

$$0 = \frac{1}{r_p} \Delta V_b + g_m \, \Delta V_c$$

$$g_m r_p = -\frac{\Delta V_b}{\Delta V_c} \qquad (A1\text{-}11)$$

The ratio $-\Delta V_b/\Delta V_c$ is just the definition of the amplification factor. The minus sign signifies that the plate voltage decreases when the grid potential increases, as discussed previously. Therefore, Eq. (A1-11) is written

$$\mu = r_p g_m \qquad (A1\text{-}12)$$

This relation is useful in obtaining one of the three parameters if the other two are known.

According to Eq. (A1-10) small changes in plate current around the operating point caused by changes in grid voltage and plate voltage can be calculated if suitable values of the small-signal parameters μ, r_p, and g_m appropriate to the operating point are known. The plate resistance is just the reciprocal of the slope of the plate characteristic at the operating point and g_m is the slope of the transfer characteristic at the operating point so that quantitative values can be obtained from the current-voltage characteristics of the tube. Typical values of small-signal parameters for several triodes are given in Table A1-1.

The magnitudes of the small-signal parameters depend upon the operating point. For this reason it is usually necessary to evaluate μ, r_p, and g_m graphically at the point determined by the dc potentials and tube characteristics. To a first

Table A1-1 Triode small-signal parameters

Type	μ	r_p, $10^3 \, \Omega$	g_m, 10^{-3} mho
6C4	20	6.3	3.1
6CW4	68	5.4	12.5
6SL7	70	44	1.6
12AT7	55	5.5	10
12AU7	17	7.7	2.2
12AX7	100	62	1.6
7895	74	7.3	10.9

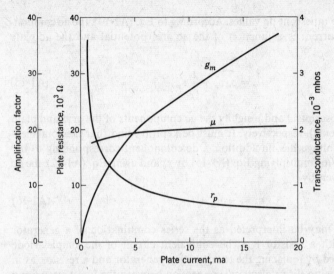

Figure A1-12 Variation of triode small-signal parameters with plate current.

approximation, the amplification factor is independent of the operating point, since it depends only upon the ratio of the inter-electrode capacitances, Eq. (A1-4). The plate resistance depends upon the plate current, according to (A1-7) and (A1-6),

$$r_p = \tfrac{2}{3}A^{-2/3}I_b^{-1/3} \tag{A1-13}$$

Similarly for the mutual transconductance, using (A1-8) and (A1-9),

$$g_m = \tfrac{3}{2}A^{2/3}\mu I_b^{1/3} \tag{A1-14}$$

According to (A1-13) and (A1-14), the plate resistance decreases slowly as the plate current increases while the transconductance increases.

Since Child's law is not useful as a quantitative description of triode behavior, it is not expected that Eqs. (A1-13) and (A1-14) exactly represent the variation of the small-signal parameters with tube current. Nevertheless, experimental data for a practical triode, Fig. A1-12, are in surprisingly good agreement with the analytical results. Note that the amplification factor is sensibly constant, except at the smallest plate currents, while g_m and r_p respectively increase and decrease with increasing plate current. According to Fig. A1-12, it is possible to obtain considerable variation in the small-signal parameters by selecting different operating points.

Triode Equivalent Circuit

The small-signal parameters of a triode are used to calculate the performance of the tube in any circuit, so long as the signal voltages and currents are small compared with the quiescent dc values. According to Eq. (A1-10), the ac component of the plate current is a function of the ac grid potential and the ac plate voltage so that

$$i_p = g_m v_g + \frac{1}{r_p} v_p \qquad (A1\text{-}15)$$

Henceforth the subscripts g and p signify the ac components of the grid and plate voltages (and currents), respectively. It must be kept in mind that the total electrode current or voltage has, in addition, a dc component corresponding to the operating point. After multiplying Eq. (A1-15) by r_p and using Eq. (A1-12), the ac plate voltage is given by

$$v_p = -\mu v_g + i_p r_p \qquad (A1\text{-}16)$$

Equation (A1-16) may be interpreted as the series combination of a generator $-\mu v_g$ in series with a resistor r_p. The equivalent circuit of the simple triode amplifier is obtained by replacing the triode with a generator and a resistor, as in Fig. A1-13. There is no direct electrical connection between the input circuit and the output circuit. This illustrates the valve action of the grid in controlling the plate current with essentially zero power expenditure. Note that the minus sign associated with the generator accounts for the 180° phase difference between the grid voltage and the plate voltage.

Figure A1-13 (a) Triode amplifier and (b) its ac equivalent circuit.

The output voltage of the amplifier circuit is calculated immediately from the equivalent circuit.

$$v_o = i_p R_L = \frac{-\mu v_g}{r_p + R_L} R_L = \frac{-\mu v_g}{1 + r_p/R_L} \qquad (A1\text{-}17)$$

The gain of the amplifier is

$$a = \frac{v_o}{v_i} = \frac{-\mu}{1 + r_p/R_L} \qquad (A1\text{-}18)$$

According to Eq. (A1-18), the maximum gain of the circuit is equal to the amplification factor of the triode. This value is achieved when the load resistance is much greater than the plate resistance.

It is not always desirable to make the load resistance much larger than the triode plate resistance in a practical amplifier. A large load resistance introduces considerable dc power loss and requires a large plate supply voltage V_{bb} to put the tube at its optimum operating point. The amplification factor of most tubes is sufficiently large to yield appreciable gains with lower values of load resistance. It is also possible to use a transformer in the plate circuit. The impedance-matching properties of a transformer reflect a large ac impedance from the secondary, as explained in Chap. 3. In this case the dc resistance corresponds to the primary-winding resistance, which may be quite small. The expense and limited frequency range of transformers restrict their use to rather special applications.

The magnitude of the grid resistor R_g in this circuit is limited only by the voltage drop caused by residual grid current. The corresponding IR drop across R_g can change the grid bias from the value set by the grid bias battery V_{cc}. Since grid current is normally an uncontrolled quantity and varies considerably even between tubes of the same type, this is an undesirable condition. Grid resistances in the 0.5- to 10-MΩ range are satisfactory for most triodes. These values are large enough that the loading effect of the grid resistor on the input voltage source may be safely ignored in most applications.

Pentodes may be represented by the same equivalent circuit as triodes so long as appropriate values of the small-signal parameters are used. These parameters depend upon the operating point and therefore vary with the screen and suppressor potentials, in addition to the grid bias and plate voltage. Because of the constant-current properties of pentodes, it is generally more useful to use the Norton representation in preference to the Thévenin circuit. According to Eq. (1-56), the equivalent current generator is given by the ratio of the Thévenin equivalent voltage generator divided by the equivalent internal resistance,

$$I_{eq} = \frac{-\mu v_g}{r_p} = -g_m v_g \qquad (A1\text{-}19)$$

A pentode amplifier stage, Fig. A1-11, is represented by the equivalent circuit shown in Fig. A1-14. The output voltage is simply the constant-current source

Figure A1-14 Equivalent circuit of pentode amplifier.

times the parallel combination of the plate resistance and load resistance,

$$v_o = -g_m v_i \frac{r_p R_L}{r_p + R_L} \tag{A1-20}$$

and the gain is

$$a = \frac{v_o}{v_i} = -g_m \frac{R_L}{1 + R_L/r_p} \tag{A1-21}$$

In most cases, the plate resistance is much greater than the load resistance so that a satisfactory approximation to Eq. (A1-21) is just

$$a = -g_m R_L \tag{A1-22}$$

Plate load resistors much in excess of $0.1 \text{ M}\Omega$ or so are inconvenient, because the excessive dc voltage drop caused by the plate current requies a very large plate-voltage source. Nevertheless, amplifications as great as 1000 are obtainable from a single pentode amplifier stage.

Index

Index